HANDBOOK OF HYPERGEOMETRIC INTEGRALS

THEORY, APPLICATIONS, TABLES, COMPUTER PROGRAMS

Series: Mathematics and its Applications

MATHEMATICS & ITS APPLICATIONS

Series Editor: Professor G. M. Bell
Chelsea College, University of London

Mathematics and its applications are now awe-inspiring in their scope, variety and depth. Not only is there rapid growth in pure mathematics and its applications to the traditional fields of the physical sciences, engineering and statistics, but new fields of application are emerging in biology, ecology and social organisation. The user of mathematics must assimilate subtle new techniques and also learn to handle the great power of the computer efficiently and economically.

The need of clear, concise and authoritative texts is thus greater than ever and our series will endeavour to supply this need. It aims to be comprehensive and yet flexible. Works surveying recent research will introduce new areas and up-to-date mathematical methods. Undergraduate texts on established topics will stimulate student interest by including applications relevant at the present day. The series will also include selected volumes of lecture notes which will enable certain important topics to be presented. earlier than would otherwise be possible.

In all these ways it is hoped to render a valuable service to those who learn, teach, develop and use mathematics.

The Foundation Programme includes:

Mathematical Models in Social Life and Management Sciences
David N. Burghes and Alistair D. Wood, Cranfield Institute of Technology.

Modern Introduction to Classical Mechanics and Control
David Burghes, Cranfield Institute of Technology and Angela Downs, University of Sheffield.

Vector & Tensor Methods
Frank Chorlton, University of Aston, Birmingham.

Lecture Notes on Queueing Systems
Brian Conolly, Chelsea College, London University.

Mathematics for the Biosciences
G. Eason, C. W. Coles and G. Gettinby, University of Strathclyde.

Handbook of Hypergeometric Integrals: Theory, Applications, Tables, Computer Programs
Harold Exton, The Polytechnic, Preston.

Multiple Hypergeometric Functions
Harold Exton, The Polytechnic, Preston.

Computational Geometry for Design and Manufacture
Ivor D. Faux and Michael J. Pratt, Cranfield Institute of Technology.

Applied Linear Algebra
Ray J. Goult, Cranfield Institute of Technology.

Generalised Functions: Theory, Applications
Roy F. Hoskins, Cranfield Institute of Technology.

Mechanics of Continuous Media
S. C. Hunter, University of Sheffield.

Using Computers
Brian Meek and Simon Fairthorne, Queen Elizabeth College, University of London.

Environmental Aerodynamics
R. S. Scorer, Imperial College of Science and Technology, University of London.

Physics of the Liquid State: A Survey for Scientists and Technologists
H. N. V. Temperley, University College of Swansea, University of Wales and
H. D. Trevena, University of Wales, Aberystwyth.

HANDBOOK OF HYPERGEOMETRIC INTEGRALS

THEORY, APPLICATIONS, TABLES, COMPUTER PROGRAMS

HAROLD EXTON, B.Sc., M.sc., Ph.D.
The Polytechnic, Preston

ELLIS HORWOOD LIMITED
Publishers Chichester

Halsted Press: a division of
JOHN WILEY & SONS
Chichester · New York · Brisbane · Toronto

The publisher's colophon is reproduced from James Gillison's drawing of the ancient Market Cross, Chichester.

First published in 1978 by

ELLIS HORWOOD LIMITED
Market Cross House, 1 Cooper Street, Chichester, Sussex, England

Distributors:

Australia, New Zealand, South-east Asia:
Jacaranda-Wiley Ltd., Jacaranda Press,
JOHN WILEY & SONS INC.,
G.P.O. Box 859, Brisbane, Queensland 4001, Australia.

Canada:
JOHN WILEY & SONS CANADA LIMITED
22 Worcester Road, Rexdale, Ontario, Canada.

Europe, Africa:
JOHN WILEY & SONS LIMITED
Baffins Lane, Chichester, Sussex, England.

North and South America and the rest of the world:
HALSTED PRESS, a division of
JOHN WILEY & SONS
605 Third Avenue, New York, N.Y. 10016, U.S.A.

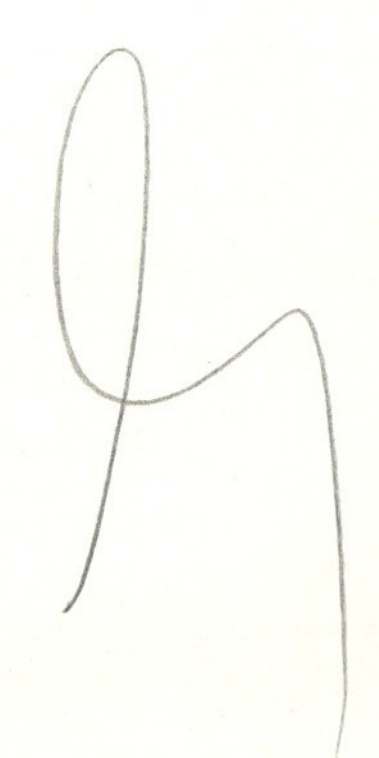

British Library Cataloguing in Publication Data

Exton, Harold,
Handbook of hypergeometric integrals. –
(Mathematics and its applications).
1. Functions – Hypergeometric – Tables
I. Title II. Series

515'.55 QA353.H9 78-40120

ISBN 0-85312-122-2 (Ellis Horwood Ltd., Publishers)
ISBN 0-470-26342-3 (Halsted)

Printed in Great Britain by Biddles of Guildford

Table of Contents

Author's Preface

This book has been designed with the object of providing tables of integrals of hypergeometric functions of many different types. It is believed that, as a result of a re-ordering of the material, a number of the integrals included here have not previously appeared explicitly in the literature. At the same time, a number of computer programs to evaluate numerically some of the hypergeometric integrals concerned are given. These programs have been set up in such a way that they may be extended to cover wider results with a minimum of essential modification; and the international computer language FORTRAN IV has been used so that the programs have the widest possible applicability. It is believed that this is the first time that programs of this type have been included in a book of tables of integrals.

Handbooks of mathematical formulae very often include little or no theoretical background to the results listed. In an attempt to fill this gap, the first part of this book consists of seven chapters on the general theory of hypergeometric integrals. Additionally, an important aspect of this work is the concluding bibliography. The breadth of these aims, combined with the necessity for keeping the size of the book within bounds, has made it necessary to be as concise as is compatible with intelligibility. It may be observed that an important part of the fascination of working with integrals of hypergeometric functions is that the results themselves are of hypergeometric form.
The wide variety of applications of hypergeometric integrals is exemplified in Chapter 7 where a number of instances of the appearance of such expressions in statistics and mathematical physics are cited. One reason why such integrals are of importance is that the majority of the classical special functions, such as those of Laguerre, Legendre, Hermite and Bessel, are particular cases of hypergeometric functions.

In the case of an analytical work such as this, it is almost certain that some mistakes, clerical or otherwise, have remained undetected. The number of such errors has been considerably diminished by the great care taken by Dr. Lucy J. Slater of the Department of Applied Economics, University of Cambridge, in her reading of the manuscript. For this and constant advice and encouragement so freely given, I offer my most sincere thanks. Grateful acknowledgement is also due to the Director and Staff of Preston Polytechnic for the use of computer facilities.

Harold Exton
The Polytechnic
Preston, January 1978

Foreword

A hypergeometric integral is any integral which has a hypergeometric function in its integrand. Such hypergeometric integrals have frightened mathematicians for a long time, because they look complicated. This is mainly the fault of the notation which is necessary to exhibit the underlying structure of the integrals. The functions in these integrands range from simple Bessel and Whittaker functions up to generalised hypergeometric functions of several variables with many parameters.

This handbook supplies an ordered list of the main analytical formulae of these integrals, together with some simple computer programs, in FORTRAN, which will evaluate these integrals numerically provided that we keep away from values of the parameters which are near any poles of the integrands.

The book is in two parts. The first part consists of seven chapters which give the general theory of these integrals, and examples of their uses, particularly in the study of statistical distributions, and in various branches of physics. The second part consists of about 100 pages of tables of the formulae and about 50 pages of computer programs to evaluate the integrals. There is also an extensive bibliography.

There are many types of hypergeometric functions. The simple Gauss series ${}_2F_1(a,b;c;x)$ with one variable and three parameters contains as special cases most of the ordinary functions of mathematical physics, such as exponental, Bessel and Whittaker functions. The first step in generalisation is the ordinary generalised hypergeometric function ${}_AF_C((a);(c);x)$, which has one variable, A numerator parameters and C denominator parameters.

The next step in generalisation is to introduce another variable to produce the hypergeometric function of two variables:

$$F^{A:B;B'}_{C:D;D'}\left(\begin{matrix}(a):(b);(b');\\(c):(d);(d');\end{matrix}x,y\right).$$

This may be further extended to give rise to the multiple hypergeometric functions of many variables. Certain special cases of these are the Lauricella functions and the generalised Kampé de Fériet functions. the theory of all these functions forms the subject matter of Chapter 1.

In Chapter 2, we meet the various types of Euler integrals which have these functions in their integrands. Chapter 3 covers the theory of definite integrals and repeated integrals which involve all these functions. In Chapter 4, there is a discussion of the contour integrals which occur in this theory, that is the Pochhammer and Barnes type integrals. Chapter 5 covers the infinite integrals and their connections with these contour integrals and Chapter 6 discussed the multiple Euler, Laplace and Barnes type integrals of hypergeometric functions.

The seventh chapter contains several examples of some very important applications of this theory to various branches of statistics and mathematical physics. In statistics, probability distributions such as the well-known Beta and Gamma distributions, density functions and cumulative distributions can all be expressed in terms of hypergeometric integrals, and so their theory can be extended and unified. Also, once they have been expressed in terms of such integrals, the numerical evaluation of these functions becomes a comparatively simple matter. In physics, hypergeometric integrals occur in the theories of vibration of thin elastic plates, heat conduction in a cylinder, the motion of a viscous fluid, and some types of electrical networks. They also have applications in the theory of the flow of heat in solids moving in conducting media and in the theory of certain types of non-linear oscillations. In quantum chemistry, they occur in some molecular integrals, and in communications engineering where Gaussian noise is a well-known phenomenon. Other applications which are discussed are in the theory of electrical impedences and in combinatorial analysis.

Once again, I am happy to recommend Dr. Exton's work to all those mathematicians, statisticians, engineers, physicists, chemists and computer systems analysts, who study the problem of the evaluation of hypergeometric integrals.

L. J. Slater Sc.D.
Cambridge, January 1978

Part One: Theory and Applications

Chapter 1

Hypergeometric Functions of One or More Variables

INTRODUCTION

The term 'hypergeometric' was first used by Wallis in his work *Arithmetica Infinitorum* published in the year 1655 when he was a professor at the University of Oxford. This expression was employed to denote any series which may be regarded as a generalisation of the ordinary geometric series

$$1 + z + z^2 + z^3 + \dots ,$$

where each pair of adjacent terms has a common multiplicative ratio.

It will be seen that the term 'hypergeometric series' literally means 'beyond the geometric series'. Exactly why a geometric series should be so named remains rather obscure.

The present-day use of the term 'hypergeometric' in connection with hypergeometric functions leads to some considerable confusion among many readers. At a first glance, it might seem that some connection with, and, in fact, extension of some branch of higher geometry in many dimensions is implied. Nothing could be further from the truth. It would, perhaps, be desirable to refer to hypergeometric series by some such term as 'gamma-product series' which would be more in keeping with their actual structure. Since the currently used term has been in use for such a long time, however, it seems to the author to be undesirable even to attempt such a change of this nature.

The notation of hypergeometric functions frequently presents a somewhat complicated appearance. Initially this may seem to be rather alarming, but this type of notation is necessary on account of the number of parameters and variables involved in a hypergeometric function of higher order. This scheme of nomenclature is simple in essence, however, and readily displays the rich inherent structure of a given hypergeometric function once the reader has familarised himself with it.

Many special functions of mathematical physics and chemistry, statistics and number theory are of hypergeometric form, so that any general results may be used as master formulae from which expressions applicable to such functions as those of Legendre and Bessel may easily be deduced.

When considering the numerical evaluation of hypergeometric integrals, it is interesting to note that any integral of a hypergeometric function is itself a hypergeometric function. This may, of course, take the form of an elementary function such as an exponential or a binomial function. Within their regions of definition, hypergeometric integrals may be evaluated numerically by fast present-day computers provided that we avoid those values of the parameters coinciding with the poles of the integrand. A number of computer programs is given in Part II.

1.1 The Gauss Hypergeometric Function

Gauss in his thesis presented at Göttingen in 1812 systematically discussed the series

$$1 + \frac{a\ b}{1.c} x + \frac{a(a+1)\ b(b+1)}{1.2\ c(c+1)} x^2 + \ldots \qquad (1.1.1)$$

which is of great importance to mathematicians. This series is known as the ordinary *hypergeometric series* and may be regarded as a generalisation of the geometric series

$$\sum_n x^n \quad . \qquad (1.1.2)$$

In terms of the Pochhammer symbol (a,r) defined by the relations

$$(a,r) = \Gamma(a+r)/\Gamma(a)$$

$$= a(a+1)\ldots(a+r-1), \qquad (1.1.3)$$

$$(a,0) = 1 \qquad (1.1.4)$$

and

$$(a,-r) = (-1)^r/(1-a,r) \quad , \qquad (1.1.5)$$

where r is an integer, positive, negative or zero, the series (1.1.1) may be written

$$\sum_{n=0}^{\infty} \frac{(a,n)(b,n)}{(c,n)\ n!} x^n \quad . \qquad (1.1.6)$$

This series may be regarded as a representation of the Gauss *hypergeometric function* which we denote by the symbol

$${}_2F_1(a,b;c;x) \quad . \qquad (1.1.7)$$

In the hypergeometric series, the three elements a,b and c are described as the parameters and the element x is called the variable of the series. All four of these quantities may be any numbers, real or complex. There is one exception, namely that the series representation (1.1.6) is not defined if c, the denominator parameter, is a non-positive integer, as then the numerical value of the series becomes infinite, unless one of the numerator parameters a or b is also a negative integer, such that $-c < -a$, say. In general, if either of the numerator parameters is a negative integer, the series terminates.

1.1.1 The Convergence of the series $_2F_1$

From the series representation (1.1.6), the general term of the Gaussian series is given by

$$u_n = \frac{(a,n)(b,n)}{(c,n)\quad n!} x^n , \qquad (1.1.1.1)$$

and so, the ratio of succesive terms is

$$u_{n+1}/u_n = \frac{(a+n)(b+n)}{(c+1)(1+n)} x , \qquad (1.1.1.2)$$

from which it follows from d'Alembert's ratio test (Bromwich[1931] page 39), that the series converges for all x, real or complex, such that $|x| < 1$ and diverges if $|x| > 1$. When $x = 1$, (1.1.6) converges absolutely if $\mathrm{Re}(c-a-b) > 0$ and diverges if $\mathrm{Re}(c-a-b)$ is less than or equal to zero. At other points on the circle of convergence $|x| = 1$, more delicate tests are necessary to discuss the convergence of the series in question. See Slater (1966) page 5.

1.1.2 Confluent Forms of the Gauss Function

If we replace x by x/b in Gauss's hypergeometric series, a power series in x whose radius of convergence is $|b|$ is obtained; and if we let $b \to \infty$, taking into account that

$$\lim_{b \to \infty} \frac{(b,n)}{b^n} = 1, \qquad (1.1.2.1)$$

we obtain the series

$${}_1F_1(a;c;x) = \sum_{n=0}^{\infty} \frac{(a,n)x^n}{(c,n)n!} , \qquad (1.1.2.2)$$

which possesses only one numerator parameter. This is known as the confluent hypergeometric function. This is because the process of taking the limit $b \to \infty$ results in the coalescence of the singularities at $x=1$ and $x=\infty$ of the function ${}_2F_1(a,b;c;x)$; see Section 1.1.3.

The function defined by (1.1.2.2) was first discussed in detail by Kummer (1836), and it is often named after him. If we repeat the same limiting process, we obtain the result

$$ {}_0F_1(-;c;x) = \sum_{n=0}^{\infty} \frac{x^n}{(c,n)n!} \quad . \tag{1.1.2.3} $$

Kummer himself gave the interesting realtion

$$ {}_1F_1(a;2a;2x) = e^x {}_0F_1(-;a+1/2;x^2/4), \tag{1.1.2.4} $$

which is sometimes referred to as Kummer's second theorem.

The functions ${}_2F_1$, ${}_1F_1$ and ${}_0F_1$ and many other functions of hypergeometric form possess very many applications in mathematics and mathematical physics, and the reader is referred to such texts as Erdélyi et al. (1953) and the author's work (1976), for example.

1.1.3 The Hypergeometric Differential Equation

The formula

$$ \frac{d^p}{dx^p} {}_2F_1(a,b;c;x) = \frac{(a,p)(b,p)}{(c,p)} {}_2F_1(a+p,b+p;c+p;x) \tag{1.1.3.1} $$

may be established from (1.1.6) by direct differentiation, which is readily justified when $|x| < 1$. By direct substitution, or otherwise, it follows that the function ${}_2F_1(a,b;c;x)$ is a particular integral of the hypergeometric differential equation

$$ x(1-x)y'' + [c-(a+b+1)x]y' - aby = 0. \tag{1.1.3.2} $$

This equation possesses three singularities, all regular, at the points $x=0,1$ and ∞, relative to which we have the following exponents: $(0,1-c)$, $(0,c-a-b)$ and (a,b), respectively. This differential equation may be integrated completely by using the method of Fröbenius as is well-known; see Whittaker and Watson (1952), Chapter 10.

A much more meaningful approach to the discussion of the complete solution of (1.1.3.2), for non-exceptional values of the parameters, is through the medium of the Euler integral formula for the function ${}_2F_1$. Consider the integral

$$ I = \int_0^1 t^{a-1}(1-t)^{c-a-1}(1-xt)^{-b}\, dt. \tag{1.1.3.3} $$

If $|x| < 1$, we may expand the factor $(1-xt)^{-b}$ in the integrand of (1.1.3.3) and integrate term-by-term, when we find that

$$ I = \sum_{n=0}^{\infty} \frac{(b,n)}{n!} x^n \int_0^1 t^{a+n-1}(1-t)^{c-a-1}\, dt. \tag{1.1.3.4} $$

The inner integral of (1.1.3.4) may be expressed as the beta function $B(c+n,c-a)$ provided that the real parts of a and c-a are positive. It follows immediately that

$$ {}_2F_1(a,b;c;x) = \frac{\Gamma(c)}{\Gamma(a)\Gamma(c-a)}\int_0^1 t^{a-1}(1-t)^{c-a-1}(1-xt)^{-b}dt, \tag{1.1.3.5}$$

provided that $\mathrm{Re}(c)$, $\mathrm{Re}(c-a) > 0$. The above expression is known as the Euler integral for the hypergeometric function and was first given by Euler in his work *Introductio in Analysis Infinitorum* published in Lausanne in the year 1748.

This integral suggests that a trial solution of the hypergeometric differential equation is of the form

$$ z\int_g^h u^{b-c}(u-1)^{c-a-1}(u-x)^{-b}\,du. \tag{1.1.3.6}$$

Direct substitution into (1.1.3.2) shows that (1.1.3.6) is indeed a solution, provided that the path of integration is either a simple contour joining any two of the singularities of the integrand, or a contour closed on the Riemann surface of the integrand, see Erdélyi (1950) for example.

In the first case, we see that g and h must be any two of the quantities 0, 1, x or ∞, and this gives the following six possibilities:

$$ z_1 = \int_1^\infty V du,\quad z_2 = \int_0^x V du,\quad z_3 = \int_{-\infty}^0 V du,\quad z_4 = \int_1^x V du,\quad z_5 = \int_0^1 V du $$

and $z_6 = \int_x^\infty V du$, where $V = u^{b-c}(u-1)^{c-a-1}(u-x)^{-b}$.

Each of the integrals z_1 to z_6 may be reduced to the form (1.1.3.5) by means of the transformations of the variable of integration: $u=1/v$, $u=xv$, $u=1-1/v$, $u=1-v(1-x)$, $u=v$ and $u=x/v$ respectively, so that we may write

$$ z_1 = x^{-b}\,{}_2F_1(b,b+1-c;b+1-a;1/x),\quad z_2 = x^{1-c}\,{}_2F_1(b+1-c,a+1-c;2-c;x), $$

$$ z_3 = (1-x)^{c-a-b}\,{}_2F_1(c-a,c-b;c+1-a-b;1-x),\quad z_4 = {}_2F_1(a,b;a+b-c;1-x), $$

$$ z_5 = {}_2F_1(a,b;c;x) \text{ and } z_6 = x^{-a}\,{}_2F_1(a,a+1-c;a+1-b;1/x). $$

Furthermore, the integral (1.1.3.6) is unchanged in form by means of the substitutions $u=1-v$, $u=v/(1-x-xv)$ and $u=(1-v)/(1-xv)$, and we have the Euler transformations

$$ \begin{aligned} {}_2F_1(a,b;c;x) &= (1-x)^{-a}\,{}_2F_1(a,c-b;c;x/[x-1]) \\ &= (1-x)^{-b}\,{}_2F_1(c-a,b;c;x/[x-1]) \\ &= (1-x)^{c-a-b}\,{}_2F_1(c-a,c-b;c;x), \end{aligned} \tag{1.1.3.7}$$

from which we may obtain the twenty-four well-known solutions of the hypergeometric differential equation first given by Kummer (1836).

1.1.4 Special Cases

A large number of the functions of mathematical physics may be written in terms of the Gauss hypergeometric function and its confluent forms. A few examples of such cases are given in the list which follows; for further information, the reader should consult such works as Erdélyi et al. (1953), for instance.

Special Cases of the Function ${}_2F_1$

The binomial function

$${}_2F_1(a,b;b;x) = (1-x)^{-a},$$ Slater (1966) page 17,

the Legendre polynomial

$${}_2F_1(n+1,-n;1;\tfrac{1-x}{2}) = P_n(x),$$ Whittaker and Watson (1952) page 312,

the associated Legendre polynomial

$${}_2F_1(-m-n,1-m+n;1-m;\tfrac{1-x}{2}) = \Gamma(1-m)2^{-m}(x^2-1)^{m/2}\,P_n^m(x),$$ Slater (1966) page 18,

the Tchebcheff polynomial

$${}_2F_1(-n,n;1/2;\tfrac{1-x}{2}) = T_n(x),$$ Sneddon (1966) page 100,

the Gegenbauer polynomial

$${}_2F_1(n+2p,-n;p+1/2;\tfrac{1-x}{2}) = \frac{n!\Gamma(2p)}{n+2p}\,C_n^p(x),$$ Sneddon (1966) page 101,

the Jacobi polynomial

$${}_2F_1(-n,a+n;c;x) = F_n^{a,c}(x),$$ Appell and Kampé de Fériet (1926) page 11

and the complete elliptical integrals

$${}_2F_1(1/2,1/2;1;k^2) = K(k)$$

and $${}_2F_1(-1/2,1/2;1;k^2) = E(k),$$ MacRobert (1962) page 174.

Special Cases of the Function ${}_1F_1$

The exponential function

$${}_1F_1(a;a;x) = e^x,$$ Sneddon (1966) page 46,

the Bessel function

$${}_1F_1(n+1/2;2n+1;2x) = n!e^x(2/x)^nI_n(x),$$ MacRobert (1962) page 347,

the associated Laguerre polynomial

$$ {}_1F_1(-n;m+1;x) = \frac{m!n!}{(m+n)!} L_n^m(x), \text{ Sneddon (1966) page 166,} $$

the Toronto function

$$ {}_1F_1(m/2+1/2;1+n;x^2) = \frac{n!x^{m-2n-1}}{\Gamma(m/2+1/2)} T(m,n,x), $$

the incomplete gamma function

$$ {}_1F_1(a;a+1;-x) = a\, x^{-a}\gamma(a,x), $$

the Charlier polynomial

$$ {}_1F_1(-n;1+p-n;x) = \frac{\sqrt{(n!)}x^{n/2}}{(1+p-n,n)} \rho_n(p,x), $$

the parabolic cylinder function

$$ {}_1F_1(-p/2;1/2;x^2/2) = 2^{-1/2} \exp(x^2/4)E_p^{(0)}(x) $$

and the hyperbolic function

$$ {}_1F_1(1;2;2x) = x^{-1}e^x \sinh x; $$

see Abramowitz et al. (1965) page 509.

We also note that the Bessel function is given by

$$ J_p(x) = (x/2)^p {}_0F_1(-;p+1;-x^2/4)/\Gamma(p+1), \text{ Erdélyi et al (1953) Vol. II page 4.} $$

1.2 THE GENERALISED HYPERGEOMETRIC FUNCTION

A natural generalisation of the Gauss hypergeometric function is to increase the number of numerator and denominator parameters. The generalised hypergeometric function is then obtained and its series representation is given by

$$ {}_AF_C[(a);(c);x] = \sum_{n=0}^{\infty} \frac{((a),n)x^n}{((c),n)n!}, \tag{1.2.1} $$

where the symbol (a) denotes the sequence of parameters $a_1,..,a_A$, etc., a notation first adopted by Slater, see Slater (1966) page 41, for example.

The generalised function ${}_AF_C$ is a special solution of the differential equation

$$ [x \frac{d}{dx}(x \frac{d}{dx}+c_1-1)..(x\frac{d}{dx}+c_C-1)-x(x\frac{d}{dx}+a_1)..(x\frac{d}{dx}+a_A)]y = 0. \tag{1.2.2} $$

Near the origin, equation (1.2.2) possesses C convergent series solutions of the form

$$x^{1-c_p}\,{}_AF_C\left[\begin{matrix}1+a_1-c_p, 1+a_2-c_p, \,..\,, 1+a_A-c_p & ; \\ 1+c_1-c_p, 1+c_2-c_p, .\,, *, .\,, 1+c_C-c_p, 2-c_p; & \end{matrix} x\right] \qquad (1.2.3)$$

for p=1,2,.,C. It will be seen that the expression $1+c_p-c_p$ is omitted from the sequence of denomiantor parameters of (1.2.3). The order of the function ${}_AF_C$ is the order of the corresponding differential equation (1.2.2), and it is the greater of the two numbers A and C+1.

1.2.1 Convergence

As in the case of the series ${}_2F_1$, the convergence of ${}_AF_C(x)$ may be investigated by means of the d'Alembert ratio test, and we have the following general results:

(i) If $A \leq C$, the series converges for all finite values of its argument, real or complex.

(ii) If $A = C+1$, the series converges for $|x| < 1$ for all values of its parameters, and also if $x = 1$ and $x = -1$ provided that

$$\mathrm{Re}\left(\sum_{j=1}^{C} c_j - \sum_{j=1}^{A} a_j\right) > 0$$

and

$$\mathrm{Re}\left(\sum_{j=1}^{C} c_j - \sum_{j=1}^{A} a_j\right) > -1, \quad \text{respectively.}$$

(iii) If $A > C+1$, the series only converges in the trivial case when $x = 0$ and is only a formal representation of the generalised hypergeometric function concerned. If one or more of the numerator parameters is a negative integer, then the series terminates and the question of convergence does not arise.

1.2.2 The E, G and H Functions

In an attempt to give a meaning to the symbol ${}_pF_q$ when $p > q+1$, MacRobert (1962) page 352 defined the E-function by means of the multiple integral

$$\frac{\Gamma(c_1-a_1)\Gamma(c_2-a_2)..\Gamma(c_q-a_q)}{\Gamma(a_{q+1})}\, E(p;a_r:q;c_s;x)$$

$$= \prod_{m=1}^{q}\int_0^\infty y_m^{c_m-a_m-1}(1+y_m)^{-c_m}\,dy_m \times \prod_{n=2}^{p-q-1}\int_0^\infty e^{-y_{q+n}}\, y_{q+n}^{a_{q+n}-1}\,dy_{q+n} \times$$

(continued) (1.2.2.1)

(continued)

$$\times \int_0^\infty e^{-y_p} y_p^{a_p-1} \left[1+ \frac{y_{q+2}\, y_{q+3} \cdots y_p}{(1+y_1)(1+y_2)\cdots(1+y_q)x} \right]^{-a_{q+1}} dy_p, \quad (1.2.2.1)$$

where $|\arg x| < \pi$, when $p \geq q+1$, and the a_r and c_s are such that the integrals are convergent.

MacRobert also showed that the above definition is equivalent to

(i)

$$E(p;a_r:q;c_s;x) = \frac{\Gamma(a_1)\cdots\Gamma(a_p)}{\Gamma(c_1)\cdots\Gamma(c_q)}\, {}_pF_q(a_1,\ldots,a_p;c_1,\ldots,c_q;-1/x), \quad (1.2.2.2)$$

when $p \leq q+1$; $x \neq 0$ if $p < q$ and $|x| > 1$ if $p=q+1$,

and (ii)

$$E(p;a_r:q;c_s;x)$$

$$= \sum_{r=1}^{p} \frac{\prod_{s=1}^{p}{}' \Gamma(a_s-a_r)\Gamma(a_r)x^{a_r}}{\prod_{t=1}^{q} \Gamma(c_t-a_r)}\, {}_{q+1}F_{p+1}\left[\begin{matrix} a_r, a_r-c_1+1, \ldots, a_r-c_q+1; \\ a_r-a_1, \ldots, *, \ldots, a_r-a_p+1\ ; \end{matrix} (-1)^{p+q}\, x\right].$$

$$(1.2.2.3)$$

Zero and negative integer values of $c_1,\ldots,c_q$ or $a_1,\ldots,a_p$ are tacitly excluded in (1.2.2.2) and (1.2.2.3) respectively.

An interpretation of the symbol ${}_pF_q$ when $p > q+1$ has also been given by Meijer (1946) which accords completely with the one given by MacRobert's E-function. This is given by

$$2\pi i G^{m,n}_{p,q}\left(x \middle| \begin{matrix} a_1,\ldots,a_p \\ b_1,\ldots,b_q \end{matrix}\right)$$

$$= \int_{-i\infty}^{i\infty} \frac{\prod_{j=1}^{m} \Gamma(b_j-s) \prod_{j=1}^{n} \Gamma(1-a_j+s)}{\prod_{j=m+1}^{q} \Gamma(1-b_j+s) \prod_{j=n+1}^{p} \Gamma(a_j-s)}\, x^s\, ds\ . \quad (1.2.2.4)$$

See Section 5.2.1 for a more detailed description of this function.

A more general class of function is the H-function of Fox (1961). This is defined in a manner similar to the G-function, that is

$$2\pi i\, H^{m,n}_{p,q}\left(x \middle| \begin{matrix} (a_1,A_1),\ldots,(a_p,A_p) \\ (b_1,B_1),\ldots,(b_q,B_q) \end{matrix}\right) = \quad (1.2.2.5)$$

(continued)

(continued)

$$= \int_C \frac{\prod_{j=1}^{m} \Gamma(b_j - B_j s) \prod_{j=1}^{n} \Gamma(1-a_j+A_j s)}{\prod_{j=m+1}^{q} \Gamma(1-b_j+B_j s) \prod_{j=n+1}^{p} \Gamma(a_j - A_j s)} x^s \, ds, \qquad (1.2.2.5)$$

where an empty product is interpreted as unity, $0 \leq m \leq p$, $0 \leq n \leq q$, the A_j and B_j are all real and positive and the poles of the integrand of (1.2.2.5) are simple. The path of integration C is a suitable contour of Barnes type which runs from $-i\infty$ to $i\infty$, indented if necessary such that all the poles of $\Gamma(b_j - B_j s)$, $j=1,..,m$ are to the right, and those of $\Gamma(1-a_j+A_j s)$, $j=1,..,n$ to the left of the contour C. The integral (1.2.2.5) converges if

$|\arg x| < \frac{\pi}{2} D$, where $D = \sum_{j=1}^{n} A_j - \sum_{j=n+1}^{p} A_j + \sum_{j=1}^{m} B_j - \sum_{j=m+1}^{q} B_j > 0$.

1.2.3 Special Cases

A short list of special cases of the generalised hypergeometric function of one variable is now given.

The Bateman polynomials

$${}_3F_2(-n,n+1,1/2+x/2;1,1;1) = F_n(x), \qquad \text{Bateman (1933)},$$

$${}_2F_2(-n,n+1;1,1;x) = Z_n(x)$$

and $${}_1F_2(-n;u+1,v+1+u/2;z^2) = \frac{n!\,\Gamma(u+1)\Gamma(v+1+u/2)}{\Gamma(v+n+1+u/2)} z^{-u} J_n^{u,v}(z),$$

Bateman (1936),

the Rice polynomials

$${}_3F_2(-n,n+1,k;1,p;v) = H_n(k,p,v), \qquad \text{Rice (1939)},$$

the Fasenmeyer polynomials

$${}_{p+2}F_{q+2}(-n,n+1,a_1,..,a_p;1/2,1,b_1,..,b_q;x) = f_n(a_i;b_i;x),$$

Fasenmeyer (1947),

the product of two Bessel functions

$${}_2F_3(1/2+n/2+m/2,1+n/2+m/2;1+n,1+m,1+n+m;-x^2)$$

$$=\Gamma(n+1)\Gamma(m+1)(2/x)^{n+m} J_n(x)J_m(x),$$

Erdélyi et al. (1953) Vol. II page 11,

the Lommel polynomial

$$ {}_2F_3(1/2-m/2,-m/2;n,-m,1-n-m;-x^2) = \frac{(x/2)^m}{(n,m)} R_{m,n}(x), $$

Erdelyi et al. (1953) Vol. II page 35,

and the Struve function

$$ {}_1F_2(1;3/2+p,3/2;-x^2/4) = \sqrt{\pi}/2\ (2/x)^{1+p}\ \underline{H}_p(x), $$

Erdélyi et al. (1953) Vol. II page 38.

1.3 HYPERGEOMETRIC FUNCTIONS OF TWO VARIABLES

In addition to increasing the number of parameters, hypergeometric functions may be generalised along the lines of increasing the number of variables. Appell (1880) was the first author to treat hypergeometric functions of two variables on a systematic basis, and he defined the four functions which follow:

$$ F_1(a,b,b';c;x,y) = \sum_{m,n=0}^{\infty} \frac{(a,m+n)(b,m)(b',n)}{(c,m+n)m!n!} x^m y^n, \tag{1.3.1} $$

$$ F_2(a,b,b';c,c';x,y) = \sum_{m,n=0}^{\infty} \frac{(a,m+n)(b,m)(b',n)}{(c,m)(c',n)\ m!n!} x^m y^n, \tag{1.3.2} $$

$$ F_3(a,a',b,b';c;x,y) = \sum_{m,n=0}^{\infty} \frac{(a,m)(a',n)(b,m)(b',n)}{(c,m+n)\ m!n!} x^m y^n, \tag{1.3.3} $$

and

$$ F_4(a,b;c,c';x,y) = \sum_{m,n=0}^{\infty} \frac{(a,m+n)(b,m+n)}{(c,m)(c',n)m!n!} x^m y^n. \tag{1.3.4} $$

The new functions are all generalisations of the Gauss function. Appell derived them by considering, first of all the simple product of two Gauss functions

$$ {}_2F_1(a,b;c;x)\,{}_2F_1(a',b';c';y) = \sum_{m,n=0}^{\infty} \frac{(a,m)(a',n)(b,m)(b',n)}{(c,m)(c',n)\ m!n!} x^m y^n \tag{1.3.5} $$

and replacing, in turn, each pair of products (a,m)(a',n), for example, by the composite product (a,m+n), where both indices of summation m and n are involved.

In addition to the functions F_1, F_2, F_3 and F_4 given above, it would appear that we may also have the double series

$$ \sum_{m,n=0}^{\infty} \frac{(a,m+n)(b,m+n)}{(c,m+n)\ m!n!} x^m y^n. \tag{1.3.6} $$

A straightforward application of the binomial theorem, however, reduces (1.3.6) to the function ${}_2F_1(a,b;c;x+y)$; see Exton (1976) page 24.

1.3.1 The Horn Functions

Other hypergeometric functions of two variables were investigated by Horn in a long series of papers extending over the fifty year period 1889 to 1939. Here all the double hypergeometric functions of the second order and of two independent variables were systematically studied. In this work, as well as products of the type $(a,m+n)$, we also encounter the types $(a,m-m)$, $(a,2m+n)$ and $(a,2m-n)$. Typical examples of the functions listed are

$$G_2(a,a',b,b';x,y) = \sum_{m,n=0}^{\infty} \frac{(a,m)(a',n)(b,n-m)(b',m-n)}{m!n!} x^m y^n \quad (1.3.1.1)$$

and

$$H_4(a,b;c,d;x,y) = \sum_{m,n=0}^{\infty} \frac{(a,2m+n)(b,n)}{(c,m)(d,n)m!n!} x^m y^n \; . \quad (1.3.1.2)$$

A list of all these functions is given in Erdélyi et al. (1953) Vol. I page 224.

1.3.2 The Kampé de Fériet Function

It is possible to generalise the Appell functions F_1 to F_4 so as to obtain the double hypergeometric function of higher order by increasing the number of parameters in a manner similar to the generalisation of the single hypergeometric function in Section 1.2. This function was first defined and studied by Kampé de Fériet (1921) and is named after him; it has the following series representation:

$$F^{A:B;B'}_{C:D;D'}\left({(a):(b);(b'); \atop (c):(d);(d');} x,y\right) = \sum_{m,n=0}^{\infty} \frac{\prod_{j=1}^{A}(a_j,m+n) \prod_{j=1}^{B}(b_j,m) \prod_{j=1}^{B'}(b'_j,n) x^m y^n}{\prod_{j=1}^{C}(c_j,m+n) \prod_{j=1}^{D}(d_j,m) \prod_{j=1}^{D'}(d'_j,n) m!n!} . \quad (1.3.2.1)$$

The notation used on the left of (1.3.2.1) due to Burchanll end Chaundy (1941), is more compact than that used originally by Kampé de Fériet.

The Appell functions are special cases of the Kampé de Fériet function as follows:

$$F^{1:1;1}_{1:0;0} = F_1,\; F^{1:1;1}_{0:1;1} = F_2,\; F^{0:2;2}_{1:0;0} = F_3 \text{ and } F^{2:0;0}_{0:1;1} = F_4 . \quad (1.3.2.2)$$

Also certain Kampé de Fériet functions are immediately reducible to generalised hypergeometric functions of one variable, such as

$$F^{A:0;0}_{C:0;0}\left({(a):-;-;\atop (c):-;-;}x,y\right) = {}_AF_C\left({(a);\atop (c);}x+y\right), \qquad (1.3.2.3)$$

$$F^{0:B;B'}_{0:D;D'}\left({-:(b);(b');\atop -:(d);(d');}x,y\right) = {}_BF_D\left({(b)\atop (d)};x\right){}_{B'}F_{D'}\left({(b');\atop (d');}y\right), \qquad (1.3.2.4)$$

and

$$F^{A:1;1}_{C:0;0}\left({(a):b;b';\atop (c):-;-\ ;}x,x\right) = {}_{A+1}F_C\left({(a),b+b';\atop (c)\quad ;}x\right). \qquad (1.3.2.5)$$

These functions are discussed at greater length in Appell et Kampé de Fériet (1926) Chapter 9.

1.3.3 Convergence

As described in Appell et Kampé de Fériet (1926) page 396, Horn gave for the first time the general definition of hypergeometric functions of two variables in which he stated that the double power series

$$F(x,y) = \sum_{m,n=0}^{\infty} A_{m,n}x^m y^n, \qquad (1.3.3.1)$$

where the coefficients satisfy the conditions

$$\frac{A_{m+1,n}}{A_{m,n}} = \frac{P(m,n)}{R(m,n)} \quad \text{and} \quad \frac{A_{m,n+1}}{A_{m,n}} = \frac{Q(m,n)}{S(m,n)}, \qquad (1.3.3.2)$$

is of hypergeometric type; P,Q,R and S denote polynomials in the indices of summation m and n of degree p,q,r and s respectively. Apart from the compatibility relation

$$\frac{P(m,n+1)Q(m,n)}{R(m,n+1)S(m,n)} = \frac{Q(m+1,n)P(m,n)}{S(m+1,n)R(m,n)}, \qquad (1.3.3.3)$$

P,Q,R and S may be chosen arbitrarily.

In order to investigate the region of convergence of (1.3.3.1), Horn puts

$$\Phi(m,n) = \lim_{\varepsilon\to\infty}\frac{P(\varepsilon m,\varepsilon n)}{R(\varepsilon m,\varepsilon n)} \quad \text{and} \quad \Psi(m,n) = \lim_{\varepsilon\to\infty}\frac{Q(\varepsilon m,\varepsilon n)}{S(\varepsilon m,\varepsilon n)}, \qquad (1.3.3.4)$$

when it is clear that $\Phi(m,n)$ is infinite if $p > r$, identically zero if $p < r$, and is a rational function of m and n if $p = r$. Let $\xi = |x|$ and $\eta = |y|$, and denote by D the rectangle in the positive quadrant of the plane $O\xi\eta$ bounded by the coordinate axes and the straight lines parallel to the coordinate axes

$$\xi = 1/|\Phi(1,0)| \quad \text{and} \quad \eta = 1/|\Psi(0,1)|. \qquad (1.3.3.5)$$

We also take C to be the curve whose parametric equations are

$$\xi = 1/\Phi(m,n) \quad \text{and} \quad \eta = 1/\Psi(m,n). \tag{1.3.3.6}$$

In so far as the convergence of (1.3.3.1) is concerned, it now remains to consider the following five possibilities:

(i) If $p > r$ and $q > s$, the region of convergence consists only of the origin.

(ii) If $p < r$ and $q < s$, the region of convergence consists of the whole positive quadrant.

(iii) If $p < r$ and $q = s$, the region of convergence consists of the strip between the axis Ox and the straight line $\eta = 1/|\Psi(0,1)|$.

(iv) If $p = r$ and $q < s$, the region of convergence consists of the strip between the axis Oy and the straight line $\xi = 1/|\Phi(1,0)|$.

(v) If $p = r$ and $q = s$, (1.3.3.1) converges in that region of the plane $O\xi\eta$ common to D and C and which contains the origin O. This last case is the most interesting in that it deals with the complete (non-confluent) double hypergeometric functions.

The general theory outlined above is now applied to the Kampé de Fériet function. For this, with reference to the definition (1.3.2.1), suppose, for convenience, that $B'=B$ and $D'=D$. It is evident that if $A+B < C+D+1$, then $p < r$ and $q < s$, and so the series converges for all finite values of the variables x and y. If $A+B > C+D+1$, then $p > r$ and $q > s$ and the region of convergence reduces to the origin only in the Oxy plane.

The greatest interest attaches to those functions where $p=q=r=s$, or $A+B=C+D+1$, and then

$$\Phi(m,n) = (m+n)^{A-C}\, m^{B-D-1} \quad \text{and} \quad \Psi(m,n) = (m+n)^{A-C}\, n^{B-D-1},$$

from which, as before, D is the unit square with one corner coinciding with the origin and with two sides lying along the positive coordinate axes.

Three different possibilities now arise:

(i) $A=C$, giving $\Phi(m,n) = \Psi(m,n) = 1$, and the region of convergence is $|x| < 1$ and $|y| < 1$.

(ii) $A-C= -k < 0$, when $\Phi(m,n) = m^k/(m+n)^k$ and $\Psi(m,n)=n^k/(m+n)^k$, and the Cartesian equation of C is $\xi^{-1/k}+\eta^{-1/k} = 1$. Thus C lies entirely outside of D, so that the series again converges for $|x| < 1$ and $|y| < 1$.

(iii) Finally, $A-C = k > 0$, when $\Phi(m,n) = (m+n)^k\, m^{-k}$ and $\Psi(m,n) = (m+n)^k\, n^{-k}$, and the region of convergence is now such that $|x^{1/k}| + |y^{1/k}| < 1$.

1.3.4 Special Cases

Although hypergeometric functions of two variables occur in a number of applications, the situations where they have been presented in terms of functions with individual notation occur comparatively rarely. We now give a few cases.

The Appell polynomials,

$$F_2(c+c'-a-m-n,c+m,c'+n;c,c';x,y) = (1-x-y)^{a-c-c'}F_{m,n}(a,c,c';x,y)$$

and

$$F_2(a+m+n,-m,-n;c,c';x,y) = E_{m,n}(a,c,c';x,y),$$

Appell et Kampé de Fériet (1926) Chapter 6,

the incomplete elliptical integral of the second kind,

$$F_1(1/2,1/2,1/2;3/2;\sin^2\phi,k^2\sin^2\phi) = \text{Cosec}\phi\ F(\phi|k)$$

Carlson (1961),

and the random flight probability function in two dimensions,

$$F_4(1-p/4,1-3p/4;p/2,p/2;a_1^2/r^2,a_2^2/r^2)$$

$$= \frac{\Gamma(p/2)\Gamma(3/2-p/4)\Gamma(3p/4-1)2(1-p)}{r^{n-2}(a_1a_2)^{1-p/2}\Gamma(2-p/2)}P_2(r;a_1,a_2|p),$$

Watson (1944) page 421.

1.4 MULTIPLE HYPERGEOMETRIC FUNCTIONS

The process of generalising the Gauss function and its confluent forms by both increasing the number of parameters and also increasing the number of variables may be carried on to any desired extent. To this end, Srivastava and Daoust (1969) have given the multiple series

$$\sum \frac{\prod_{j=1}^{A}\Gamma[a_j+\sum_{i=1}^{n}m_i\theta_j^{(i)}]\prod_{j=1}^{B'}\Gamma[b_j^{(1)}+m_1\phi_j^{(1)}]..\prod_{j=1}^{B^{(n)}}\Gamma[b_j^{(n)}+m_n\phi_j^{(n)}]x_1^{m_1}..x_n^{m_n}}{\prod_{j=1}^{C}\Gamma[c_j+\sum_{i=1}^{n}m_i\psi_j^{(i)}]\prod_{j=1}^{D'}\Gamma[d_j^{(1)}+m_1\delta_j^{(1)}]..\prod_{j=1}^{D^{(n)}}\Gamma[d_j^{(n)}+m_n\delta_j^{(n)}]m_1!..m_n!} \tag{1.4.1}$$

Here and in what follows, it is taken that all indices of summation run from zero to infinity unless otherwise indicated. This extremely general multiple hypergeometric series is denoted by either of the two symbols which follow:

$$S^{A:B';..;B^{(n)}}_{C:D';..;D^{(n)}}\begin{bmatrix} x_1 \\ \vdots \\ x_n \end{bmatrix}$$

$$= S^{A:B';..;B^{(n)}}_{C:D';..;D^{(n)}}\begin{bmatrix} [(a):\theta',..,\theta^{(n)}]:[(b'):\phi'];.;[(b^{(n)}):\phi^{(n)}]; & \\ & x_1,.,x_n \\ [(c):\psi',..,\psi^{(n)}]:[(d'):\delta'];.;[(d^{(n)}):\delta^{(n)}]: & \end{bmatrix}.$$

The alternative notation is

$$\frac{\prod_{j=1}^{A}\Gamma(a_j)\prod_{j=1}^{B'}\Gamma(b'_j)..\prod_{j=1}^{B^{(n)}}\Gamma(b_j^{(n)})}{\prod_{j=1}^{C}\Gamma(c_j)\prod_{j=1}^{D'}\Gamma(d'_j)..\prod_{j=1}^{D^{(n)}}\Gamma(d_j^{(n)})}$$

$$\times F^{A:B';.;B^{(n)}}_{C:D';.;D^{(n)}}\begin{bmatrix} [(a):\theta',..,\theta^{(n)}]:[(b'):\phi'];.;[(b^{(n)}):\phi^{(n)}]; & \\ & x_1,.,x_n \\ [(c):\psi',..,\psi^{(n)}]:[(d'):\delta'];.;[(d^{(n)}):\delta^{(n)}]; & \end{bmatrix},$$

where the θ's,ϕ's,ψ's and δ's are positive constants or zero. It is clear that, if these positive constants are all taken to equal unity, then, for example, $F^{1:1;.;1}_{0:1;.;1}$ corresponds to the Lauricella function $F_A^{(n)}$, see (1.4.1.1) below.

While (1.4.1) is certainly a useful generalisation, it seems that for some purposes, multiple hypergeometric functions which are of a less general nature are of more immediate value. To this end, we consider the generalised Kampé de Fériet function first given by Karlsson (1973) and defined as

$$F^{A:B}_{C:D}\begin{bmatrix} (a):(b_1);.;(b_n); & \\ & x_1,.,x_n \\ (c):(d_1);.;(d_n); & \end{bmatrix}$$

$$= \sum\frac{((a),m_1+.+m_n)((b_1),m_1)..((b_n),m_n)x_1^{m_1}..x_n^{m_n}}{((c),m_1+.+m_n)((d_1),m_1)..((d_n),m_n)m_1!..m_n!}\ . \qquad (1.4.2)$$

A more general form of this function is sometimes employed:

$$F^{A:B';.;B^{(n)}}_{C:D';.;D^{(n)}}\begin{bmatrix} (a):(b');..;(b^{(n)}); & \\ & x_1,..,x_n \\ (c):(d');..;(d^{(n)}); & \end{bmatrix}$$

$$= \sum\frac{((a),m_1+.+m_n)((b'),m_1)..((b^{(n)}),m_n)x_1^{m_1}..x_n^{m_n}}{((c),m_1+.+m_n)((d'),m_1)..((d^{(n)}),m_n)m_1!..m_n!}\ . \qquad (1.4.3)$$

1.4.1 The Lauricella Functions and their Confluent Forms

By far the most important hypergeometric functions of several variables are the Lauricella functions:

$$F_A^{(n)}(a,b_1,..,b_n;c_1,..,c_n;x_1,..,x_n)$$

$$= \sum \frac{(a,m_1+.+m_n)(b_1,m_1)..(b_n,m_n)x_1^{m_1}..x_n^{m_n}}{(c_1,m_1)..(c_n,m_n)\quad m_1!..m_n!}, \qquad (1.4.1.1)$$

$$F_B^{(n)}(a_1,..,a_n,b_1,..,b_n;c;x_1,..,x_n)$$

$$= \sum \frac{(a_1,m_1)..(a_n,m_n)(b_1,m_1)..(b_n,m_n)x_1^{m_1}..x_n^{m_n}}{(c,m_1+.+m_n)\quad m_1!..m_n!}, \qquad (1.4.1.2)$$

$$F_C^{(n)}(a,b;c_1,..,c_n;x_1,..,x_n)$$

$$= \sum \frac{(a,m_1+.+m_n)(b,m_1+.+m_n)x_1^{m_1}..x_n^{m_n}}{(c_1,m_1)...(c_n,m_n)\quad m_1!..m_n!} \qquad (1.4.1.3)$$

and

$$F_D^{(n)}(a,b_1,..,b_n;c;x_1,..,x_n)$$

$$= \sum \frac{(a,m_1+.+m_n)(b_1,m_1)..(b_n,m_n)x_1^{m_1}..x_n^{m_n}}{(c,m_1+.+m_n)\quad m_1!..m_n!} \qquad (1.4.1.4)$$

These four functions were first defined and studied by Lauricella (1893). If n, the number of variables, is made equal to two, these four functions reduce to the Appell functions F_2, F_3, F_4 and F_1 respectively: and if n=1, all four functions become the Gauss function ${}_2F_1$.

By means of appropriate limiting processes, a large number of possible confluent forms of the Lauricella functions arise. The most important of these are

$$\Phi_2^{(n)}(b_1,..,b_n;c;x_1,..,x_n) = \sum \frac{(b_1,m_1)..(b_n,m_n)x_1^{m_1}..x_n^{m_n}}{(c,m_1+.+m_n)\quad m_1!..m_n!}, \qquad (1.4.1.5)$$

$$\Psi_2^{(n)}(b;c_1,..,c_n;x_1,..,x_n) = \sum \frac{(b,m_1+.+m_n)\quad x_1^{m_1}..x_n^{m_n}}{(c_1,m_1)..(c_n,m_n)m_1!..m_n!} \qquad (1.4.1.6)$$

See Exton (1976) Chapter 2.

and $\Phi_D^{(n)}(a,b_1,..,b_{n-1},-;c;x_1,..,x_n)$

$$= \sum \frac{(a,m_1+.+m_n)(b_1,m_1)..(b_{n-1},m_{n-1})x_1^{m_1}..x_n^{m_n}}{(c,m_1+.+m_n)\quad m_1!..m_n!}. \qquad (1.4.1.7)$$

1.4.2 Convergence

By means of a straightforward generalisation of Horn's general theory of convergence as outlined in Section 1.3.3, the series representations of the Lauricella functions are found to be convergent within the following regions:

$$F_A^{(n)} \qquad |x_1|+.+|x_n| < 1,$$

$$F_B^{(n)} \qquad |x_1|,.,|x_n| < 1,$$

$$F_C^{(n)} \qquad |\sqrt{x_1}|+.+|\sqrt{x_n}| < 1$$

and $$F_D^{(n)} \qquad |x_1|,.,|x_n| < 1.$$

The series representations of the functions $\Phi_2^{(n)}$ and $\Psi_2^{(n)}$ converge for all finite values of their variables, and in the case of $\Phi_D^{(n)}$, we have convergence when $|x_1|,.,|x_{n-1}| < 1$, and x_n may assume any finite value.

1.4.3 Systems of Partial Differential Equations

We have seen in Section 1.1.3 that the Gauss hypergeometric function of one variable satisfies a certain differential equation. In a similar way, the four Lauricella functions of several variables defined in Section 1.4.1 are particular solutions of certain systems of partial differential equations. There are four of these systems, one associated with each of the four functions $F_A^{(n)}$, $F_B^{(n)}$, $F_C^{(n)}$ and $F_D^{(n)}$ respectively.

These four systems are particularly important in that they enable these four functions to be defined for all possible values of the independent variables $x_1,.,x_n$, real or complex. The need for a means of carrying out such a complete definition of the four functions in question is the prime motivation of the investigation into the general integration of the four systems. This also has an important bearing upon the application of the Lauricella functions. In fact, the above remarks may be applied to all hypergeometric functions of one or more variables, but we confine ourselves to the Lauricella functions here.

These systems of partial differential equations are:-

for the function $F_A^{(n)}$, $$x_j(1-x_j)\frac{\partial^2 F}{\partial x_j^2} - x_j\sum_{\substack{k=1\\k\neq j}}^{n} x_k \frac{\partial^2 F}{\partial x_k \partial x_j} + \qquad (1.4.3.1)$$

(continued)

(continued)

$$+ [c_j-(a+b_j+1)x_j]\frac{\partial F}{\partial x_j} - b_j\sum_{\substack{k=1\\k\neq j}}^{n} x_k\frac{\partial F}{\partial x_k} - ab_jF = 0, \qquad (1.4.3.1)$$

for the function $F_B^{(n)}$, $x_j(1-x_j)\frac{\partial F}{\partial x_j^2} + \sum_{\substack{k=1\\k\neq j}}^{n} x_k\frac{\partial^2 F}{\partial x_k\partial x_j}$

$$+ [c-(a_j+b_j+1)x_j]\frac{\partial F}{\partial x_j} - a_jb_jF = 0, \qquad (1.4.3.2)$$

for the function $F_C^{(n)}$, $x_j(1-x_j)\frac{\partial^2 F}{\partial x_j^2} - \sum_{r=1}^{n} x_r \sum_{s=1}^{n} x_s\frac{\partial^2 F}{\partial x_r\partial x_s}$ ($s\neq j$ when $r=j$)

$$+ [c_j-(a+b+1)x_j]\frac{\partial F}{\partial x_j} - (a+b+1)\sum_{\substack{k=1\\k\neq j}}^{n} x_k\frac{\partial F}{\partial x_k} - abF = 0 \qquad (1.4.3.3)$$

and for the function $F_D^{(n)}$, $x_j(1-x_j)\frac{\partial^2 F}{\partial x_j^2} + (1-x_j)\sum_{\substack{k=1\\k\neq j}}^{n} x_k\frac{\partial^2 F}{\partial x_k\partial x_j}$

$$+ [c-(a+b_j+1)x_j]\frac{\partial F}{\partial x_j} - b_j\sum_{\substack{k=1\\k\neq j}}^{n} x_k\frac{\partial F}{\partial x_k} - ab_jF = 0. \qquad (1.4.3.4)$$

In the above four systems, $j = 1,.,n$.

It has been shown by Lauricella (1893), that the general integrals of the systems satisfied by the functions $F_A^{(n)}$, $F_B^{(n)}$ and $F_C^{(n)}$ depend linearly upon 2^n arbitrary constants, while that satisfied by $F_D^{(n)}$ depends linearly upon only $(n+1)$ arbitrary constants.For discussions of these partial differential systems in more detail, the reader is referred to Lauricella (1893), Appell et Kampé de Fériet (1926) page 117,Erdélyi (1950) and Exton (1976) Chapter 5.

The rest of this book is devoted to a study of various types of integrals of hypergeometric functions. These integrals are of importance because of the fact that a large number of special functions of applied mathematics are of hypergeometric form, and integrals which involve special functions are of frequent occurrence in such fields as mathematical physics and statistics. A few examples are given in Chapter 7.

Chapter 2

Integrals of Euler Type

2.1 GENERAL EULER INTEGRALS

Euler integrals constitute an important class of finite integrals and the general integral of this type may be written in the form

$$\int_0^1 x^{a-1}(1-u)^{b-1}\ f(x)\ dx. \qquad (2.1.1)$$

If the function f(x) in the integrand is capable of expansion in a power series such as

$$f(x) = \sum_{n=0}^{\infty} c_n x^n, \qquad (2.1.2)$$

then, provided that the radius of convergence of (2.1.2) is not greater than unity, we have

$$\int_0^1 x^{a-1}(1-x)^{b-1}\ f(x)\ dx = \sum_{n=0}^{\infty} c_n \int_0^1 x^{a+n-1}(1-x)^{b-1}\ dx. \qquad (2.1.3)$$

The inner integral on the right of (2.1.3) may be evaluated as a beta function [Erdélyi et al. (1953) Vol. I page 9]

$$B(a+n,b), \qquad (2.1.4)$$

provided that Re(a) and Re(b) are both positive.

The Beta function (2.1.4) may be written as

$$\frac{\Gamma(a+n)\Gamma(b)}{\Gamma(a+b+n)} = \frac{\Gamma(a)\Gamma(b)}{\Gamma(a+b)} \cdot \frac{(a,n)}{(a+b,n)}, \qquad (2.1.5)$$

so that

$$\int_0^1 x^{a-1}(1-x)^{b-1} f(x)\ dx = \frac{\Gamma(a)\Gamma(b)}{\Gamma(a+b)} \sum_{n=0}^{\infty} \frac{(a,n)c_n}{(a+b,n)}. \qquad (2.1.6)$$

This general result may easily be extended to cover multiple series representations of f(x), and has very many special cases involving special functions.

2.1.1 The Gauss Function and its Confluent Forms

If f(x) is of the form of a Gauss function ${}_2F_1$, then

$$f(x) = {}_2F_1(c,d;f;sx^k) = \sum_{n=0}^{\infty} \frac{(c,n)(d,n)}{(f,n)n!}(sx^k)^n. \qquad (2.1.1.1)$$

Hence, $\int_0^1 x^{a-1}(1-x)^{b-1}{}_2F_1\left({c,d; \atop f\;;}sx^k\right)dx$

$$= \frac{\Gamma(a)\Gamma(b)}{\Gamma(a+b)}\;{}_{2+k}F_{1+k}\left[{c,d,a/k,(a+1)/k,..,(a+k-1)/k\;; \atop f,(a+b)/k,(a+b+1)/k,..,(a+b+k-1)/k;}s\right], \tag{2.1.1.2}$$

where k is a positive integer and Re(a) and Re(b) > 0.

This result may be expressed in closed form when the hypergeometric function on the right is summable. If we put c=f, the Gauss function takes the form of a binomial function giving the result

$$\int_0^1 x^{a-1}(1-x)^{b-1}(1-sx^k)^{-d}\,dx = \frac{\Gamma(a)\Gamma(b)}{\Gamma(a+b)}\;{}_{k+1}F_k\left[{d,\frac{a}{k},..,\frac{a+k-1}{k}\;; \atop \frac{a+b}{k},..,\frac{a+b+k-1}{k};}\;s\right]$$

Re(a), Re(b) > 0. (2.1.1.3)

If, further, we put k=1, we obtain the well-known Euler integral formula for the Gauss function, see (1.1.3.5),

$$\int_0^1 x^{a-1}(1-x)^{b-1}(1-sx)^{-d}\,dx = \frac{\Gamma(a)\Gamma(b)}{\Gamma(a+b)}\;{}_2F_1\left({d,a; \atop a+b;}s\right), \tag{2.1.1.4}$$

Re(a), Re(b) > 0.

A number of special cases of the ${}_{2+k}F_{1+k}$ function on the right of (2.1.1.2) may be expressed in closed form using the various summation theorems for the generalised single hypergeometric function. See Slater (1966) Appendix III. We now give a few examples:

$$\int_0^1 x^{a-1}(1-x)^{b-1}{}_2F_1(c,d;f;sx)dx = \frac{\Gamma(a)\Gamma(b)}{\Gamma(a+b)}\;{}_3F_2\left({c,d,a; \atop f,a+b;}s\right). \tag{2.1.1.5}$$

The Clausen function of the previous expression may be reduced in a number of cases, some of which are (i) a=f, b=d-f, (ii) a=f, s=1 and (iii) a=-n, b=d-f+1-n, s=1. In case (i), the right-hand member of (2.1.1.5) reduces to a binomial function, and we have

$$\int_0^1 x^{f-1}(1-x)^{d-f-1}{}_2F_1(c,d;f;sx)\,dx = \frac{\Gamma(f)\Gamma(d-f)}{\Gamma(d)}(1-s)^{-f} \tag{2.1.1.6}$$

Re(f), Re(d-f) > 0, $|s| < 1$.

In case (ii), the integral concerned may be evaluated by applying Gauss's summation theorem [Slater (1966) Appendix III], and we have

$$\int_0^1 x^{f-1}(1-x)^{b-1}{}_2F_1\left({c,d; \atop f\;;}x\right)dx = \frac{[\Gamma(f)]^2\Gamma(b)\Gamma(f-c-d)}{\Gamma(f+b)\Gamma(f-c)\Gamma(f-d)}, \tag{2.1.1.7}$$

Re(f), Re(b), Re(f-c-d) > 0.

Saalschütz's theorem is now applied to the hypergeometric function on the right of (2.1.1.5), when case (iii) yields the result

$$\int_0^1 x^{a-1}(1-x)^{d-f+1-n}{}_2F_1(c,d;f;x)\,dx = \quad\text{(cont.)} \tag{2.1.1.8}$$

(continued)

$$= \frac{\Gamma(a)\Gamma(d-f+1-n)\Gamma(f-d+n)\Gamma(f-a+n)\Gamma(f)\Gamma(f-d-a)}{\Gamma(a+d-f+1-n)\Gamma(f-d)\Gamma(f-a)\Gamma(f+n)\Gamma(f-d-a+n)}, \qquad (2.1.1.8)$$

$$\mathrm{Re}(a),\ \mathrm{Re}(d-f+1-n) > 0.$$

By applying the same method as was employed in deducing (2.1.1.2), we have the formulae

$$\int_0^1 x^{a-1}(1-x)^{b-1}\,{}_1F_1\left(\begin{matrix}d;\\f;\end{matrix}sx^k\right)dx = \frac{\Gamma(a)\Gamma(b)}{\Gamma(a+b)}\,{}_{1+k}F_{1+k}\left[\begin{matrix}d,\frac{a}{k},..,\frac{a+k-1}{k} & ;\\ f,\frac{a+b}{k},..,\frac{a+b+k-1}{k}; & \end{matrix} s\right] \qquad (2.1.1.9)$$

and

$$\int_0^1 x^{a-1}(1-x)^{b-1}\,{}_0F_1(-;f;sx^k)dx = \frac{\Gamma(a)\Gamma(b)}{\Gamma(a+b)}\,{}_kF_{k+1}\left[\begin{matrix}\frac{a}{k},...,\frac{a+k-1}{k} & ;\\ f,\frac{a+b}{k},..,\frac{a+b+k-1}{k}; & \end{matrix} s\right], \qquad (2.1.1.10)$$

where, as before, Re(a) and Re(b) > 0. If we put k=1,a=f and b=d-f, (2.1.1.9) reduces to

$$\int_0^1 x^{f-1}(1-x)^{d-f-1}\,{}_1F_1(d;f;sx)dx = \frac{\Gamma(f)\Gamma(f-a)}{\Gamma(d)}\,e^s, \qquad (2.1.1.11)$$

$$\mathrm{Re}(f),\mathrm{Re}(d-f) > 0.$$

2.1.2 Generalised Hypergeometric Function

A straightforward generalisation of (2.1.1.4) may be written

$$\int_0^1 x^{a-1}(1-x)^{b-1}\,{}_CF_D[(c);(d);sx^k]dx. \qquad (2.1.2.1)$$

This integral may be evaluated using the same technique as was employed in the previous section, and it becomes

$$\frac{\Gamma(a)\Gamma(b)}{\Gamma(a+b)}\,{}_{C+k}F_{D+k}\left[\begin{matrix}(c),a/k\ ,...,\ (a+k-1)/k & ;\\ (d),(a+b)/k,..,(a+b+k-1)/k; & \end{matrix} s\right], \qquad (2.1.2.2)$$

provided that $p \leqq q+1$, k is a positive integer and that Re(a) and Re(b) > 0, in order to ensure convergence.

Many summation theorems may be employed to sum the ${}_{C+k}F_{D+k}$ function on the right of (2.1.2.2). Two examples are now given using the binomial theorem and the summation theorem for a well-poised ${}_5F_4$ series of unit argument [see Slater (1966) App. III]:

$${}_5F_4\left[\begin{matrix}a,1+\frac{a}{2}, & b,\ c,\ d & ;\\ \frac{a}{2},1+a-b,1+a-c,1+a-d; & & \end{matrix} 1\right] = \frac{\Gamma(1+a-b)\Gamma(1+a-c)\Gamma(1+a-d)\Gamma(1+a-b-c-d)}{\Gamma(1+a)\Gamma(1+a-b-c)\Gamma(1+a-b-d)\Gamma(1+a-c-d)},$$

$$\mathrm{Re}(b+c+d-a-1) > 0. \qquad (2.1.2.3)$$

Firstly, we have

$$\int_0^1 x^{a-1}(1-x)^{b-1}\,{}_{k+1}F_k\left[\begin{matrix} c,\frac{a+b}{k},..,\frac{a+b+k-1}{k}; \\ \frac{a}{k},..,\frac{a+k-1}{k} \quad ; \end{matrix} sx^k\right]dx = \frac{\Gamma(a)\Gamma(b)}{\Gamma(a+b)}(1-s)^{-a}, \tag{2.1.2.4}$$

where Re(a), Re(b) > 0 and $|s| < 1$.

Secondly, if the parameters of (2.1.2.2) are suitably specialised, (2.1.2.3) can be applied, and we have

$$\int_0^1 x^{d-1}(1-x)^{a-2d}\,{}_4F_3\left[\begin{matrix} a,1+\frac{a}{2}, \ b, \ \ c \ ; \\ \frac{a}{2},1+a-b,1+a-c; \end{matrix} x\right]dx$$

$$= \frac{\Gamma(d)\Gamma(1+a-2d)\Gamma(1+a-b)\Gamma(1+a-c)\Gamma(1+a-b-c-d)}{\Gamma(1+a)\Gamma(1+a-b-c)\Gamma(1+a-b-d)\Gamma(1+a-c-d)}, \tag{2.1.2.5}$$

provided that Re(d),Re(a-2d) > -1 and Re(b+c+d-a) > -1. If either b or c is a negative integer, the third condition of convergence of (2.1.2.5) may be relaxed because the hypergeometric functions involved are terminating.

2.1.3 Double Hypergeometric Function

We now consider integrals of the form

$$\int_0^1 x^{a-1}(1-x)^{b-1}\,F^{C:D;D'}_{F:G;G'}\left(\begin{matrix}(c):(d);(d');\\(f):(g);(g');\end{matrix} rx^k,sx^k\right)dx, \tag{2.1.3.1}$$

$$\int_0^1 x^{a-1}(1-x)^{b-1}\,F^{C:D;D'}_{F:G;G'}\left(\begin{matrix}(c):(d);(d');\\(f):(g);(g');\end{matrix} r,sx^k\right)dx \tag{2.1.3.2}$$

and
$$\int_0^1 x^{a-1}(1-x)^{b-1}\,F^{C:D;D'}_{F:G;G'}\left(\begin{matrix}(c):(d);(d');\\(f):(g);(g');\end{matrix} rx^k,s[1-x]^k\right)dx. \tag{2.1.1.3}$$

If the Kampé de Fériet function inside the integrand of (2.1.3.1) is expanded as a double series, then this integral may be expressed in the form

$$\sum\frac{((c),m+n)((d),m)((d'),n)r^m s^n}{((f),m+n)((g),m)((g'),n)m!n!}\int_0^1 x^{a+k(m+n)-1}(1-x)^{b-1}dx. \tag{2.1.3.4}$$

For the process of interchanging the operations of double summation and integration in the previous expression to be justified, the Kampé de Fériet function concerned must converge for $|x|\leq 1$. The inner integral of (2.1.3.4) may be written in the form of a beta function, so that (2.1.3.1) becomes

$$\frac{\Gamma(a)\Gamma(b)}{\Gamma(a+b)}F^{C+k:D;D'}_{F+k:G;G'}\left[\begin{matrix}(c),\frac{a}{k},...,\frac{a+k-1}{k} \quad :(d);(d');\\(f),\frac{a+b}{k},..,\frac{a+b+k-1}{k}:(g);(g');\end{matrix} r,s\right], \tag{2.1.3.5}$$

where Re(a), Re(b) > 0, k is a positive integer and all the series concerned are either convergent or terminating.

If we treat the integrals (2.1.3.2) and (2.1.3.3) similarly, they may be written, respectively, as

$$\frac{\Gamma(a)\Gamma(b)}{\Gamma(a+b)} F^{C:D;D'+k}_{F:G;G'+k}\left[\begin{matrix}(c):(d);(d'),\frac{a}{k},\ldots,\frac{a+k-1}{k};\\(f):(g);(g'),\frac{a+b}{k},..,\frac{a+b+k-1}{k};\end{matrix}r,s\right] \qquad (2.1.3.6)$$

$$\frac{\Gamma(a)\Gamma(b)}{\Gamma(a+b)} F^{C:D+k;D'+k}_{F+k:G;G'}\left[\begin{matrix}(c)&:(d),\frac{a}{k},..,\frac{a+k-1}{k};&(d'),\frac{b}{k},..,\frac{b+k-1}{k};\\(f),\frac{a+b}{k},..,\frac{a+b+k-1}{k}:&(g);&(g');\end{matrix}r,s\right].$$

(2.1.3.7)

A number of simpler forms of (2.1.3.1) to (2.1.3.3) may be deduced. Suppose that we let $C=F=k$, $c_1=(a+b)/k,..,c_k=(a+b+k-1)/k$ and $f_1=a/k,..,f_k=(a+k-1)/k$ in the first of these expressions. The function (2.1.3.5) becomes

$$\frac{\Gamma(a)\Gamma(b)}{\Gamma(a+b)} F^{0:D;D'}_{0:G;G'}\left(\begin{matrix}-:(d);(d');\\-:(g);(g');\end{matrix}r,s\right), \qquad (2.1.3.8)$$

which splits up into the product of a pair of hypergeometric functions of one variable giving the result

$$\int_0^1 x^{a-1}(1-x)^{b-1} F^{k:D;D'}_{k:G;G'}\left[\begin{matrix}\frac{a+b}{k},..,\frac{a+b+k-1}{k}:(d);(d');\\\frac{a}{k},\ldots,\frac{a+k-1}{k}:(g);(g');\end{matrix}rx^k,sx^k\right]dx$$

$$= \frac{\Gamma(a)\Gamma(b)}{\Gamma(a+b)} {}_DF_G\left(\begin{matrix}(d);\\(g);\end{matrix}r\right) {}_{D'}F_{G'}\left(\begin{matrix}(d');\\(g');\end{matrix}s\right). \qquad (2.1.3.9)$$

In (2.1.3.6), put $r=s$, $D=1, G=0, D =k+1, D'=k$ and obtain the formula

$$\int_0^1 x^{a-1}(1-x)^{b-1} F^{C:1;k+1}_{F:0;\ k}\left[\begin{matrix}(c):d;d',\frac{a+b}{k},..,\frac{a+b+k-1}{k};\\(f):-;\ \frac{a}{k},\ldots,\frac{a+k-1}{k}\ ;\end{matrix}s,sx^k\right]dx$$

$$= \frac{\Gamma(a)\Gamma(b)}{\Gamma(a+b)} {}_{C+1}F_G((c),d+d';(g);s) \qquad (2.1.3.10)$$

Compare Exton (1976) page 132.

The expression (2.1.3.7) gives

$$\int_0^1 x^{a-1}(1-x)^{b-1} F^{C:0;0}_{F:k;k}\left[\begin{matrix}(c):&-&;&-&;\\(f):\frac{a}{k},..,\frac{a+k-1}{k};&\frac{b}{k},..,\frac{b+k-1}{k};\end{matrix}rx^k,s(1-x)^k\right]dx$$

$$= \frac{\Gamma(a)\Gamma(b)}{\Gamma(a+b)} {}_CF_{G+k}\left((c);(f),\frac{a+b}{k},..,\frac{a+b+k-1}{k};r+s\right). \qquad (2.1.3.11)$$

Results of the type (2.1.3.9) to (2.1.3.11) may often be expressed in closed form if the parameters and variables are suitably specialised; the various summation theorems of the generalised hypergeometric functions of one variable are employed. See Slater (1966) Appendix III.

2.1.4 Multiple Hypergeometric Function

In this section, Euler integrals of the generalised Kampé de Fériet function [Karlsson (1973)] are considered. The following integrals will be evaluated:

$$\int_0^1 x^{a-1}(1-x)^{b-1}\, F^{C:D}_{F:G}\left[\begin{matrix}(c):(d');.;(d^{(n)});\\(f):(g');.;(g^{(n)});\end{matrix}\, s_1x^k,.,s_nx^k\right] dx, \quad (2.1.4.1)$$

$$\int_0^1 x^{a-1}(1-x)^{b-1}\, F^{C:D}_{F:G}\left[\begin{matrix}(c):(d');.;(d^{(n)});\\(f):(g');.;(g^{(n)});\end{matrix}\, s_1x^k[1-x]^k,.,s_nx^k[1-x]^k\right] dx \quad (2.1.4.2)$$

and $$\int_0^1 x^{a-1}(1-x)^{b-1}\, F^{C:D}_{F:G}\left[\begin{matrix}(c):(d');.;(d^{(n)});\\(f):(g');.;(g^{(n)});\end{matrix}\, s_1,.,s_{n-1},s_nx^k\right] dx. \quad (2.1.4.3)$$

As in the previous sections of this chapter, k is a positive integer and the real parts of a and b are both taken to be positive. The integrals (2.1.4.1) and (2.1.4.3) are straightforward generalisations of (2.1.3.5) and (2.1.3.6) respectively, and so we may state immediately that (2.1.4.1) and (2.1.4.3) are equal to, respectively

$$\frac{\Gamma(a)\Gamma(b)}{\Gamma(a+b)}\, F^{C+k:D}_{F+k:G}\left[\begin{matrix}(c),\frac{a}{k},\dots,\frac{a+b-1}{k}:(d');.;(d^{(n)});\\(f),\frac{a+b}{k},.,\frac{a+b+k-1}{k}:(g');.;(g^{(n)});\end{matrix}\, s_1,.,s_n\right] \quad (2.1.4.4)$$

and

$$\frac{\Gamma(a)\Gamma(b)}{\Gamma(a+b)}\, F^{C:D+k}_{F:G+k}\left[\begin{matrix}(c):(d'),h,.,h;.;(d^{(n-1)}),h,.,h;\\(f):(g'),h,.,h;.;(g^{(n-1)}),h,.,h;\end{matrix}\right.$$
$$\left.\begin{matrix}(d^{(n)}),\frac{a}{k},\dots,\frac{a+k-1}{k};\\(g^{(n)}),\frac{a+b}{k},.,\frac{a+b+k-1}{k};\end{matrix}\, s_1,.,s_k\right]. \quad (2.1.4.5)$$

In the expression (2.1.4.5), the dummy parameters have been introduced in order to preserve the compact notation of the generalised Kampé de Fériet function.

The integral (2.1.4.2) is slightly different from (2.1.4.1) and (2.1.4.2), and will be discussed in a little more detail. As in the previous section, it is assumed that the multiple series are either convergent over the appropriate ranges of their variables or that they are terminating. The multiple hypergeometric function of the integrand is expanded in series of its arguments and (2.1.4.2) may then be written

$$\sum\frac{((c),m_1+.+m_n)((d'),m_1)..((d^{(n)}),m_n)s_1^{m_1}..s_n^{m_n}}{((f),m_1+.+m_n)((g'),m_1)..((g^{(n)}),m_n)m_1!..m_n!}$$

$$\times\int_0^1 x^{a+km_1+.+km_n-1}(1-x)^{b+km_1+.+km_n-1}\,dx. \qquad (2.1.4.6)$$

Once again, we have a beta integral involved in the expression under consideration. This integral may be written in the form

$$\frac{\Gamma(a)\Gamma(b)}{\Gamma(a+b)}\ \frac{(a,km_1+.+km_n)(b,km_1+.+km_n)}{(\frac{a+b}{2},km_1+.+km_n)(\frac{a+b+1}{2},km_1+.+km_n)}\,4^{-km_1-.-km_n}, \qquad (2.1.4.7)$$

provided that Re(a) and Re(b) are both positive. The required evaluation then follows:

$$\frac{\Gamma(a)\Gamma(b)}{\Gamma(a+b)}F^{C+2k:D}_{F+2k:G}\left[\begin{matrix}(c),\ \frac{a}{k},\ldots,\frac{a+k-1}{k}:(d');.;(d^{(n)});\\(f),\frac{a+b}{k},..,\frac{a+b+k-1}{k}:(g');.;(g^{(n)});\end{matrix}\ \frac{s_1}{4^k},..,\frac{s_n}{4^k}\right].$$

$$(2.1.4.8)$$

2.1.5 Special Cases

A large number of Euler integrals involving special functions may be deduced from the expressions given in Sections 2.1.1 to 2.1.4. Certain representative examples are now obtained.

Replace the Gauss function of the integrand of (2.1.1.5) by the corresponding Jacobi polynomial and the following result is obtained:

$$\int_0^1 x^{a-1}(1-x)^{b-1}\,P_n^{c,d}(1-2xy)\,dx = \frac{\Gamma(n+c+1)\Gamma(a)\Gamma(b)}{n!\Gamma(c+1)\Gamma(a+b)}\,{}_3F_2\left(\begin{matrix}-n,n+c+d+1,a;\\c+1,a+b\ \ ;\end{matrix}y\right)$$

$Re(a),\ Re(b) > 0.$ (2.1.5.1)

From (2.1.1.6), we obtain a result involving a Legendre polynomial

$$\int_0^1(1-x)^{n+1}\,P_n(1-2xy)\,dx = n^{-1}(1-y)^{-n}\ . \qquad (2.1.5.2)$$

Formula (2.1.1.10) may be expressed in terms of an integral of a Bessel function, when it takes the form

$$\int_0^1 x^{a-1}(1-x)^{b-1}\,J_c(sx)\,dx = \frac{2^{-c}\Gamma(a+c)\Gamma(b)}{\Gamma(c+1)\Gamma(a+b+c)}\,{}_2F_3\left[\begin{matrix}\frac{a+c}{2},\frac{a+c+1}{2}\ ; & -s^2/4\\\frac{a+b+c}{2},\frac{a+b+c+1}{2},c+1; & \end{matrix}\right]$$

$Re(a+c), Re(b) > 0.$ (2.1.5.3)

If b=1 and a=c+2, we have the simpler result

$$\int_0^1 x^{c-1} J_c(sx)\,dx = \frac{2^{(c+1)/2}\Gamma([3c+5]/2)}{\Gamma(c+2)\; s^{3(c+1)/2}} J_{3(c+1)/2}(s), \qquad (2.1.5.4)$$

provided that Re(c) > 0.

A few examples of special cases of integrals of Kampé de Fériet functions are now considered. From (2.1.3.1), we have, on putting k=1,

$$\int_0^1 x^{a-1}(1-x)^{b-1}\; F_1(c,d,d';f;rx,sx)\,dx \qquad |r|,|s| < 1,$$

$$= \frac{\Gamma(a)\Gamma(b)}{\Gamma(a+b)} F^{2:1;1}_{2:0;0}\left(\begin{matrix}c,\ a:d;d';\\ f,a+b:-;-\ ;\end{matrix} r,s\right),\ \mathrm{Re}(a),\mathrm{Re}(b)>0. \quad (2.1.5.5)$$

If we let a=f and b=c-f, the above result reduces to

$$\int_0^1 x^{f-1}(1-x)^{c-f-1}\; F_1(c,d,d';f;rx,sx) = \frac{\Gamma(f)\Gamma(c-f)}{\Gamma(c)}(1-r)^{-d}(1-s)^{-d'},$$

$$\mathrm{Re}(f),\ \mathrm{Re}(c-f) > 0. \qquad (2.1.5.6)$$

Now, suppose that C=F=0 in (2.1.3.5), so that, if k=1, we have

$$\int_0^1 x^{a-1}(1-x)^{b-1}\; F^{0:D;D'}_{0:G;G'}\left(\begin{matrix}-:(d);(d');\\ -:(g);(g');\end{matrix} rx,sx\right)dx \qquad \mathrm{Re}(a),\ \mathrm{Re}(b) > 0.$$

$$= \frac{\Gamma(a)\Gamma(b)}{\Gamma(a+b)} F^{1:D;D'}_{1:G;G'}\left(\begin{matrix}a\ :(d);(d');\\ a+b:(g);(g');\end{matrix} r,s\right). \qquad (2.1.5.7)$$

The Kampé de Fériet function on the left of (2.1.5.7) may be written as the product of a pair of hypergeometric functions of one variable. Many special forms may be deduced from this, for example, we may write

$$\int_0^1 x^{a-1}(1-x)^{b-1}\; K(rx)\;K(sx)\,dx = \frac{\pi^2\Gamma(a)\Gamma(b)}{4\;\Gamma(a+b)} F^{1:2;2}_{1:1;1}\left[\begin{matrix}a\ :\frac{1}{2},\frac{1}{2};\frac{1}{2},\frac{1}{2};\\ a+b;\ 1\ ;\ 1\ ;\end{matrix} r,s\right],$$

$$\mathrm{Re}(a),\ \mathrm{Re}(b) > 0, \qquad (2.1.5.8)$$

and K(m) is the complete elliptical integral.

This section is concluded by mentioning a few special cases of Euler integrals of the generalised Kampé de Fériet function (2.1.4.1) to (2.1.4.3). The Lauricella function $F_D^{(n)}$ is of frequent occurrence in a number of applications, see Exton (1976) Chapters 7 and 8, so that we give three different types of integral formulae which involve this function. If we put C=D=F=k=1 and G=0 in (2.1.4.4), we have

$$\int_0^1 x^{a-1}(1-x)^{b-1}\; F_D^{(n)}(c,d_1,..,d_n;f;s_1x,..,s_nx)\,dx = \qquad (2.1.5.9)$$

(continued)

(continued) $= \frac{\Gamma(a)\Gamma(b)}{\Gamma(a+b)} F^{2:1}_{1:1}\left(\begin{matrix} c, & a & ;d_1;.;d_n; \\ f, & a+b: & - \quad ; \end{matrix} s_1,.,s_n\right),$ (2.1.5.9)

$\mathrm{Re}(a), \mathrm{Re}(b) > 0, |s_1|,.,|s_n| < 1.$

Two special cases of this result now follow. Firstly, suppose that $f=c$, when the $F_D^{(n)}$ function on the left of (2.1.5.9) splits up into the product of n binomial factors and we obtain the Picard integral for the Lauricella function $F_D^{(n)}$ itself. See Lauricella (1893).

$$\int_0^1 x^{a-1}(1-x)^{b-1}(1-s_1x)^{-d_1}..(1-s_nx)^{-d_n}\,dx$$
$$= \frac{\Gamma(a)\Gamma(b)}{\Gamma(a+b)} F_D^{(n)}(a,d_1,.,d_n;a+b;s_1,.,s_n). \quad (2.1.5.10)$$

The right-hand member of (2.1.5.9) also reduces to an $F_D^{(n)}$ function if $a=f$ and $b=c-f$, and so

$$\int_0^1 x^{f-1}(1-x)^{c-f-1}\ F_D^{(n)}(c,d_1,.,d_n;f;s_1x,.,s_nx)\,dx$$
$$= \frac{\Gamma(a)\Gamma(b)}{\Gamma(a+b)} F_D^{(n)}(f,d_1,.,d_n;c;s_1,.,s_n). \quad (2.1.5.11)$$

Similarly, (2.1.4.5) gives the expression

$$\int_0^1 x^{a-1}(1-x)^{b-1}\ F_D^{(n)}(c,d_1,.,d_n;f;s_1,.,s_{n-1},s_nx)\,dx$$
$$= \frac{\Gamma(a)\Gamma(b)}{\Gamma(a+b)} F^{1:2}_{1:1}\left[\begin{matrix} c:d_1,h;.;d_{n-1},h;d_n,a; \\ f: \ \ h \ \ ;.; \ \ h \ \ ;a+b \ ; \end{matrix} s_1,.,s_n\right] \quad (2.1.5.12)$$

and its special case

$$\int_0^1 x^{a-1}(1-x)^{b-1} F_D^{(n)}(c,d_1,.,d_{n-1},a+b;f;s_1,.,s_{n-1},s_nx)\,dx$$
$$= \frac{\Gamma(a)\Gamma(b)}{\Gamma(a+b)} F_D^{(n)}(c,d_1,.,d_{n-1},a;f;s_1,.,s_n), \quad (2.1.5.13)$$

$\mathrm{Re}(a), \mathrm{Re}(b) > 0$ and $|s_1|,.,|s_n| < 1.$

Turning to (2.1.4.8), we have

$$\int_0^1 x^{a-1}(1-x)^{b-1}\ F_D^{(n)}(c,d_1,.,d_n;f;s_1x[1-x],.,s_nx[1-x])\,dx$$
$$= \frac{\Gamma(a)\Gamma(b)}{\Gamma(a+b)} F^{3:1}_{3:0}\left[\begin{matrix} c, \ a \ , \ b \ :d_1;.;d_n; \\ f, \frac{a+b}{2}, \frac{a+b+1}{2}: \ — \ ; \end{matrix} \frac{s_1}{4},.,\frac{s_n}{4}\right], \quad (2.1.5.14)$$

$\mathrm{Re}(a), \mathrm{Re}(b) > 0$ and $|s_1|,.,|s_n| < 4.$

Finally, the formula (2.1.4.4) is specialised so that an integral of the product of several Gauss functions is obtained:-

$$\int_0^1 x^{a-1}(1-x)^{b-1}\,{}_2F_1(c_1,d_1;f_1;s_1x)\,..\,{}_2F_1(c_n,d_n;f_n;s_nx)\,dx$$

$$= \frac{\Gamma(a)\Gamma(b)}{\Gamma(a+b)} F^{1:2}_{1:1}\left(\begin{matrix} a\ :c_1,d_1;.;c_n,d_n; \\ a+b:\ f_1\ \ ;.;\ f_n\ \ ; \end{matrix} s_1,.,s_n\right), \qquad (2.1.5.15)$$

$Re(a), Re(b) > 0$ and $|s_1|,.,|s_n| < 1$.

2.2 EULER INTEGRALS ASSOCIATED WITH THE CONFLUENT HYPERGEOMETRIC FUNCTION

Many integrals of Euler type which are of interest involve the Kummer function ${}_1F_1$ or the Bessel function, which may be written as a ${}_0F_1$ series. We begin by considering two general integrals of the type

$$\int_0^1 x^{a-1}(1-x)^{b-1}\,{}_1F_1(c;f;sx^k)f(x)dx \qquad (2.2.1)$$

and

$$\int_0^1 x^{a-1}(1-x)^{b-1}\,{}_0F_1(-;f;sx^k)f(x)dx, \qquad (2.2.2)$$

where $Re(a)$, $Re(b) > 0$ and k is a positive integer.

It is now assumed that $f(x)$ can be expanded as a power series of the type (2.1.2) whose radius of convergence is not less than 1. The integral (2.2.1) may thus be written

$$\sum_{n=0}^{\infty} h_n \int_0^1 x^{a+n-1}(1-x)^{b-1}\,{}_1F_1(c;f;sx^k)dx, \qquad (2.2.3)$$

where the process of interchanging the summation and integration is valid on account of the assumed convergence of the series expansion of $f(x)$. The integral of (2.2.3) may be evaluated by means of (2.1.1.9). The integral (2.2.1) may then be written as

$$\frac{\Gamma(a)\Gamma(b)}{\Gamma(a+b)} \sum_{n=0}^{\infty} \frac{(a,n)h_n}{(a+b,n)}\,{}_{1+k}F_{2+k}\left[\begin{matrix} c,\frac{a+n}{k},.,\frac{a+n+k-1}{k} & ; \\ f,\frac{a+b+n}{k},.,\frac{a+b+n+k-1}{k} & ; \end{matrix} s\right]. \qquad (2.2.4)$$

If $k=1$, we have the simpler result

$$\frac{\Gamma(a)\Gamma(b)}{\Gamma(a+b)} \sum \frac{(a,m+n)(c,n)h_m s^n}{(a+b,m+n)(f,n)n!}. \qquad (2.2.5)$$

Similarly, using (2.1.1.10), the integral (2.2.2) may be evaluated in the form

$$\frac{\Gamma(a)\Gamma(b)}{\Gamma(a+b)} \sum_{n=0}^{\infty} \frac{(a,n)h_n}{(a+b,n)}\,{}_kF_{k+2}\left[\begin{matrix} \frac{a+n}{k},\dots,\frac{a+n+k-1}{k} & ; \\ f,\frac{a+b+n}{k},.,\frac{a+b+n+k-1}{k} & ; \end{matrix} s\right]. \qquad (2.2.6)$$

The expressions (2.2.5) and (2.2.6) are important because they include many integrals involving special functions.

2.2.1 Generalised Hypergeometric Function

If f(x) of the previous section is of the form of a generalised hypergeometric function of one variable, we have the two results which follow:

$$\int_0^1 x^{a-1}(1-x)^{b-1}{}_1F_1(c;f;sx^k)\,{}_DF_G((d);(g);rx^k)dx$$
$$= \frac{\Gamma(a)\Gamma(b)}{\Gamma(a+b)}F^{k:1;D}_{k:1;G}\left[\begin{matrix}\frac{a}{k},..,\frac{a+k-1}{k}:c;(d);\\ \frac{a+b}{k},..,\frac{a+b+k-1}{k}:f;(g);\end{matrix}\,s,r\right] \quad (2.2.1.1)$$

and

$$\int_0^1 x^{a-1}(1-x)^{b-1}{}_0F_1(-;f;sx^k)\,{}_DF_G((d);(g);rx^k)dx$$
$$= \frac{\Gamma(a)\Gamma(b)}{\Gamma(a+b)}F^{k:0;D}_{k:1;G}\left[\begin{matrix}\frac{a}{k},..,\frac{a+k-1}{k}:-;(d);\\ \frac{a+b}{k},..,\frac{a+b+k-1}{k}:f;(g);\end{matrix}\,s,r\right], \quad (2.2.1.2)$$

where Re(a) and Re(b) > 0.

Simpler forms of the two previous results occur if r takes certain particular values. For example, if we put k and r both equal to unity in (2.2.1.1), we obtain the result

$$\int_0^1 x^{a-1}(1-x)^{b-1}{}_1F_1(c;f;sx)\,{}_DF_G((d);(g);x)dx$$
$$= \frac{\Gamma(a)\Gamma(b)}{\Gamma(a+b)}\sum_{m=0}^{\infty}\frac{(a,m)(c,m)s^m}{(a+b,m)(f,m)m!}\,{}_{D+1}F_{G+1}[a+m,(d);a+b+m,(g);1], \quad (2.2.1.3)$$

and further specialisation of the ${}_DF_G$ function may result in a form of the inner ${}_{D+1}F_{G+1}$ series of (2.2.1.3) which can be expressed in closed form. Let $D=2, G=1, d_2=-N$ and $g=d_1-N-b+1$, where N is a positive integer. Saalschütz's theorem may now be used to sum the inner series in the form

$$\frac{(a+b-d_1,N)(a+b-d_1+N,m)(a+b,m)}{(a+b,N)(a+b-d_1,m)(a+b+N,m)}. \quad (2.2.1.4)$$

Hence,

$$\int_0^1 x^{a-1}(1-x)^{b-1}{}_1F_1(c;f;sx)\,{}_2F_1(d_1,-N;d_1-b-N+1;x)dx$$
$$= \frac{\Gamma(a)\Gamma(b)}{\Gamma(a+b)}\frac{(a+b-d_1,N)}{(a+b,N)}\,{}_3F_3(a,c,a+b-d_1+N;f,a+b-d_1,a+b+N;s). \quad (2.2.1.5)$$

2.2.2 Double Hypergeometric Function

Consider the integral

$$\int_0^1 x^{a-1}(1-x)^{b-1}\,{}_1F_1(c;f;sx)F^{D:E;E'}_{G:H;H'}\left({(d):(e);(e'); \atop (g):(h);(h');}rx,tx\right)dx. \quad (2.2.2.1)$$

A result similar to (2.2.5) where f(x) is expanded as a double series holds, so that (2.2.2.1) takes the form

$$\frac{\Gamma(a)\Gamma(b)}{\Gamma(a+b)}\sum\frac{(a,m+n+p)((d),n+p)(c,m)((e),n)((e'),p)s^m r^n t^p}{(a+b,m+n+p)((g),n+p)(f,m)((h),n)((h'),p)m!n!p!}. \quad (2.2.2.2)$$

This result may be expressed in terms of Srivastava's triple hypergeometric function [Srivastava (1967b)] as follows:-

$$\frac{\Gamma(a)\Gamma(b)}{\Gamma(a+b)}F^{(3)}\left({a\ ::-:(d):-:c;(e);(e'); \atop a+b::-:(g):-:f;(h);(h');}s,r,t\right). \quad (2.2.2.3)$$

We also have

$$\int_0^1 x^{a-1}(1-x)^{b-1}\,{}_1F_1(c;f;sx)F^{D:E;E'}_{G:H;H'}\left({(d):(e);(e'); \atop (g):(h);(h');}rx,t\right)dx =$$

$$\frac{\Gamma(a)\Gamma(b)}{\Gamma(a+b)}F^{(3)}\left({-::\ a\ :(d):-:c;(e);(e'); \atop -::a+b:(g):-:f;(h);(h');}s,r,t\right) \quad (2.2.2.4)$$

and

$$\int_0^1 x^{a-1}(1-x)^{b-1}\,{}_1F_1(c;f;sx)F^{D:E;E'}_{G:H;H'}\left({(d):(e);(e'); \atop (g):(h);(h');}rx,t[1-x]\right)dx$$

$$= \frac{\Gamma(a)\Gamma(b)}{\Gamma(a+b)}F^{(3)}\left({-\ ::a:(d):-:c;(e);(e'); \atop a+b::-:(g):-:f;(h);(h');}s,r,t\right). \quad (2.2.2.5)$$

Other results of this type may be deduced. Special cases where, for example, Appell functions are involved will be discussed in Section 2.2.4.

2.2.3 Multiple Hypergeometric Function

The vast majority of the special functions of mathematical physics and chemistry may be expressed in hypergeometric form. The most convenient generalisation of the hypergeometric function of several variables is the generalised Kampé de Fériet function of Karlsson (1973). We recall that this function possesses the following multiple series representation:-

$$F^{A:B}_{C:D}\left({(a):(b');.;(b^{(n)}); \atop (c):(d');.;(d^{(n)});}x_1,.,x_n\right)$$

$$= \sum\frac{((a),m_1+.+m_n)((b'),m_1)..((b^{(n)}),m_n)x_1^{m_1}..x_n^{m_n}}{((c),m_1+.+m_n)((d'),m_1)..((d^{(n)}),m_n)m_1!..m_n!}. \quad (2.2.3.1)$$

If A=C=0, this function reduces to the product of n generalised hypergeometric functions of one variable ${}_BF_D$. On the other hand, if B=D=0, we have a single function ${}_AF_C$ whose argument consists of the sum of the arguments of the generalised Kampé de Fériet function from which it is obtained. Furthermore, if B=1 and D=0, a reducible form of (2.2.3.1) occurs when all its arguments are made equal to each other. It then takes the form

$$ {}_{A+1}F_C[(a),b'+.+b^{(n)};(c);x]. $$

Three general types of Euler integral involving the generalised Kampé de Fériet function and a confluent hypergeometric function are considered. These are

$$ \int_0^1 x^{a-1}(1-x)^{b-1}{}_1F_1(c;f;sx)F^{D:E}_{G:H}\binom{(d):(e');.;(e^{(n)});}{(g):(h');.;(h^{(n)});}r_1x,.,r_nx)dx, \tag{2.2.3.2} $$

$$ \int_0^1 x^{a-1}(1-x)^{b-1}{}_1F_1(c;f;sx)F^{D:E}_{G:H}\binom{(d):(e');.;(e^{(n)});}{(g):(h');.;(h^{(n)});}r_1x,r_2,.,r_n)dx \tag{2.2.3.3} $$

and

$$ \int_0^1 x^{a-1}(1-x)^{b-1}{}_1F_1(c;f;sx)F^{D:E}_{G:H}\left[\begin{matrix}(d):(e');.;(e^{(n)});\\(g):(h');.;(h^{(n)});\end{matrix}\, r_1x,.,r_kx,\right. $$

$$ \left. r_{k+1}[1-x],.,r_n[1-x]\right]dx. \tag{2.2.3.4} $$

These integrals can be evaluated by using the multi-dimensional extension of (2.2.6). Hence, (2.2.3.2) may be written as

$$ \frac{\Gamma(a)\Gamma(b)}{\Gamma(a+b)}\sum\frac{(a,m+m_1+.+m_n)((d),m_1+.+m_n)(c,m)((e'),m_1)..((e^{(n)},m_n)}{(a+b,m+m_1+.+m_n)((g),m_1+.+m_n)(f,m)((h),m_1)..((h^{(n)}),m_n)} \times\frac{s^m r_1^{m_1}..r_n^{m_n}}{m!m_1!..m_n!}. \tag{2.2.3.5} $$

The integrals (2.2.3.3) and (2.2.3.4) become, respectively

$$ \frac{\Gamma(a)\Gamma(b)}{\Gamma(a+b)}\sum\frac{((d),m_1+.+m_n)(c,m)(a,m)(b,m_1)((e'),m_1)..((e^{(n)}),m_n)}{((g),m_1+.+m_n)(a+b,m+m_1)(f,m)((h'),m_1)..((h^{(n)}),m_n)} \times\frac{s^m r_1^{m_1}..r_n^{m_n}}{m!m_1!..m_n!} \tag{2.2.3.6} $$

and

$$ \frac{\Gamma(a)\Gamma(b)}{\Gamma(a+b)}\sum\frac{((d),m_1+.+m_n)(a,m+m_1+.+m_k)(b,m_{k+1}+.+m_n)(c,m)}{(a+b,m+m_1+.+m_n)((g),m_1+.+m_n)(f,m)}\times \tag{2.2.3.7} $$

(continued)

(continued)

$$\times \frac{((e'),m_1)..((e^{(n)}),m_n)s^m r_1^{m_1}..r_n^{m_n}}{((h'),m_1)..((h^{(n)}),m_n)m!m_1!..m_n!}. \qquad (2.2.3.7)$$

The results (2.2.3.5) to (2.2.3.7) require that Re(a) and Re(b) are both positive for convergence of the integrals concerned. It is also understood that all the multiple series concerned are either convergent over the range of integration or that they are terminating. Some of the special cases of the various Euler integrals involving the confluent hypergeometric function and other types of hypergeometric functions will be discussed in the next section.

2.2.4 Special Cases

A large number of Euler integrals involving special functions may be obtained by appropriate specialisation of previous formulae in Section 2.2 and its subsections. The generalised Laguerre polynomial and the Bessel function of the first kind may be expressed as hypergeometric functions by means of the the following formulae respectively [see Erdélyi et al. (1953) Vol. II pages 189 and 4]:-

$$L_n^a(x) = \frac{(a+1,n)}{n!}\,{}_1F_1(-n;a+1;x) \qquad (2.2.4.1)$$

and

$$J_c(x) = (x/2)^c/\Gamma(c+1)\,{}_0F_1(-;c+1;-\frac{x^2}{4}). \qquad (2.2.4.2)$$

If these expressions are substituted into (2.2.1.1), we have

$$\int_0^1 x^{a-1}(1-x)^{b-1}L_n^d(sx)J_c[\sqrt{(rx)}]dx$$

$$= \frac{(d+1,n)(r/4)^{c/2}\Gamma(a+c/2)\Gamma(b)}{n!\,\Gamma(c+1)\,\Gamma(a+b+c/2)} F^{1:1;0}_{1:1;1}\left(\begin{matrix} a+c/2 :-n ; - ; \\ a+b+c/2:d+1;c+1; \end{matrix} s,-r/4\right). \qquad (2.2.4.3)$$

Now replace the Legendre polynomial by the corresponding hypergeometric function[Erdélyi et al. (1953) Vol. II page 180], again using (2.2.1.1) in a slightly different form. We then have

$$\int_0^1 x^{a-1}(1-x)^{b-1}L_n^d(sx)P_m(1-2r+2rx)dx$$

$$= \frac{(d+1,n)\Gamma(a)\Gamma(b)}{n!\,\Gamma(a+b)} F^{0:2;3}_{1:1;1}\left(\begin{matrix} - :a,-n;b,-m,m+1; \\ a+b:d+1 ; \quad 1 \quad ; \end{matrix} s,r\right). \qquad (2.2.4.4)$$

The results (2.2.2.3) to (2.2.2.5) may be specialised so that integrals involving the confluent hypergeometric function and the Appell are evaluated. This gives the formulae

$$\int_0^1 x^{a-1}(1-x)^{b-1}\,{}_1F_1(c;f;sx)F_2(d,e,e';g,g';rx,tx)dx$$

$$=\frac{\Gamma(a)\Gamma(b)}{\Gamma(a+b)}F^{(3)}\left(\begin{matrix}a ::-:d:-;c;e;e';\\ a+b::-:-:-;f;g;g';\end{matrix}s,r,t\right), \qquad (2.2.4.5)$$

$$\int_0^1 x^{a-1}(1-x)^{b-1}\,{}_1F_1(c;f;sx)F_3(d,d',e,e';g;rx,t)dx$$

$$=\frac{\Gamma(a)\Gamma(b)}{\Gamma(a+b)}F^{(3)}\left(\begin{matrix}-:: a :-:-;c;d,e;d',e';\\ -::a+b:g:-;f; - ; - ;\end{matrix}s,r,t\right) \qquad (2.2.4.6)$$

and

$$\int_0^1 x^{a-1}(1-x)^{b-1}\,{}_1F_1(c;f;sx)F_4(d,e;g,g';rx,t[1-x])dx$$

$$=\frac{\Gamma(a)\Gamma(b)}{\Gamma(a+b)}F^{(3)}\left(\begin{matrix}- ::a:d,e:-;c;-;- ;\\ a+b::-: - :-;f;g;g';\end{matrix}s,r,t\right), \qquad (2.2.4.7)$$

$Re(a),\ Re(b) > 0.$

We now evaluate an integral involving the incomplete gamma function and the incomplete elliptical integral of the second. This is

$$\int_0^1 x^{a-1}(1-x)^{b-1}\gamma(c,sx^2)F(r[1-x],t)dx. \qquad (2.2.4.8)$$

Now the two special functions of the integrand may be expressed in hypergeometric form by means of the formulae

$$\gamma(a,x)= x^a a^{-1}\,{}_1F_1(a;a+1;-x) \qquad (2.2.4.9)$$

and

$$F(x,k) = xF_1(1/2,1/2,1/2;3/2;x^2,k^2x^2). \qquad (2.2.4.10)$$

See Erdélyi et al. (1953) Vol. II page 133 and Carlson (1961) respectively.

If we make a quadratic change of the variable of integration, $z=x^2$, (2.2.4.8) assumes a standard form and it may be evaluated as

$$\frac{\Gamma(\frac{a+c+1}{2})\Gamma(b)}{2c\Gamma(\frac{a+c+1}{2}+b)}F^{(3)}\left[\begin{matrix}- & ::-:\frac{1}{2},b:-:c,\frac{a+c+1}{2};\frac{1}{2};\frac{1}{2}; -s,r,rt\\ \frac{a+c+1}{2}+b & ::-:\frac{3}{2} \quad :-:c+1 \quad ;-;-;\end{matrix}\right], \qquad (2.2.4.11)$$

$Re(a+c) > -1,\ Re(b) > 0.$

This section is concluded by giving certain cases of the integrals under consideration where products of several hypergeometric functions of one variable and the Lauricella functions are involved. Such results as (2.2.3.5), (2.2.3.6) and (2.2.3.7) may be used. For example, we have

$$\int_0^1 x^{a-1}(1-x)^{b-1}\, L_n^c(rx)L_{m_1}^{d_1}(s_1x)\ldots L_{m_k}^{d_k}(s_kx)\,dx$$

$$= \frac{(c+1,n)(d_1+1,m_1)\ldots(d_k+1,m_k)\Gamma(a)\Gamma(b)}{n!\qquad m_1!\ldots\ldots m_k!\qquad \Gamma(a+b)}$$

$$\times F^{1:1}_{1:1}\left[\begin{matrix} a:-n;-m_1;\ldots;-m_k;r,s_1,\ldots,s_k \\ a+b:c+1;d_1+1;\ldots;d_k+1; \end{matrix}\right], \qquad (2.2.4.13)$$

$\mathrm{Re}(a)$, $\mathrm{Re}(b) > 0$.

Now we consider a finite Laplace transform of the incomplete elliptical integral of the third kind.This last function may be expressed as a Lauricella function of three variables by means of the formula [Carlson(1961)]

$$\Pi(x,b,m) = xF_D^{(3)}(\tfrac{1}{2},\tfrac{1}{2},\tfrac{1}{2},1;\tfrac{3}{2};x^2,m^2x^2,-bx^2). \qquad (2.2.4.14)$$

We then have

$$\int_0^1 x^{a-1}(1-x)^{b-1}e^{-sx}\Pi(\sqrt{x},c,n)\,dx$$

$$= \int_0^1 x^{a-1/2}(1-x)^{b-1}e^{-sx}F_D^{(3)}(\tfrac{1}{2},\tfrac{1}{2},\tfrac{1}{2},1;\tfrac{3}{2};x,n^2x,-cx) \qquad (2.2.4.15)$$

This is a special case of (2.2.3.2) which yields a confluent hypergeometric function of four variables of higher order

$$\frac{\Gamma(a+1/2)\Gamma(b)}{\Gamma(a+b+1/2)}\sum\frac{(a+1/2,m_1+.+m_4)(1/2,m_2+m_3+m_4)(1/2,m_2)(1/2,m_3)}{(a+b+1/2,m_1+.+m_4)\;m_1!m_2!m_3!}$$

$$\times(-s)^{m_1}n^{2m_3}(-c)^{m_4}, \qquad (2.2.4.16)$$

$\mathrm{Re}(a) > -1/2$, $\mathrm{Re}(b) > 0$.
The function (2.2.4.16) is related to the hypergeometric functions studied by the author. See Exton (1972b) and (1973a).

Finally, we give an integral of Euler type which includes a Lauricella function $F_C^{(n)}$ in its integrand. This is obtained by specialising (2.2.3.4). Take D=2, E=G=0 and H=1, when we have

$$\int_0^1 x^{a-1}(1-x)^{b-1}{}_1F_1(c;f;sx)F_C^{(n)}(d_1,d_2;h_1,\ldots,h_n;r_1x,\ldots,r_kx,$$

$$r_{k+1}[1-x],\ldots,r_n[1-x])\,dx$$

$$= \frac{\Gamma(a)\Gamma(b)}{\Gamma(a+b)}\sum\frac{(d_1,m_1+.+m_n)(d_2,m_1+.+m_n)(a,m+m_1+.+m_k)(b,m_{k+1},\ldots,m_n)}{(a+b,m+m_1+.+m_n)(f,m)\qquad m!m_1!\ldots m_n!}$$

$\mathrm{Re}(a),\mathrm{Re}(b)>0.$ $\quad\times(c,m)(h_1,m_1)\ldots(h_n,m_n)\;s^m r_1^{m_1}\ldots r_n^{m_n}. \qquad (2.2.4.17)$

2.3 EULER INTEGRALS ASSOCIATED WITH THE GAUSS FUNCTION

In this section Euler integrals involving a Gauss function as well as another hypergeometric function are discussed. Such integrals may be written in the form

$$\int_0^1 x^{a-1}(1-x)^{b-1}\,{}_2F_1(c,d;f;sx^k)f(x)dx, \tag{2.3.1}$$

where Re(a) and Re(b) > 0, and k is a positive integer. As in the previous section, it is assumed that f(x) can be expanded as a power series of the form (2.1.2) whose radius of convergence is not less than unity. The integral (2.3.1) may thus be integrated term-by-term and may thus be written

$$\sum_{n=0}^{\infty} h_n \int_0^1 x^{a+n-1}(1-x)^{b-1}\,{}_2F_1(c,d;f;sx^k)dx. \tag{2.3.2}$$

Suppose that k=1 when the preceding integral is of the same form as (2.1.1.5) and, on evaluation, this integral becomes

$$\frac{\Gamma(a+n)\Gamma(b)}{\Gamma(a+b+n)}\,{}_3F_2\left({c,d,a+n; \atop f,a+b+n;}s\right). \tag{2.3.3}$$

Hence, (2.3.1) takes the form

$$\frac{\Gamma(a)\Gamma(b)}{\Gamma(a+b)}\sum_{n=0}^{\infty}\frac{(a,n)h_n}{(a+b,n)}\,{}_3F_2\left({c,d,a+n; \atop f,a+b+n;}s\right), \tag{2.3.4}$$

provided that Re(a) and Re(b) are both positive. If the ${}_2F_1$ and ${}_3F_2$ series are not terminating, it is sufficient that $|s| < 1$. If s=1, then we must have the condition Re(f-c-d) > 0 as well. If either c or d is a negative integer, then we need only retain the condition upon a and b for the convergence of (2.3.4).

Let us now suppose that c = -N, a negative integer, and that b = d-f-N+1. Saalschütz's theorem [Slater (1966) appendix III] may now be used to sum the Clausen function of (2.3.4) if s=1 also. This sum is

$$\frac{(f-d,N)(f-a,N)(1+a-f,n)(1+a+d-f-N,n)}{(f,N)(f-a-d,N)(1+a-f-N,n)(1+a+d-f,n)}. \tag{2.3.5}$$

The expression (2.3.4) now takes on a more elegent form, that is

$$\frac{\Gamma(a)\Gamma(d-f-N+1)(f-d,N)(f-a,N)}{\Gamma(a+d-f-N+1)(f,N)(f-a-d,N)}\sum_{n=0}^{\infty}\frac{(a,n)(1+a-f,n)h_n}{(1+a-f-N,n)(1+a+d-f,n)}. \tag{2.3.6}$$

If we let k=2, then (2.3.1) becomes

$$\frac{\Gamma(a)\Gamma(b)}{\Gamma(a+b)}\sum_{n=0}^{\infty}\frac{(a,n)}{(a+b,n)}\,{}_4F_3\left[{c,d,a/2+n,a/2+1/2+n; \atop f,\frac{a+b}{2}+n,\frac{a+b+1}{2}+n\ ;}s\right]. \tag{2.3.7}$$

If b=1, then the previous ${}_4F_3$ series reduces to

$$ {}_3F_2(c,d,a/2+n;f,a/2+1+n;s) \tag{2.3.8}$$

and on letting a = -N and f = d-N, Saalschütz's theorem may be applied once again if s=1. The function (2.3.8) may then be written

$$\frac{(a/2+1-d,N)(1,N)(a/2+1-d+N,n)(a/2+1,n)}{(a/2+1,N)(1-d,N)(a/2+1-d,n)(a/2+1+N,n)} . \tag{2.3.9}$$

Hence, (2.3.1) is now equal to

$$\frac{(a/2+1-d,N)(1,N)}{a(a/2+1,N)(1-d,N)}\sum_{n=0}^{\infty}\frac{(a,n)(a/2+1-d+N,n)(a/2+1,n)h_n}{(a+1,n)(a/2+1-d,n)(a/2+1+N,n)} \tag{2.3.10}$$

The expressions (2.3.4), (2.3.7), (2.3.6) and (2.3.10), perticularly the two latter, readily lend themselves to the evaluation of integrals involving the Gauss function.

2.3.1 Generalised Hypergeometric Function

The integrals whose general form is

$$\int_0^1 x^{a-1}(1-x)^{b-1}\,{}_2F_1(c,d;f;sx^k)\,{}_GF_H((g);(h);rx^k)dx \tag{2.3.1.1}$$

and

$$\int_0^1 x^{a-1}(1-x)^{b-1}\,{}_2F_1(c,d;f;sx^k)\,{}_GF_H((g);(h);r[1-x]^k)dx \tag{2.3.1.2}$$

where k is a positive integer, are now investigated. It is assumed that, for convergence, Re(a) and Re(b) are both positive. Also, unless the generalised hypergeometric function is terminating, we take it that $G \leqq H+1$.

If the inner generalised hypergeometric function of (2.3.1.1) is expanded as a series in its argument, this integral becomes

$$\sum_{n=0}^{\infty}\frac{((g),n)r^n}{((h),n)n!}\int_0^1 x^{a+kn-1}(1-x)^{b-1}\,{}_2F_1(c,d;f;sx^k)dx. \tag{2.3.1.3}$$

The inner integral of the previous expression may be written

$$\frac{\Gamma(a+kn)\Gamma(b)}{\Gamma(a+b+kn)}\;{}_{2+k}F_{1+k}\left[\begin{matrix}c,d,\frac{a+kn}{k},..,\frac{a+kn+n-1}{k} & ;\\ f,\frac{a+b+kn}{k},..,\frac{a+b+kn+k-1}{k} & ;\end{matrix}s\right] , \tag{2.3.1.4}$$

so that (2.3.1.3) becomes

$$\frac{\Gamma(a)\Gamma(b)}{\Gamma(a+b)}F^{k:2;G}_{k:1;H}\left[\begin{matrix}a/k,.......,[a+k-1]/k & :c,d;(g); \\ [a+b]/k,..,[a+b+k-1]/k: & f\ ;(h);\end{matrix}s,r\right] . \tag{2.3.1.5}$$

Similarly, the integral (2.3.1.2) may be shown to be equal to

$$\frac{\Gamma(a)\Gamma(b)}{\Gamma(a+b)} F^{0:k+2;k+G}_{k:\ 1\ ;\ H}\left[\begin{matrix} \text{—} & :c,d,\frac{a}{k},..,\frac{a+k-1}{k};(g),\frac{b}{k},..,\frac{b+k-1}{k}; \\ \frac{a+b}{k},\frac{a+b+k-1}{k}: & f \qquad ;(h) \qquad ; \end{matrix} s,r\right]. \tag{2.3.1.6}$$

A number of simpler forms of these two last results may be obtained, and we give the following example:-
Let $a = -N$, $b=c-d-N+1$ and $k=s=1$. The inner integral of (2.3.1.3) may then be expressed as a Saalschützian Clausen series of unit argument, which when summed takes the form

$$\frac{\Gamma(a)\Gamma(c-d-N+1)(d-c,N)(d-a,N)(a,n)(1+a-d,n)}{\Gamma(a+c-d-N+1)(d,N)(d-a-c,N)(1+a-d-N,n)(1+a+c-d,n)}. \tag{2.3.1.7}$$

Hence, (2.3.1.2) may now be written

$$\frac{\Gamma(a)\Gamma(c-d-N+1)(d-c,N)(d-a,N)}{\Gamma(a+c-d-N+1)(d,N)(d-a-c,N)}\ {}_{G+2}F_{H+2}\left(\begin{matrix} a \quad , \quad 1+a-d,(g); \\ 1+a+c-d,1+a-d-N,(h); \end{matrix} r\right). \tag{2.3.1.8}$$

The generalised hypergeometric function associated with (2.3.1.8) may be summed in many cases for special values of its parameters and variable. Suppose that $G=2$, $H=1$, $g_1=1+a-d-N, g_2=1+a+c-d$ and $h=a$. The binomial theorem may be applied to express the resulting ${}_1F_0$ series in closed form, and we obtain the expression

$$\int_0^1 x^{a-1}(1-x)^{c-d-N}\,{}_2F_1(-N,c;d;x)\,{}_2F_1(1+a-d-N,1+a+c-d;a;rx)dx$$

$$= \frac{\Gamma(a)\Gamma(c-d-N+1)(d-c,N)(d-a,N)}{\Gamma(a+c-d-N+1)(d,N)(d-a-c,N)}(1-r)^{d-a-1}, \tag{2.3.1.9}$$

where $\mathrm{Re}(a) > 0$, $\mathrm{Re}(c-d-N) > -1$ and $|r| < 1$. If $1+a-d$ is also a negative integer, then the condition on r is not necessary, since all the series involved are then terminating.

2.3.2 Double Hypergeometric Function

Three Euler integrals, each of which involves a combination of a Gauss function and a Kampé de Fériet function are

$$\int_0^1 x^{a-1}(1-x)^{b-1}\,{}_2F_1(c,d;f;rx^k)F^{G:H;H'}_{P:Q;Q'}\left(\begin{matrix}(g):(h);(h'); \\ (p):(q);(q');\end{matrix} sx^k,tx^k\right)dx, \tag{2.3.2.1}$$

$$\int_0^1 x^{a-1}(1-x)^{b-1}\,{}_2F_1(c,d;f;rx^k)F^{G:H;H'}_{P:Q;Q'}\left(\begin{matrix}(g):(h);(h'); \\ (p):(q);(q');\end{matrix} sx^k,t\right)dx \tag{2.3.2.2}$$

$$\int_0^1 x^{a-1}(1-x)^{b-1}\,{}_2F_1(c,d;f;rx^k)F^{G:H;H'}_{P:Q;Q'}\left(\begin{matrix}(g):(h);(h'); \\ (p):(q);(q');\end{matrix} s[1-x]^k,tx^k\right)dx, \tag{2.3.2.3}$$

where k is a positive integer. It is understood that Re(a) and Re(b) are both positive and that all the series are either convergent over the range of integration or terminating. Other similar integrals also exist which may be dealt with in the same way as will be used in evaluating (2.3.2.1) to (2.3.2.3).

On expanding the Kampé de Fériet function and interchanging the operations of integration and summation, (2.3.2.1) becomes

$$\sum\frac{((g),m+n)((h),m)((h'),n)s^m t^n}{((p),m+n)((q),m)((q'),n)m!n!}\int_0^1 x^{a+km+kn-1}(1-x)^{b-1}{}_2F_1\left(\begin{matrix}c,d;\\ f\;;\end{matrix}rx^k\right)dx. \tag{2.3.2.4}$$

The inner integral may be written

$$\frac{\Gamma(a+km+kn)\Gamma(b)}{\Gamma(a+b+km+kn)}\;{}_{2+k}F_{1+k}\left(\begin{matrix}[a+km+kn]/k,.,[a+km+kn+k-1]/k,c,d\;;\\ [a+b+km+kn]/k,.,[a+b+km+kn+k-1]/k,f;\end{matrix}r\right), \tag{2.3.2.5}$$

see (2.3.4), for example. Hence, (2.3.2.1) takes the form

$$\frac{\Gamma(a)\Gamma(b)}{\Gamma(a+b)}F^{(3)}\left(\begin{matrix}a/k,....,[a+k-1]/k\quad ::-:(g):-;c,d;(h);(h');\\ [a+b]/k,.,[a+b+k-1]/k::-:(p):-;\;f\;;(q);(q');\end{matrix}r,s,t\right). \tag{2.3.2.6}$$

Using the same type of process, it may be shown that (2.3.2.2) and (2.3.2.3) become, respectively,

$$\frac{\Gamma(a)\Gamma(b)}{\Gamma(a+b)}F^{(3)}\left(\begin{matrix}-::a/k,.....,[a+k-1]/k\quad :(g):-;c,d;(h);(h');\\ -::[a+b]/k,.,[a+b+k-1]/k:(p):-;\;f\;;(q);(q');\end{matrix}r,s,t\right) \tag{2.3.2.7}$$

and

$$\frac{\Gamma(a)\Gamma(b)}{\Gamma(a+b)}F^{(3)}\left[\begin{matrix}\text{---}\quad ::-:(g):a/k,.,[a+k-1]/k;\\ [a+b]/k,.,[a+b+k-1]/k::-:(p):\quad\text{---}\quad ;\end{matrix}\right.$$
$$\left.\begin{matrix}c,d;b/k,.,[b+k-1]/k,(h);(h');\\ f\;;\qquad (q)\qquad ;(q');\end{matrix}r,s,t\right] \tag{2.3.2.8}$$

Suppose that k=r=1, c = -N and b = d-f-N+1, then, by Saalschütz's theorem, we may express (2.3.2.5) in closed form:-

$$\frac{\Gamma(a)\Gamma(d-f-N+1)\;\;(f-d,N)(f-a,N)(a,m+n)(1+a-f,m+n)}{\Gamma(a+d-f-N+1)(f,N)(f-d-a,N)(1+a-f-N,m+n)(1+a+d-f,m+n)}\;. \tag{2.3.2.9}$$

A special case of (2.3.2.1) may now be evaluated as the more compact expression which follows:

$$\frac{\Gamma(a)\Gamma(d-f-N+1)(f-d,N)(f-a,N)}{\Gamma(a+d-f-N+1)(f,N)(f-d-a,N)}$$
$$\times\;F^{G+2:B;B'}_{P+2:Q;Q'}\left(\begin{matrix}(g),a,1+a-f\qquad :(h);(h');\\ (p),1+a-f-N,1+a+d-f:(q);(q');\end{matrix}s,t\right). \tag{2.3.2.10}$$

Similar results follow from (2.3.2.7) and (2.3.2.8).

2.3.3 Multiple Hypergeometric Function

The results given above may be generalised still further and we give a brief discussion of the integral

$$\int_0^1 x^{a-1}(1-x)^{b-1}{}_2F_1(c,d;f;rx^k)F^{G:H}_{P:Q}\left[\begin{matrix}(g):(h');.;(h^{(n)});\\(p):(q');.;(q^{(n)});\end{matrix}s_1x^k,.,s_nx^k\right]dx \qquad (2.3.3.1)$$

as an example. The generalised Kampé de Fériet function in the integrand is expanded as a multiple series, and if this series is convergent over the range of integration, the integral (2.3.3.1) takes the form

$$\sum\frac{((g),m_1+.+m_n)((h'),m_1)..((h^{(n)}),m_n)s_1^{m_1}..s_n^{m_n}}{((p),m_1+.+m_n)((q'),m_1)..((q^{(n)}),m_n)m_1!..m_n!}$$
$$\times\int_0^1 x^{a+km_1+.+km_n-1}(1-x)^{b-1}{}_2F_1\left(\begin{matrix}c,d;\\f\ ;\end{matrix}rx^k\right)dx, \qquad (2.3.3.2)$$

and the inner integral is equal to

$$\frac{\Gamma(a+km_1+.+km_n)\Gamma(b)}{\Gamma(a+b+km_1+.+km_n)}\,{}_3F_2(a+km_1+.+km_n,c,d;a+b+km_1+.+km_n,f;r) \qquad (2.3.3.3)$$

Hence, (2.3.3.1) becomes

$$\frac{\Gamma(a)\Gamma(b)}{\Gamma(a+b)}\sum\frac{(a,km+km_1+.+km_n)((g),m_1+.+m_n)((h'),m_1)..((h^{(n)}),m_n)}{(a+b,km+km_1+.+km_n)((p),m_1+.+m_n)((q'),m_1)..((q^{(n)}),m_n)}$$
$$\times\frac{r^m s_1^{m_1}..s_n^{m_n}}{m!m_1!..m_n!}. \qquad (2.3.3.4)$$

It will be seen that, in its general form, the integral (2.3.3.1) cannot be evaluated in terms of the generalised Kampé de Fériet function in its compact form. Two results may be deduced from (2.3.3.4) which are more elegent, however. If G=P=0 and H=Q+1, we have the formula

$$\int_0^1 x^{a-1}(1-x)^{b-1}{}_2F_1(c,d;f;rx^k)F^{0:Q+1}_{0:\ Q}\left[\begin{matrix}-:(h');.;(h^{(n)});\\-:(q');.;(q^{(n)});\end{matrix}s_1x^k,.,s_nx^k\right]dx$$

$$=\frac{\Gamma(a)\Gamma(b)}{\Gamma(a+b)}F^{k:Q+1}_{k:\ Q}\left[\begin{matrix}a/k,.....,[a+k-1]/k\ :c,d,u,.,u;(h');.;(h^{(n)});\\ [a+b]/k,.,[a+b+k-1]/k:\ f\ \ u,.,u;(q');.;(q^{(n)});\end{matrix}r,s_1,.,s_n\right], \qquad (2.3.3.5)$$

where $|r|$, $|s_1|,\ldots,$ $|s_n|$ < 1 and Re(a) and Re(b) > 0.

Now, suppose that k=1 and c = -N, when the Clausen function of (2.3.3.3) may be summed by Saalschütz's theorem if we also put r=1 and b = d-f-N+1. Hence, (2.3.3.3) may be written

$$\frac{\Gamma(a)\Gamma(d-f-N+1)(f-d,N)(f-a,N)(a,m_1+.+m_n)(1+a-f,m_1,.,m_n)}{\Gamma(a+d-f-N+1)(f,N)(f-d-a,N)(1+a-f-N,m_1+.+m_n)(1+a+d-f,m_1+.+m_n)}.$$

and so, (2.3.3.6)

$$\int_0^1 x^{a-1}(1-x)^{b-1}{}_2F_1\left({-N,d; \atop f\ ;}x\right)F^{G:H}_{P:Q}\left[{(g):(h');.;(h^{(n)}); \atop (p):(q');.;(q^{(n)});}s_1x,.,s_nx\right]dx$$

$$= \frac{\Gamma(a)\Gamma(d-f-N+1)(f-d,N)(f-a,N)}{\Gamma(a+d-f-N+1)(f,N)(f-d-a,N)}$$

$$\times F^{G+2:H}_{P+2:Q}\left[{(g),\ a\ ,\ 1+a-f\ :(h');.;(h^{(n)}); \atop (p),1+a-f-N,1+a+d-f:(q');.;(q^{(n)});}s_1,.,s_n\right]. \quad (2.3.3.7)$$

Here, the integral is evaluated in terms of a generalised Kampé de Fériet function with the same number of variables as that in the integrand, but with its order augumented by two. A number of special cases of Euler integrals associated with the Gauss function are discussed in the next section.

2.3.4 Special Cases

Let us consider the Euler integral of a product of Legendre polynomials

$$\int_0^1 x^{a-1}(1-x)^{b-1}P_n(1-2rx)P_m(1-2sx)dx, \quad (2.3.4.1)$$

which may be written in the form

$$\int_0^1 x^{a-1}(1-x)^{b-1}{}_2F_1(-n,n+1;rx)\,{}_2F_1(-m,m+1;1;sx)dx, \quad (2.3.4.2)$$

where, for convergence Re(a) and Re(b) must both be positive. The integral (2.3.4.2) is clearly a special case of (2.3.1.3), and so it may be evaluated as

$$\frac{\Gamma(a)\Gamma(b)}{\Gamma(a+b)}F^{1:2;2}_{1:1;1}\left({a\ :-n,n+1;-m,m+1; \atop a+b:\ \ 1\ \ ;\ \ 1\ \ ;}r,s\right). \quad (2.3.4.3)$$

Other convergence conditions are unnecessary because all the hypergeometric functions involved are terminating. If, in addition, r=b=1, a special case of (2.3.1.8) arises, so that (2.3.4.1) now becomes

$$\frac{(-1)^n(1-a,n)}{a(-a-n,n)}{}_3F_2\left({a-n,m+1,-m; \atop a+n+1,\ 1\ ;}s\right). \quad (2.3.4.4)$$

Now let a=n+1 and r=1, when we have the further simplification using Gauss's theorem

$$\frac{n!(2n-m+1,m)}{(n+1)(-2n-1,m)(2n+2,m)}. \qquad (2.3.4.5)$$

If we now take a=n+1 and m=2n+1 in (2.3.4.4), this takes the form

$$\frac{n!(1-r)^{2n+1}}{(n+1)(-2n-1,n)}. \qquad (2.3.4.6)$$

We now investigate an Euler integral of a Legendre polynomial and a Gegenbauer polynomial. This is

$$\int_0^1 x^{a-1}(1-x)^{b-1}P_n(1-2rx)C_m^c(1-2sx)dx. \qquad (2.3.4.7)$$

We replace the Legendre and Gegenbauer polynomials by their corresponding hypergeometric representations. The formula (2.3.1.3) may then be applied. It is clear that (2.3.4.7) is now equal to

$$\frac{m!\Gamma(a)\Gamma(b)}{(2c,m)\Gamma(a+b)}F^{1:2;2}_{1:1;1}\left(\begin{matrix}a & :-n,n+1;-m,m+2c; \\ a+b: & 1 \quad ; \; c+1/2 \; ;\end{matrix} r,s\right). \qquad (2.3.4.8)$$

Two special cases of this result are now considered. Firstly, let r=b=1, when we may apply (2.3.1.8) and (2.3.4.7) becomes

$$\frac{m!(-n,n)(1-a,n)}{(2c,m)(1,n)a(-a-n,n)}\;{}_3F_2\left(\begin{matrix}a-n,-m,m+2c; \\ 1+a+n, \quad 1 \quad ;\end{matrix} s\right). \qquad (2.3.4.9)$$

If we now put c=n+1 and s=1, then the Clausen function of (2.3.4.9) may be summed by Saalschütz's theorem giving a yet simpler special case of (2.3.4.7):

$$\frac{(1-a,m+n)(-3n-1,m)m!(-1)^n}{a(-a-n,n)(-a-m-1,m)n!(2n+2,m)}, \qquad (2.3.4.10)$$

provided that Re(a) > 0.

A number of interesting results are obtainable by specialising (2.3.2.6). As examples, we investigate two integrals involving the complete elliptical integral of the first kind and the incomplete integral of the second kind and the complete elliptical integral of the third kind respectively. Elliptical integrals, and also hyperelliptical integrals may be conveniently be expressed in hypergeometric form. See Carlson (1961), for example. We may thus write

$$\frac{2\Gamma(a+b+1/2)}{\pi s\Gamma(a+1/2)\Gamma(b)}\int_0^1 x^{a-1}(1-x)^{b-1}\; K(rx)\; F(s\sqrt{x},k\sqrt{x})dx \qquad (2.3.4.11)$$

$$= F^{(3)}\left(\begin{matrix}a+1/2 \; ::-:1/2:-;1/2,1/2;1/2;1/2; \\ a+b+1/2::-:3/2:-: \quad 1 \quad ; \; - \; ; \; - \; ;\end{matrix} r,s^2,s^2k^2\right).$$

$$\frac{4\Gamma(a+b)}{\pi^2\Gamma(a)\Gamma(b)}\int_0^1 x^{a-1}(1-x)^{b-1}\ K(rx)\ \Pi(cx,k\sqrt{x})dx$$

$$= F^{(3)}\left(\begin{matrix}a\ ::-:1/2:-;1/2,1/2;1/2;1;\\ a+b::-:\ 1\ :-;\quad 1\quad ;\ -\ ;-;\end{matrix}r,k^2,-c\right). \tag{2.3.4.12}$$

In (2.3.4.11), Re(a) > 1/2 and Re(b) > 0, and in (2.3.4.12), Re(a), Re(b) > 0.

Finally, we mention a special case of (2.3.3.4):-

$$\frac{2\Gamma(a+b+1/2)}{\pi s\Gamma(a+1/2)\Gamma(b)}\int_0^1 x^{a-1}(1-x)^{b-1}\ K(rx)\ \Pi(\sqrt{x},s,k)dx$$

$$= \sum\frac{(a+1/2,m+m_1+m_2+m_3)(1/2,m_1+m_2+m_3)(\frac{1}{2},m)(\frac{1}{2},m)(\frac{1}{2},m_1)(\frac{1}{2},m_2)(\frac{1}{2},m_3)}{(a+b+1/2,m+m_1+m_2+m_3)(3/2,m_1+m_2+m_3)(1,m)m!m_1!m_2!m_3!}$$

$$\times\ r^m s^{2m_1}(sk)^{2m_2}\ . \tag{2.3.4.13}$$

This gives an Euler integral of a complete elliptical integral of the first kind and an incomplete elliptical integral of the third kind.

Finally, we conclude this chapter by giving an integral which involves the fourth kind of Lauricella function. This result may be written down directly as a special case of (2.3.3.7),

$$\int_0^1 x^{a-1}(1-x)^{b-c-N}{}_2F_1(-N;b;c;x)F_D^{(n)}(d,f_1,..,f_n;g;s_1x,..,s_nx)dx$$

$$= \frac{\Gamma(a)\Gamma(b-c-N+1)(c-b,N)(c-a,N)}{\Gamma(a+b-c-N+1)(c,N)(c-b-a,N)}$$

$$\times\ F^{3:1}_{3:0}\left(\begin{matrix}d,a,1+a-c:f_1;.;f_n;\\ g,1+a-c-N:\ \ \text{---}\end{matrix}s_1,..,s_n\right), \tag{2.3.4.14}$$

where Re(a) > 0 and Re(b-c-N) > -1.

Chapter 3

Definite Integrals and Repeated Integrals

3.1 DEFINITE INTEGRALS INVOLVING ONE HYPERGEOMETRIC FUNCTION

Hypergeometric integrals with variable limits of integration and which are of frequent occurrence, may be reduced to the following two types:-

$$\int_0^z u^{a-1}(1-u)^{b-1} f(u)\, du \tag{3.1.1}$$

and

$$\int_0^z e^{-bu}\, u^{a-1} f(u)\, du, \tag{3.1.2}$$

where Re(a), Re(b) > 0.
It is evident that the first of these two integrals is similar to the Euler integral and the second to the Laplace integral. See Exton (1976) pages 16 and 17 respectively.

Suppose that f(u) may be expanded in the form

$$f(u) = \sum_{n=0}^{\infty} h_n\, u^n, \tag{3.1.3}$$

where the radius of convergence does not exceed the absolute value of z. From (3.1.1), we have

$$\int_0^z u^{a-1}(1-u)^{b-1} f(u)du = \sum_{n=0}^{\infty} h_n \int_0^z u^{n+a-1}(1-u)^{b-1}du, \tag{3.1.4}$$

and from (3.1.2), we obtain the corresponding formula

$$\int_0^z e^{-bu}\, u^{a-1} f(u)du = \sum_{n=0}^{\infty} h_n \int_0^z e^{-bu}\, u^{a+n-1}du. \tag{3.1.5}$$

The interchanging of the summation and integration in these two results is justified on account of the convergence conditions imposed upon the series (3.1.3).

It will be seen that the integral on the right of (3.1.4) is an incomplete beta function. See Erdélyi et al. (1953) Vol. I page 87. This may be expressed as a Gauss function as follows:-

$$B_z(a+n,b) = \frac{z^{a+n}}{a+n}\; {}_2F_1(a+n,1-b;a+1+n;z). \tag{3.1.6}$$

Now, because $\quad a+n = a(a+1,n)/(a,n), \quad$ (3.1.7)

it is possible to evaluate the integral on the right of (3.1.4) as a series involving the Gauss function together with gamma products expressed as Pochhammer symbols. See Sneddon (1961) page 19. Hence, the integral (3.1.1) may be written as a double series, namely

$$\frac{z^a}{a}\sum \frac{h_m\,(a,m+n)(1-b,n)z^{m+n}}{(1+a,m+n)\quad n!}\,. \tag{3.1.8}$$

Similarly, it will be observed that the integral on the right of (3.1.5) is an incomplete gamma function

$$b^{a+n}\,\gamma(a+n,z/b) = \frac{z^{a+n}\quad (a,n)}{a\ (a+1,n)}\,{}_1F_1(a+n;a+1+n;-z/b). \tag{3.1.9}$$

This takes the place of the Gauss function when considering (3.1.4). Hence, (3.1.2) becomes

$$\frac{z^a}{a}\sum \frac{h_m\,(a,m+n)z^m(-z/b)^n}{(a+1,m+n)n!}\,. \tag{3.1.10}$$

This last result is a limiting form of (3.1.8). The formulae (3.1.8) and (3.1.10) form the basis of much of the work on definite hypergeometric integrals.

3.1.1 Generalised Hypergeometric Function

We now take f(u) to be of the form

$$f(u) = {}_CF_D((c);(d);ru), \tag{3.1.1.1}$$

where ${}_CF_D$ is the generalised hypergeometric function of one variable. The expression (3.1.8) now becomes

$$\frac{z^a}{a}\sum\frac{((c),m)(a,m+n)(1-b,n)(zr)^m z^n}{((d),m)(a+1,m+n)\qquad m!n!} \tag{3.1.1.2}$$

which may be expressed as the following Kampé de Fériet function:

$$z^a/a\ F^{1:C;1}_{1:D;0}\left(\begin{matrix} a\ :(c);1-b;\\ a+1:(d);\ -\ ;\end{matrix}zr,z\right). \tag{3.1.1.3}$$

On the other hand, (3.1.10) gives

$$z^a/a\ F^{1:C;0}_{1:D;0}\left(\begin{matrix} a\ :(c);-;\\ a+1:(d);-;\end{matrix}zr,\,-z\right). \tag{3.1.1.4}$$

which is a limiting form of (3.1.1.3).

3.1.2 Double Hypergeometric Function

The results (3.1.8) and (3.1.10) may be extended in a straightforward fashion to double and multiple series. Hence, we are led to consider the integrals which follow.

$$\int_0^z u^{a-1}(1-u)^{b-1}\, F^{C:D;D'}_{F:G;G'}\left({(c):(d);(d'); \atop (f):(g);(g');} ru, su\right) du, \tag{3.1.2.1}$$

$$\int_0^z u^{a-1}(1-u)^{b-1}\, F^{C:D;D'}_{F:G;G'}\left({(c):(d);(d'); \atop (f):(g);(g');} ru, s\right) du \tag{3.1.2.2}$$

and

$$\int_0^z u^{a-1}(1-u)^{b-1}\, F^{C:D;D'}_{F:G;G'}\left({(c):(d);(d'); \atop (f):(g);(g');} ru, s[1-u]\right). \tag{3.1.2.3}$$

Expand the integrand of (3.1.2.1) as a double power series in u, when this integral may be written

$$\sum \frac{((c),m+n)((d),m)((d'),n) r^m s^n}{((f),m+n)((g),m)((g'),n) m! n!} \int_0^z u^{a+m+n-1}(1-u)^{b-1}\, du. \tag{3.1.2.4}$$

The inner integral is an incomplete beta function which may be expressed as

$$\frac{z^{a+m+n}(a,m+n)}{a(a+1,m+n)}\, {}_2F_1(a+m+n, 1-b; a+1+m+n; z). \tag{3.1.2.5}$$

Hence, after a little re-arrangement, we find that (3.1.2.1) becomes

$$\frac{z^a}{a}\, F^{(3)}\left({a ::(c):-:-;(d);(d');1-b; \atop a+1::(f):-:-;(g);(g');\ -\ ;} rz, sz, z\right), \tag{3.1.2.6}$$

where Re(a), Re(b) > 0, $|z| \leq 1$, and the function $F^{(3)}$ is Srivastava's general triple hypergeometric function. See Exton (1976) page 108.

Similarly, the integrals (3.1.2.2) and (3.1.2.3) respectively take the form

$$\frac{z^a}{a}\, F^{(3)}\left({-::(c):-:\ a\ ;(d);(d');1-b; \atop -::(f):-:1+a;(g);(g');\ -\ ;} rz, s, z\right) \tag{3.1.2.7}$$

and

$$\frac{z^a}{a} \sum \frac{((c),m+n)(a,m+p)(1-b,p-n)(b,n)((d),m)((d'),n)}{((f),m+n)(a+1,m+p)((g),m)((g'),n)} \frac{(rz)^m}{m!} \frac{(-s)^n}{n!} \frac{z^p}{p!}. \tag{3.1.2.8}$$

We also have the following example of a definite integral of Laplace type which involves a Kampé de Fériet function:-

$$\int_0^z u^{a-1}(1-u)^{b-1} F^{C:D;D'}_{F:G;G'}\left({(c):(d);(d'); \atop (f):(g);(g');} ru, su\right) du$$

$$= \frac{z^a}{a}\, F^{(3)}\left({a ::(c):-:-;(d);(d');-; \atop a+1::(f):-:-;(g);(g');-;} rz, sz, -z\right). \tag{3.1.2.9}$$

A few special cases will be discussed in Section 3.1.4 .

3.1.3 Multiple Hypergeometric Function

If integrals involving hypergeometric functions of several variables are considered, the number of possibilities increases very rapidly with the number of independent variables of the functions concerned. We now discuss the following two integrals which lend themselves most readily to convenient treatment:-

$$\int_0^z u^{a-1}(1-u)^{b-1}\, F^{C:D}_{F:G}\left[\begin{matrix}(c):(d');.;(d^{(n)});\\(f):(g');.;(g^{(n)});\end{matrix} r_1u,.,r_nu\right] du \qquad (3.1.3.1)$$

and

$$\int_0^z u^{a-1}(1-u)^{b-1}\, F^{C:D}_{F:G}\left[\begin{matrix}(c):(d');.;(d^{(n)});\\(f):(g');.;(g^{(n)});\end{matrix} r_1u,.,r_ku,\; r_{k+1}[1-u],.,r_n[1-u]\right] du. \qquad (3.1.3.2)$$

The function $F^{C:D}_{F:G}$ is the generalised Kampé de Fériet function of Karlsson (1973), and it is assumed that Re(a) and Re(b) are both positive and that $|z|$ is sufficiently small to ensure that the hypergeometric functions in the integrands are convergent over each range of integration. If $C+D > F+G+1$, then these functions must terminate, and if $C+D \leq F+G$, the integrand in question will always converge. When $C+D = F+G+1$, each case must be considered separately.

If the inner generalised Kampé de Fériet function of (3.1.3.1) is expanded as a multiple series, term-by-term integration is allowable if the above convergence conditions are met and this integral becomes

$$\sum \frac{((c),m_1+.+m_n)((d'),m_1)..((d^{(n)}),m_n)r_1^{m_1}..r_n^{m_n}}{((f),m_1+.+m_n)((g'),m_1)..((g^{(n)}),m_n)m_1!..m_n!} B_z(a+m_1+.+m_n,b). \qquad (3.1.3.3)$$

As before, the incomplete beta function is expressed as a Gauss function and (3.1.3.1) may be written as

$$\frac{z^a}{a}\sum \frac{(a,m+m_1+.+m_n)((c),m_1+.+m_n)(1-b,m)((d'),m_1)..((d^{(n)}),m_n)}{(a+1,m+m_1+.+m_n)((f),m_1+.+m_n)((g'),m_1)..((g^{(n)}),m_n)m!m_1!..m_n!} \times (r_1z)^{m_1}..(r_nz)^{m_n}. \qquad (3.1.3.4)$$

If we investigate (3.1.3.2), we find that it is equal to

$$\frac{z^a}{a}\sum \frac{(a,m+m_1+.+m_n)((1-b,m-m_{k+1}-..-m_n)(b,m_{k+1}+.+m_n)((c),m_1+.+m_n)}{(a+1,m+m_1+.+m_n)((f),m_1+.+m_n)((g'),m_1)..((g^{(n)}),m_n)m!m_1!..m_n!}$$

$$\times((d'),m_1)..((d^{(n)}),m_n)z^m(r_1z)^{m_1}..(r_kz)^{m_k}(-r_{k+1})^{m_{k+1}}..(-r_n)^{m_n}. \qquad (3.1.3.5)$$

Next, the two similar definite integrals of Laplace type which follow are evaluated:-

$$\int_0^z u^{a-1}\ e^{-u}\ F^{C:D}_{F:G}\left[\begin{matrix}(c):(d');.;(d^{(n)});\\(f):(g');.;(g^{(n)});\end{matrix} r_1 u,.,r_n u\right] du$$

$$= \frac{z^a}{a}\sum\frac{(a+m+m_1+.+m_n)((c),m_1+.+m_n)((d'),m_1)..((d^{(n)}),m_n)}{(a+1,m+m_1+.+m_n)((f),m_1+.+m_n)((g'),m_1)..((g^{(n)}),m_n)}$$

$$\times\frac{z^m(r_1z)^{m_1}..(r_nz)^{m_n}}{m!m_1!...m_n!} \qquad (3.1.3.5)$$

and

$$\int_0^z u^{a-1}\ e^{-u}\ F^{C:D}_{F:G}\left[\begin{matrix}(c):(d');.;(d^{(n)});\\(f):(g');.;(g^{(n)});\end{matrix} r_1 u,.,r_k u,r_{k+1},.,r_n\right] du$$

$$= \frac{z^a}{a}\sum\frac{(a,m+m_1+.+m_k)((c),m_1+.+m_n)((d'),m_1)..((d^{(n)}),m_n)}{(a+1,m+m_1+.+m_k)((f),m_1+.+m_n)((g'),m_1)..((g^{(n)}),m_n)}$$

$$\times\frac{z^m(r_1z)^{m_1}..(r_kz)^{m_k}\ r_{k+1}^{m_{k+1}}..r_n^{m_n}}{m!m_1!.....m_n!}\ . \qquad (3.1.3.6)$$

In both (3.1.3.5) and (3.1.3.6), it is taken that $\mathrm{Re}(a) > 0$ and that $|z|$ is sufficiently small to ensure the convergence of the multiple hypergeometric functions of the integrand over the range of integration, unless these functions terminate.

3.1.4 Special Cases

Certain representative examples of special cases of some of the integrals of Sections 3.1.1 to 3.1.3 will now be discussed. If we express the Gegenbauer polynomial in the integrand of

$$\int_0^z u^{a-1}(1-u)^{b-1}\ C_n^c(1-2ru)du \qquad (3.1.4.1)$$

as a Gauss function, a special case of (3.1.1.2) is obtained, and so we may apply (3.1.1.3). Hence, the integral (3.1.4.1) may be evaluated to give

$$\frac{(2c,n)z^a}{n!\quad a}F^{1:2;1}_{1:1;0}\left(\begin{matrix}a\ :-n,n+2c;1-b;\\a+1:\ c+1/2\ ;\ -\ ;\end{matrix}zr,z\right). \qquad (3.1.4.2)$$

Similarly, we have the following integral involving a generalised Laguerre polynomial:-

$$\int_0^z u^{a-1}(1-u)^{b-1}L_n^c(ru)\,du = \frac{(c+1,n)z^a}{n!\;a}F^{1:1;1}_{1:1;0}\left(\begin{matrix}a\;;\;-n;1-b;\\a+1;c+1;\;-\;;\end{matrix}zr,z\right). \tag{3.1.4.3}$$

The complete elliptical integral of the third kind may be written as an Appell function of the first kind. Hence, we may evaluate the integral

$$\int_0^z u^{a-1}(1-u)^{b-1}\,\Pi(ru,k\sqrt{u})\,du$$

$$= \frac{\pi z^a}{2a}F^{(3)}\left(\begin{matrix}a\;::1/2:-:-;1/2;1;1-b;\\a+1::\;1\;:-:-;\;-\;;-;\;-\;;\end{matrix}k^2z,-rz,z\right) \tag{3.1.4.4}$$

by means of (3.1.2.6).

Also, using (3.1.2.8), we have

$$\int_0^z u^{a-1}(1-u)^{b-1}F_4(c,d;f,f';ru,s[1-u])\,du$$

$$= \frac{z^a}{a}F^{(3)}\left(\begin{matrix}-::c,d:1-b:\;a\;;-;\;-;-;\\-::\;-\;:\;-\;:a+1;c;c';-;\end{matrix}rz,s,z\right) \tag{3.1.4.5}$$

When turning to special cases of definite integrals of multiple hypergeometric functions, apart from the product of several single hypergeometric functions, very few classical special functions fall directly into the multiple hypergeometric category; even though such functions appear in divers applications, they have only recently been presented in hypergeometric form. See Exton (1976) Chapters 7 and 8, for example. B.C. Carlson (1961) has fairly recently expressed the incomplete elliptical integral of the third kind as a Lauricella function of the fourth kind in three variables. It thus follows that the integral

$$\int_0^z u^{a-1}(1-u)^{b-1}\,\Pi(r\sqrt{u},s,k)\,du \tag{3.1.4.6}$$

may be expressed as a special case of (3.1.3.4) so that it is equal to

$$\frac{rz^{a+1/2}}{a+1/2}\sum\frac{(a+1/2,m+m_1+m_2+m_3)(1/2,m_1+m_2+m_3)(1-b,m)(\frac{1}{2},m_1)(\frac{1}{2},m_2)(1,m_3)}{(a+3/2,m+m_1+m_2+m_3)(3/2,m_1+m_2+m_3)}$$

$$\times\frac{z^m(r^2z)^{m_1}(s^2r^2z)^{m_2}(-kr^2z)^{m_3}}{m!m_1!m_2!m_3!}\,. \tag{3.1.4.7}$$

If we take C=D=F=1 and G=0, C=D=G=1 and F=0 and C=D=0 respectively in (3.1.3.5), for example, we have

$$\int_0^z u^{a-1}(1-u)^{b-1}F_D^{(n)}(c,d_1,..,d_n;f;r_1u,..,r_ku,r_{k+1}[1-u],..,r_n[1-u])\,du = \tag{3.1.4.8}$$

(continued)

(continued)

$$= \frac{z^a}{a}\sum \frac{(a+m+m_1+.+m_k)(1-b,m-m_{k+1}-.-m_n)(b,m_{k+1}+.+m_n)(c,m_1+.+m_n)}{(f,m_1+.+m_n)}$$

$$\times \frac{(d_1,m_1)..(d_n,m_n)z^m(r_1z)^{m_1}..(r_kz)^{m_k}(-r_{k+1})^{m_{k+1}}..(-r_n)^{m_n}}{m!m_1!...m_n!}, \qquad (3.1.4.8)$$

$$\int_0^z u^{a-1}(1-u)^{b-1}F_A^{(n)}\Big[c,d_1,..,d_n;g_1,..,g_n;r_1u,..,r_ku, r_{k+1}[1-u],..,r_n[1-u]\Big]du$$

$$= \frac{z^a}{a}\sum \frac{(a,m+m_1+.+m_k)(1-b,m-m_{k+1}-.-m_n)(b,m_{k+1}+.+m_n)(c,m_1+.+m_n)}{(g_1,m_1)..(g_n,m_n)}$$

$$\times \frac{(d_1,m_1)..(d_n,m_n)z^m(r_1z)^{m_1}..(r_kz)^{m_k}(-r_{k+1})^{m_{k+1}}..(-r_n)^{m_n}}{m!m_1!...m_n!} \qquad (3.1.4.9)$$

and

$$\int_0^z u^{a-1}(1-u)^{b-1}{}_DF_G((d');(g');r_1u)..{}_DF_G((d^{(k)});(g^{(k)});r_ku)$$

$$\times {}_DF_G((d^{(k+1)});(g^{(k+1)});r_{k+1}[1-u])..{}_DF_G((d^{(n)});(g^{(n)});r_n[1-u])du$$

$$= \frac{z^a}{a}\sum \frac{(a,m+m_1+.+m_k)(1-b,m-m_{k+1}-.-m_n)(b,m_{k+1}+.+m_n)}{((g'),m_1)..((g^{(n)}),m_n)}$$

$$\times \frac{((b'),m_1)..((b^{(n)}),m_n)z^m(r_1z)^{m_1}..(r_kz)^{m_k}(-r_{k+1})^{m_{k+1}}..(-r_n)^{m_n}}{m!m_1!...m_n!}, \qquad (3.1.4.10)$$

3.2 DEFINITE INTEGRALS ASSOCIATED WITH THE KUMMER AND BESSEL FUNCTIONS

We now investigate integrals of the following types:-

$$\int_0^z u^{a-1}(1-u)^{b-1}{}_1F_1(c;d;ru)f(u)du, \qquad (3.2.1)$$

$$\int_0^z u^{a-1}\, e^{-u}\,{}_1F_1(c;d;ru)f(u)du, \qquad (3.2.2)$$

$$\int_0^z u^{a-1}(1-u)^{b-1}\, J_c(ru)f(u)du \qquad (3.2.3)$$

and

$$\int_0^z u^{a-1}\, e^{-u}\, J_c(ru)f(u)du. \qquad (3.2.4)$$

As before, it will be assumed that the series expansion

$$f(u) = \sum_{n=0}^{\infty} h_n u^n \tag{3.2.5}$$

has a radius of convergence which is not less than $|z|$. With this assumption, each of the above integrals may be evaluated using term-by-term integration, so that (3.2.1) to (3.2.4) become, respectively

$$\frac{z^a}{a}\sum\frac{(a,m+n+p)(c,n)(1-b,p)h_m z^m(rz)^n(-z)^p}{(a+1,m+n+p)(d,n)\quad n!p!}, \tag{3.2.6}$$

$$\frac{z^a}{a}\sum\frac{(a,m+n+p)(c,n)h_m z^m(rz)^n(-z)^p}{(a+1,m+n+p)(d,n)\quad n!p!}, \tag{3.2.7}$$

$$\frac{z^{a+c}\quad(r/2)^c}{(a+c)\Gamma(c+1)}\sum\frac{(a+c,m+2n+p)(1-b,p)h_m z^m(-z^2r^2/4)^n(-z)^p}{(a+c+1,m+2n+p)(1+c,n)\quad n!p!} \tag{3.2.8}$$

and $$\frac{z^{a+c}(r/2)^c}{(a+c)\Gamma(c+1)}\sum\frac{(a+c,m+2n+p)h_m z^m(-z^2r^2/4)^n(-z)^p}{(a+c+1,m+2n+p)(c+1,n)\quad n!p!}. \tag{3.2.9}$$

3.2.1 Generalised Hypergeometric Functions of One, Two and Several Variables

The most obvious specialisation of f(u) in the form of a single series in its argument is the generalised hypergeometric function ${}_FF_G$, see (3.1.1.1). We are thus led to consider the integral

$$\int_0^z u^{a-1}(1-u)^{b-1}\,{}_1F_1(c;d;ru)\,{}_FF_G((f);(g);su)du, \tag{3.2.1.1}$$

which is a special case of (3.2.6). Hence, (3.2.1.1) may be written as

$$\frac{z^a}{a}F^{(3)}\left(\begin{matrix}a::-:-:-;(f);c;1-b;\\a+1::-:-:-;(g);d;\ -\ ;\end{matrix}sz,rz,z\right). \tag{3.2.1.2}$$

Let f(u) take the form of a Kampé de Fériet function, so that, if applied to (3.2.2), for example, we have

$$\int_0^z u^{a-1}\ e^{-u}\,{}_1F_1(c;d;ru)F^{F:G;G'}_{J:K;K'}\left(\begin{matrix}(f):(g);(g');\\(j):(k);(k');\end{matrix}su,tu\right)du$$

$$= \frac{z^a}{a}\sum\frac{(a,m+n+p+q)((f),m+n)((g),m)((g'),n)(c,p)(sz)^m(tz)^n(ru)^p z^q}{(a+1,m+n+p+q)((j),m+n)((k),m)((k'),n)(d,p)m!n!p!q!}. \tag{3.2.1.3}$$

Similarly, integrals involving generalised Kampé de Fériet

functions of several variables may be evaluated; we have the result

$$\int_0^z u^{a-1} e^{-u^2} J_c(ru) F^{F:G}_{J:K}\left[\begin{matrix}(f):(g');.;(g^{(n)});\\(j):(k');.;(k^{(n)});\end{matrix} t_1u^2,.,t_nu^2\right] du$$

$$= -\frac{(r/2)^c z^{(a+c)/4}}{(a+c)\Gamma(c+1)} \sum \frac{([a+c]/2,p+m+m_1+.+m_n)((f),m_1+.+m_n)}{([a+c+2]/2,p+m+m_1+.+m_n)(c+1,m)((j),m_1+.+m_n)}$$

$$\times \frac{((g'),m_1)..((g^{(n)},m_n)(-z)^{p/2}(-r^2\sqrt{z}/4)^m(t_1\sqrt{z})^{m_1}..(t_n\sqrt{z})^{m_n}}{((k'),m_1)..((k^{(n)}),m_n)\ p!m!m_1!...m_n!}. \tag{3.2.1.5}$$

It appears that no general reducible forms of the integrals discussed in this section exist.

3.3 DEFINITE INTEGRALS ASSOCIATED WITH THE GAUSS FUNCTION

A number of fairly compact results may be obtained if integrals of the general type

$$\int_0^z u^{a-1} {}_2F_1(-N,b-N;b;u/z)f(u)du \tag{3.3.1}$$

are considered. As above, it is taken that

$$f(u) = \sum_{n=0}^{\infty} h_n u^n, \tag{3.3.2}$$

convergent for $|u| \leqq |z|$, and N is a non-negative integer. Hence, (3.3.1) may be written as a sum of Clausen functions. The application of Saalchutz's theorem then enables us to put the integral (3.3.1) in the form

$$\frac{(1,N)(a-b+1,N)z^a}{(1-b,N)(a+1,N)a} \sum \frac{(a,n)(a-b+1+N,n)h_n z^n}{(a-b+1,n)(a+1+N,n)}. \tag{3.3.3}$$

This last result may be employed to investigate a number of integrals involving the Gauss function together with other hypergeometric functions.

3.3.1 Generalised Hypergeometric Function

If Re(a) is positive, a number of integrals of the form

$$\int_0^z u^{a-1} {}_2F_1(-N,b;b-N;u/z) {}_CF_D((c);(d);ru)du \tag{3.3.1.1}$$

are of frequent occurrence. Here, we have specialised (3.3.1) by taking f(u) to be a generalised hypergeometric function of one variable. The formula (3.3.3) may be applied, so that the integral (3.3.1.1) may be evaluated in the form

$$\frac{N!(a-b+1,N)z^a}{a(1-b,N)(1+a,N)}\,{}_{C+2}F_{D+2}\left(\begin{matrix}a,a-b+1+N, & (c);\\ a-b+1,a+1+N,(d);\end{matrix}rz\right). \tag{3.3.1.2}$$

3.3.2 Double Hypergeometric Function

We now consider the two integrals given below, each of which involves a Gauss function and a Kampé de Fériet function.

$$\int_0^z u^{a-1}\,{}_2F_1(-N,b;b-N;u/z)F^{C:D;D'}_{F:G;G'}\left(\begin{matrix}(c):(d);(d');\\ (f):(g);(g');\end{matrix}ru,su\right)du \tag{3.3.2.1}$$

and

$$\int_0^z u^{a-1}\,{}_2F_1(-N,b;b-N;u/z)F^{C:D;D'}_{F:G;G'}\left(\begin{matrix}(c):(d);(d');\\ (f):(g);(g');\end{matrix}ru,s\right)du. \tag{3.3.2.2}$$

On expanding the Kampé de Fériet function of the integrand of (3.3.2.1), the conditions of convergence imposed upon (3.3.2) enable us to integrate term-by-term. A double series of ${}_3F_2(1)$ functions results which may be reduced by Saalschütz's theorem. Hence, (3.3.2.1) may be written as

$$\frac{z^aN!(a-b+1,N)}{a(1-b,N)(a+1,N)}F^{C+2:D;D'}_{F+2:G;G'}\left(\begin{matrix}(c),a,a-b+1+N & :(d);(d');\\ (f),a-b+1,a+1+N: & (g);(g');\end{matrix}rz,sz\right). \tag{3.3.2.3}$$

Similarly, (3.3.2.2) takes the form

$$\frac{z^aN!(a-b+1,N)}{a(1-b,N)(a+1,N)}F^{C:D+2;D'}_{F:G+2;G'}\left(\begin{matrix}(c):a,a-b+1+N, & (d) & ;(d');\\ (f):a-b+1,a+1+N, & (g) & ;(g');\end{matrix}rz,s\right). \tag{3.3.2.4}$$

3.3.3 Multiple Hypergeometric Function

The previous results may be generalised to a further degree by introducing a generalised Kampé de Fériet function of several variables into the integrand of the integral under consideration. We give the following two results as examples which may be obtained in exactly the same fashion as the expressions discussed in the previous section.

$$\int_0^z u^{a-1}\,{}_2F_1(-N,b;b-N;u/z)F^{C:D}_{F:G}\left[\begin{matrix}(c):(b');.;(b^{(n)});\\ (f):(g');.;(g^{(n)});\end{matrix}r_1u,.,r_nu\right]du$$

$$= \frac{z^aN!(a-b+1,N)}{a(1-b,N)(a+1,N)}F^{C+2:D}_{F+2:G}\left[\begin{matrix}(c),a,a-b+1+N & ;(d');.;(d^{(n)});\\ (f),a-b+1,a+1+N; & (g');.;(g^{(n)});\end{matrix}r_1z,..,r_nz\right] \tag{3.3.3.1}$$

and

$$\int_0^z u^{a-1}\,{}_2F_1(-N,b;b-N;u/z)\,F^{C:D}_{F:G}\left[\begin{matrix}(c):(b');.;(b^{(n)});\\(f):(g');.;(g^{(n)});\end{matrix}\,r_1u,r_2,.,r_n\right]du$$

$$= \frac{z^aN!(a-b+1,N)}{a(1-b,N)(a+1,N)}\,F^{C:D+2}_{F:G+2}\left[\begin{matrix}(c):a,a-b+1+N\ ,\ (d');p,q,(d'');..;\\(f):a-b+1,a+1+N,(g');p,q,(f'');..\end{matrix}\right.$$

$$\left.\begin{matrix}p,q,(d^{(n)});\\p,q,(g^{(n)});\end{matrix}\,r_1z,r_2,.,r_n\right], \qquad (3.3.3.2)$$

where p and q are dummy parameters introduced to retain the notation of the generalised Kampé de Fériet function on the right of (3.3.3.2).

3.3.4 Special Cases

The special Gauss function

$${}_2F_1(-N,b;b-N;x), \qquad (3.3.4.1)$$

which occurs in the various integrals in Sections 3.3 to 3.3.3, may be expressed in terms of a Jacobi polynomial

$$\frac{N!}{(b-N,N)}\,P_N^{b-1-N,0}(1-2x), \qquad (3.3.4.2)$$

see Erdélyi et al. (1953) Vol. I page 81, and its special case the Legendre polynomial

$$P_N(1-2x), \qquad (3.3.4.3)$$

when b = N+1.

The generalised hypergeometric function associated with (3.3.1.1) may be expressed as a variety of special functions. We consider, as examples, an integral involving a product of Legendre polynomials and an integral involving the product of a Jacobi polynomial and an error function. From (3.3.1.1) and (3.3.1.2), we have

$$\int_0^z u^{a-1}\,P_N(1-\tfrac{2u}{z})P_n(1-2ru)\,du = \frac{(-1)^N(a+N,N)z^a}{(a+1,N)\ a}\,{}_4F_3\left(\begin{matrix}a,a,-n,n+1\ ;\\a-N,a+1+N,1;\end{matrix}\,rz\right),$$

where Re(a) > 0. (3.3.4.4)

Suppose now that a=r=1, when the integral

$$\int_0^z P_N(1-2u/z)P_n(1-2u)\,du \qquad (3.3.4.5)$$

may be expressed in closed form. The right-hand member of

(3.3.4.4) becomes a Clausen function of unit argument to which Saalschütz's theorem may be applied. Hence,

$$\int_0^z P_N(1-2u/z)P_n(1-2u)du = \frac{zN(1+N)}{(N+1+n)(N-n)} . \qquad (3.3.4.6)$$

The integral

$$\int_0^z u^{a-1}P_N^{b-1-N,0}(1-2u/z)\,\mathrm{erf}(ru)du \qquad (3.3.4.7)$$

is now investigated. Replace the special functions of the integrand by their hypergeometric representations, when (3.3.4.7)takes the form

$$\frac{2(b-N,N)}{\sqrt{\pi}\ \ N!}\int_0^z u^a {}_2F_1(-N,b;b-N;u/z)\,{}_1F_1(1/2;3/2;-r^2u^2)du. \qquad (3.3.4.8)$$

In this case, it is most convenient to refer directly to (3.3.3), when (3.3.4.8) may be written as

$$\frac{2(a-b+2,N)(b-N,N)z^{a+1}}{\sqrt{\pi}(1-b,N)(a+2,N)(a+1)}\sum_{n=0}^{\infty}\frac{(a+1,2n)(a-b+2+N,2n)(1/2,n)(-rz)^2}{(a-b+2,2n)(a+2+N,2n)(3/2,n)\ \ n!}. \qquad (3.3.4.9)$$

This, in turn, may be expressed as

$$\frac{2(a-b+2,N)(b-N,N)z^{a+1}}{(1-b,N)(a+2,N)(a+1)}\ {}_5F_4\left[\begin{matrix}\frac{a+1}{2},\frac{a+2}{2},\frac{a-b+N+2}{2},\frac{a-b+N+3}{2},\frac{1}{2};-r^2z^2\\ \frac{a-b+2}{2},\frac{a-b+3}{2},\frac{a+N+3}{2},\ \ \frac{3}{2}\ \ ;\end{matrix}\right], \qquad (3.3.4.10)$$

using Legendre's duplication formula for the gamma function. See Erdélyi et al. (1953) Vol. I page 4.

Examples of special cases of integrals involving double hypergeometric functions may be furnished by the following:

$$\int_0^z u^{a-1}P_N(1-2u/z)\Pi(ru,\sqrt{[su]})du \qquad (3.3.4.11)$$

and

$$\int_0^z u^{a-1}{}_2F_1(-N,b;b-N;u/z)F_4(c,d;f,f';ru,su)du, \qquad (3.3.4.12)$$

where $\Pi(x,y)$ is a complete elliptical integral of the third kind and F_4 is an Appell function of the fourth kind. The integral (3.3.4.11) may be expressed as a special case of the standard form (3.3.2.1), and so may be evaluated to give

$$\frac{\pi z^a(a-N,N)(-1)^N}{2a\ \ (1+a,N)}F^{3:1;1}_{2:0;0}\left(\begin{matrix}1/2,a,a:1/2;1;\\1,a-N\ \ :\ -\ ;-;\end{matrix}sz,rz\right). \qquad (3.3.4.13)$$

We may also use (3.3.2.3) to show that (3.3.4.12) is equal to

$$\frac{z^aN!(a-b+1,N)}{a(1-b,N)(a+1,N)}F^{4:0;0}_{2:1;1}\left(\begin{matrix}a,a-b+1+N,c,d:-;-\ ;\\a-b+1,a+1+N\ \ :f;f';\end{matrix}rz,sz\right). \qquad (3.3.4.14)$$

If we put c=a-b+1 and d=a+N+1 in (3.3.4.14), then this simplifies to give the expression

$$\frac{z^a N!(a-b+1,N)}{a(1-b,N)(a+1,N)} F_4(a,a-b+1+N;f,f';rz,sz). \quad (3.3.4.16)$$

This section is concluded by a brief discussion of three integrals involving special cases of the generalised Kampé de Fériet function of several variables. Consider the integral

$$\int_0^z u^{a-1} {}_2F_1(-N,b;b-N;\tfrac{u}{z}) F_A^{(n)}(c,d_1,.,d_n;f_1,.,f_n;r_1u,.,r_nu)du, \quad (3.3.4.17)$$

when it will be seen that this expression is a special case of the right-hand member of (3.3.3.1). Thus the integral (3.3.4.16) may be evaluated as

$$\frac{z^a N!(a-b+1,N)}{a(1-b,N)(a+1,N)} F^{3:1}_{2:1}\left[\begin{matrix} c,a,a-b+1+N:d_1;.;d_n; \\ a-b+1,a+1+N:f_1;.;f_n; \end{matrix} r_1z,.,r_nz\right]. \quad (3.3.4.17)$$

Also, we have two further special cases of (3.3.3.1) which are worthy of note:-

$$\int_0^z u^{a-1} {}_2F_1(-N,b;b-N;\tfrac{u}{z}) F_D^{(n)}(c,d_1,.,d_n;g;r_1u,.,r_nu)du$$

$$= \frac{z^a N!(a-b+1,N)}{a(1-b,N)(a+1,N)} F^{3:1}_{3:0}\left[\begin{matrix} c,a,a-b+1+N \quad :d_1;.;d_n; \\ g,a-b+1,a+1+N: \quad - \quad ; \end{matrix} r_1z,.,r_nz\right] \quad (3.3.4.18)$$

and

$$\int_0^z u^{a-1} {}_2F_1(-N,b;b-N;\tfrac{u}{z}) {}_2F_1(c_1,d_1;f_1;r_1u)..{}_2F_1(c_n,d_n;f_n;r_nu)du$$

$$= \frac{z^a N!(a-b+1,N)}{a(1-b,N)(a+1,N)} F^{2:2}_{2:1}\left[\begin{matrix} a,a-b+1+N \quad :c_1,d_1;.;c_n,d_n; \\ a-b+1,a+1+N: \quad f_1 \;;.;\; f_n \;; \end{matrix} r_1z,.,r_nz\right]. \quad (3.3.4.19)$$

This last result leads to the following integral of a product of several Legendre polynomials:-

$$\int_0^z u^{a-1} P_N(1-2u/z) P_{m_1}(1-2r_1u)\ldots P_{m_n}(1-2r_nu)du$$

$$= \frac{z^a(a-N,N)(-1)^N}{a\;(a+1,N)} F^{2:2}_{2:1}\left[\begin{matrix} a \;,\; a \quad :-m_1,m_1+1;.;-m_n,m_n+1; \\ a-N,a+1+N: \quad 1 \quad ;.; \quad 1 \quad ; \end{matrix} r_1z,.,r_nz\right]. \quad (3.3.4.20)$$

3.4 REPEATED INTEGRALS

In discussing repeated integrals of hypergeometric functions, we first of all consider general integrals of the types

$$\int_0^z .(n). \int_0^z z^{a-1}(1-z)^{b-1} f(z)(dz)^n \tag{3.4.1}$$

and

$$\int_0^z .(n). \int_0^z z^{a-1}\, e^{-z}\, f(z)(dz)^n. \tag{3.4.2}$$

We expand the function f(z) in the form

$$f(z) = \sum_{m=0}^{\infty} h_m z^m , \tag{3.4.3}$$

where the radius of convergence of (3.4.3) is sufficiently large for the integrals (3.4.1) and (3.4.2) to be treated using term-by-term integration. We may thus write (3.4.1) and (3.4.2) in the forms

$$\sum_{m=0}^{\infty} h_m \int_0^z .(n). \int_0^z z^{a+m-1}(1-z)^{b-1}(dz)^n \tag{3.4.4}$$

and

$$\sum_{m=0}^{\infty} h_m \int_0^z .(n). \int_0^z z^{a+m-1}\, e^{-z}\, (dz)^n , \tag{3.4.5}$$

respectively. The inner integral of (3.4.4) may be written as

$$\sum_{r=0}^{\infty} \frac{(1-b,r)\, z^{a+m+n+r-1}}{r!\ (a+m+r,n)} , \tag{3.4.6}$$

which may be expressed in the form

$$\frac{z^{a+m+n-1}}{(a+m,n)}\, {}_2F_1(a+m,1-b;a+m+n;z). \tag{3.4.7}$$

Hence, the integral (3.4.1) becomes

$$\frac{z^{a+n-1}}{(a,n)} \sum_{m,r=0}^{\infty} \frac{h_m (a,m+r)(1-b,r)\, z^{m+r}}{(a+n,m+r)\quad r!} . \tag{3.4.8}$$

A similar investigation of (3.4.2) enables us to write

$$\int_0^z .(n). \int_0^z z^{a-1}\, e^{-z}\, f(z)(dz)^n$$

$$= \frac{z^{a+n-1}}{(a,n)} \sum_{m,r=0}^{\infty} \frac{h_m (a,m+r)\, z^{m+r}}{(a+n,m+r)\ r!} . \tag{3.4.9}$$

3.4.1 Generalised Hypergeometric Function

In this section, integrals of the types (3.4.1) and (3.4.2) which involve the generalised hypergeometric function of one variable are discussed. If the formula (3.4.8) is used, it follows that

$$\int_0^z .(n). \int_0^z z^{a-1}(1-z)^{b-1}\,{}_CF_D((c);(d);rz)(dz)^n$$

$$= \frac{z^{a+n-1}}{(a,n)} F^{1:A;1}_{1:C;0}\left({a\ :(c);1-b; \atop a+n:(d);\ -\ ;} rz, z\right). \qquad (3.4.1.1)$$

From (3.4.9), we have

$$\int_0^z .(n). \int_0^z z^{a-1}\ e^{-z}\,{}_CF_D((c);(d);rz)(dz)^n$$

$$= \frac{z^{a+n-1}}{(a,n)} F^{1:A;0}_{1:C;0}\left({a\ :(c);-; \atop a+n:(d);-;} rz, z\right). \qquad (3.4.1.2)$$

In the formulae (3.4.1.1) and (3.4.1.2), $\mathrm{Re}(a) > 0$, and in (3.4.1.1) $\mathrm{Re}(b) > 0$ also. If $b=1$, a simpler form of (3.4.1.1) occurs. This is

$$\int_0^z .(n). \int_0^z z^{a-1}\,{}_CF_D((c);(d);rz)(dz)^n = \frac{z^{a+n-1}}{(a,n)}\,{}_{C+1}F_{D+1}\left({(c),\ a\ ; \atop (d), a+n;} rz\right). \qquad (3.4.1.3)$$

3.4.2 Double Hypergeometric Function

The expressions (3.4.8) and (3.4.9) may readily be extended to double and multiple series. If we take $f(z)$ to be a Kampé de Fériet function, we have the following results which may be given in terms of Srivastava's triple hypergeometric function:

$$\int_0^z .(n). \int_0^z z^{a-1}(1-z)^{b-1} F^{C:D;D'}_{F:G;G'}\left({(c):(d);(d'); \atop (f):(g);(g');} rz, sz\right)(dz)^n$$

$$= \frac{z^{a+n-1}}{(a,n)} F^{(3)}\left({a\ ::(c):-:-;(d);(d');1-b; \atop a+n::(f):-:-;(g);(g');\ -\ ;} rz, sz, z\right) \qquad (3.4.2.1)$$

and

$$\int_0^z .(n). \int_0^z z^{a-1}\ e^{-z} F^{C:D;D'}_{F:G;G'}\left({(c):(d);(d'); \atop (f):(g);(g');} rz, sz\right)(dz)^n$$

$$= \frac{z^{a+n-1}}{(a,n)} F^{(3)}\left({a\ ::(c):-:-;(d);(d');-; \atop a+n::(f):-:-;(g);(g');-;} rz, sz, -z\right). \qquad (3.4.2.2)$$

Similarly, we also have other results, for example

$$\int_0^z .(n). \int_0^z z^{a-1}(1-z)^{b-1} F^{C:D;D'}_{F:G;G'}\left({(c):(d);(d'); \atop (f):(g);(g');} rz, s\right)(dz)^n = \qquad (3.4.2.3)$$

(continued)

(continued)

$$= \frac{z^{a+n-1}}{(a,n)} F^{(3)}\left({-::(c):-:\ a\ ;(d);(d');1-b; \atop -::(f):-:a+n;(g);(g');\ -\ ;} rz,s,z\right). \qquad (3.4.2.3)$$

3.4.3 Multiple Hypergeometric Function

Many further generalisations of the previous results may be obtained by the same methods as those used in the previous sections to evaluated repeated integrals involving multiple hypergeometric functions, in particular the generalised Kampé de Fériet function. The following two examples are given without proof:-

$$\int_0^z .(n). \int_0^z z^{a-1}(1-z)^{b-1} F^{C:D}_{F:G}\left[{(c):(d');.;(d^{(s)}); \atop (f):(g');.;(g^{(s)});} r_1 z,.,r_s z\right] (dz)^n$$

$$= \frac{z^{a+n-1}}{(a,n)} \sum \frac{(a,m+m_1+.+m_s)((c),m_1+.+m_s)((d'),m_1)..((d^{(s)}),m_s)}{(a+n,m+m_1+.+m_s)((f),m_1+.+m_s)((g'),m_1)..((g^{(s)}),m_s)}$$

$$\times \frac{(1-b,m)(r_1 z)^{m_1}..(r_s z)^{m_s} z^m}{m!m_1!...m_s!} \qquad (3.4.3.1)$$

and

$$\int_0^z .(n). \int_0^z z^{a-1}(1-z)^{b-1} F^{C:D}_{F:G}\left[{(c):(d');.;(d^{(s)}); \atop (f):(g');.;(g^{(s)});} r_1 z,.,r_k z, r_{k+1},.,r_s\right] \times (dz)^n$$

$$= \frac{z^{a+n-1}}{(a,n)} \sum \frac{(a,m+m_1+.+m_k)((c),m_1+.+m_s)((d'),m_1)..((d^{(s)}),m_s)}{(a+n,m+m_1+.+m_k)((f),m_1+.+m_s)((g'),m_1)..((g^{(s)}),m_s)}$$

$$\times \frac{(1-b,m)(r_1 z)^{m_1}..(r_k z)^{m_k}\, r_{k+1}^{m_{k+1}}..r_s^{m_s}}{m!m_1!..m_s!}. \qquad (3.4.3.2)$$

If b=1, the right-hand member of (3.4.3.1), for example, assumes the more compact form

$$\frac{z^{a+n-1}}{(a,n)} F^{C+1:D}_{F+1:G}\left[{(c),\ a\ :(d');.;(d^{(s)}); \atop (f),a+n:(g');.;(g^{(s)});} r_1 z,.,r_s z\right]. \qquad (3.4.3.3)$$

3.4.4 Special Cases

A number of special cases of repeated integrals each involving a single hypergeometric function are now mentioned. The Bessel function of the first kind may be represented as a ${}_0F_1$ function. Hence, from (3.4.1.3), we have the result

$$\int_0^z .(n). \int_0^z z^{a-1} J_c(rz)(dz)^n = \frac{(\frac{r}{2})^c z^{a+c+n-1}}{\Gamma(c+1)(a+c,n)} {}_2F_3\left[\begin{matrix} \frac{a+c}{2}, \frac{a+c+1}{2}; & -r^2z^2/4 \\ \frac{a+c+n}{2}, \frac{a+c+n+1}{2}, c+1; & \end{matrix}\right].$$

(3.4.4.1)

If we now consider the modified Bessel function, we have

$$\int_0^z .(n). \int_0^z z^{a-1} I_c(r\sqrt{z})(dz)^n = \frac{(-\frac{r}{2})^c z^{a+\frac{c}{2}+n-1}}{\Gamma(c+1)(a+\frac{c}{2},n)} {}_1F_2\left[\begin{matrix} a+\frac{c}{2} & ; \\ a+n+\frac{c}{2}, c+1; & \end{matrix} r^2z/4\right], \tag{3.4.4.2}$$

and, on putting a = c/2 + 1, the simpler special case given below may be obtained.

$$\int_0^z .(n). \int_0^z z^{c/2} I_c(r\sqrt{z})(dz)^n = \frac{(-2)^n}{r^n z^{(c+n)/2}} I_{c+n}(r\sqrt{z}). \tag{3.4.4.3}$$

If we replace the Laguerre polynomial by the corresponding confluent hypergeometric function, we may again use (3.4.1.3) so as to obtain the formula

$$\int_0^z .(n). \int_0^z z^{a-1} L_m^c(rz)(dz)^n = \frac{(c+1,m)z^{a+n-1}}{m!\ (a,n)} {}_2F_2\left(\begin{matrix} a\ , & -m; \\ a+n, & c+1; \end{matrix} rz\right). \tag{3.4.4.4}$$

There are also the further special cases

$$\int_0^z .(n). \int_0^z z^c L_m^c(rz)(dz)^n = \frac{z^{c+n}}{(c+1+m,n)} L_m^{c+n}(rz) \tag{3.4.4.5}$$

and

$$\int_0^z .(n). \int_0^z z^{a-1} {}_1F_1(a+n;a;rz)\ (dz)^n = \frac{z^{a+n-1}}{(a,n)} e^{rz}\ . \tag{3.4.4.6}$$

The repeated integral

$$\int_0^z .(n). \int_0^z z^{a-1} P_m(1-2rz)(dz)^n \tag{3.4.4.7}$$

is now investigated. The Legendre polynomial is written in hypergeometric form, so that (3.4.4.7) may be evaluated immediately in the form

$$\frac{z^{a+n-1}}{(a,n)} {}_3F_2\left(\begin{matrix} a, -m, m+1; \\ a+n,\ 1\ ; \end{matrix} rz\right), \tag{3.4.4.8}$$

and if a = 1, we have

$$\int_0^z .(n). \int_0^z P_m(1-2rz)(dz)^n = \frac{z^n(m!)^2}{n!(m+n)!} P_m^{n,-n}(1-2rz), \tag{3.4.4.9}$$

where the repeated integral of a Legendre polynomial is expressed in terms of a special Jacobi polynomial.

In conclusion, repeated integrals of the Lauricella functions $F_A^{(m)}$ and $F_D^{(m)}$ are evaluated by specialising the formula (3.4.3.3):

$$\int_0^z .(n). \int_0^z z^{a-1} F_A^{(m)}(c,d_1,..,d_m;f_1,..,f_m;r_1z,..,r_mz)(dz)^n$$

$$= \frac{z^{a+n-1}}{(a,n)} F^{2:1}_{1:1}\left[\begin{matrix} a,c:d_1,..,d_m; \\ a+n:f_1,..,f_m; \end{matrix} r_1z,..,r_mz\right] \tag{3.4.4.10}$$

and

$$\int_0^z .(n). \int_0^z z^{a-1} F_D^{(m)}(c,d_1,..,d_m;f;r_1z,..,r_mz)(dz)^n$$

$$= \frac{z^{a+n-1}}{(a,n)} F^{2:1}_{2:0}\left[\begin{matrix} c,\ a\ :d_1,..,d_m; \\ f,a+n:\ \ —\ \ ; \end{matrix} r_1z,..,r_mz\right] . \tag{3.4.4.11}$$

These results also yield the formulae

$$\int_0^z .(n). \int_0^z z^{a-1} F_A^{(m)}(a+n,d_1,..,d_m;f_1,..,f_m;r_1z,..,r_mz)(dz)^n$$

$$= \frac{z^{a+n-1}}{(a,n)} F_A^{(m)}(a,d_1,..,d_m;f_1,..,f_m;r_1z,..,r_mz), \tag{3.4.4.12}$$

$$\int_0^z .(n). \int_0^z z^{a-1} F_D^{(m)}(a+n,d_1,..,d_m;f;r_1z,..,r_mz)(dz)^n$$

$$= \frac{z^{a+n-1}}{(a,n)} F_D^{(m)}(a,d_1,..,d_m;f;r_1z,..,r_mz) \tag{3.4.4.13}$$

and

$$\int_0^z .(n). \int_0^z z^{a-1} F_D^{(m)}(a+n,d_1,..,d_m;a;r_1z,..,r_mz)(dz)^n$$

$$= \frac{z^{a+n-1}}{(a,n)}(1-r_1z)^{-d_1}...(1-r_mz)^{-d_m} . \tag{3.4.4.14}$$

Chapter 4

Contour Integrals

4.1 POCHHAMMER INTEGRALS

The integrals of Euler type, such as are discussed in Chapter 2, may be replaced by contour integrals, where the path of integration is a Pochhammer double-loop. See Whittaker and Watson (1952) page 256. Such a loop begins from a point P, say, between 0 and 1, encircles 0 then 1 in the positive direction and then encircles the same two points again in the negative direction, returning to the point P. Such a contour will be denoted by C unless otherwise stated. The advantage of such contour integration is that the restrictions on the parameters of Euler integrals may be lifted. Double-loop integrals are also a very powerful tool in the investigation of hypergeometric differential systems. See Erdélyi (1950) for example.

We begin by considering the general type of integral

$$\int_C u^{r-1}(1-u)^{r'-1} f(u)\, du, \qquad (4.1.1)$$

where

$$f(u) = \sum_{m=0}^{\infty} h_m u^m . \qquad (4.1.2)$$

If the contour C can be deformed, without crossing or encircling any of the singularities of the integrand of (4.1.1), so that the series representation (4.1.2) converges uniformly upon it, then the integral (4.1.1) may be replaced by

$$\sum_{m=0}^{\infty} h_m \int_C u^{r+m-1}(1-u)^{r'-1}\, du. \qquad (4.1.3)$$

The above inner integral may be evaluated by means of the useful formula [see Exton (1976) page 17]

$$\int_C (-u)^{r-1}(u-1)^{r'-1}\, du = \frac{(2\pi i)^2}{\Gamma(1-r)\Gamma(1-r')\Gamma(r+r')}, \qquad (4.1.4)$$

on the understanding that u^r is interpreted as $\exp(r \log u)$, where u is real, positive and continuous on the contour. Hence, (4.1.1) becomes

$$\frac{(-1)^{r+r'}\,(2\pi i)^2}{\Gamma(1-r)\Gamma(1-r')\Gamma(r+r')} \sum_{m=0}^{\infty} \frac{h_m\,(r,m)}{(r+r',m)} . \qquad (4.1.5)$$

Suppose that f(u) may be expanded as a double series

$$f(u) = \sum_{m,n=0}^{\infty} h_{m,n}\, u^m (1-u)^n \,, \qquad (4.1.6)$$

then (4.1.1) now takes the form

$$\frac{(-1)^{r+r'}\,(2\pi i)^2}{\Gamma(1-r)\Gamma(1-r')\Gamma(r+r')} \sum_{m,n=0}^{\infty} h_{m,n} \frac{(r,m)(r',n)}{(r+r',m+n)} \,. \qquad (4.1.7)$$

Both of the results (4.1.5) and (4.1.7) may easily be extended to multiple series and will be the basis of the following study of Pochhammer integrals.

4.1.1 Generalised Hypergeometric Function

If we let f(u) in the integral (4.1.1) take the form of a generalised hypergeometric function of one variable, we have

$$\int_C u^{r-1}(1-u)^{r'-1}\,{}_AF_B((a);(b);ux)\,du$$

$$= \frac{(-1)^{r+r'}\,(2\pi i)^2}{\Gamma(1-r)\,\Gamma(1-r')\,\Gamma(r+r')}\,{}_{A+1}F_{B+1}(r,(a);r+r',(b);x), \qquad (4.1.1.1)$$

which holds provided that the contour of integration, C, can be deformed so that the hypergeometric function of the integrand is convergent upon it. Any values of the parameters, either here, or later in the text, which make any of the associated gamma functions infinite are tacitly excluded.

If we let $r=b_1$, then the hypergeometric function on the right of (4.1.1.1) assumes a simpler form:

$${}_AF_B(a_1,..,a_A;b_1+r',b_2,..,b_B;x), \qquad (4.1.1.2)$$

and if $r'=a_1-b_1$ also, a further simplification results:

$${}_{A-1}F_{B-1}(a_2,..,a_A;b_2,..,b_B;x). \qquad (4.1.1.3)$$

For special values of x, a number of integrals of the type (4.1.1.1) may be evaluated in closed form. For example, if x is made equal to unity, and A=1 and B=0, the hypergeometric function on the right of (4.1.1.1) may be summed using Gauss's summation theorem, see Slater (1966) Appendix III. This gives the result

$$\int_C u^{b-1}(1-u)^{r-1}\,{}_2F_1\left({a,d;\atop b\ ;}u\right)du = \frac{(-1)^{b+r}\,(2\pi i)^2\,\Gamma(b+r-a-d)}{\Gamma(b+r-a)\Gamma(b+r-d)\Gamma(1-b)\Gamma(1-r)} \,. \qquad (4.1.1.4)$$

It is understood that Re(b+r-a-d) > 0, or that at least one of the parameters a or d is a non-positive integer.

A further useful result is

$$\int_C u^{r-1}(1-u)^{a-1-N}\,{}_2F_1(a,-N;b;u)\,du$$

$$= \frac{(-1)^{r+a-b+1}\,(2\pi i)^2\,(b-r,N)\,(b-a,N)}{\Gamma(1-r)\Gamma(b-a-r)\Gamma(r+a-b+1)(b,N)(b-r-a,N)}, \qquad (4.1.1.5)$$

which has been obtained in a similar fashion to (4.1.1.4) and where Saalschütz's theorem has been used in place of Gauss's theorem.

An integral similar to (4.1.1.1) is

$$\int_C u^{r-1}(1-u)^{r'-1}\,{}_AF_B((a);(b);\tfrac{ux}{1-u})\,du. \qquad (4.1.1.6)$$

If the hypergeometric function of the integrand is expanded in series, we have, after interchanging the operations of integration and summation,

$$\sum_{m=0}^{\infty} \frac{((a),m)x^m}{((b),m)m!}\int_C u^{r+m-1}(1-u)^{r'-m-1}du, \qquad (4.1.1.7)$$

and the inner integral is equal to

$$\frac{(2\pi i)^2}{\Gamma(1-r-m)\Gamma(1-r'+m)\Gamma(r+r')}, \qquad (4.1.1.8)$$

so that (4.1.1.6) becomes

$$\frac{(-1)^{r+r'}\,(2\pi i)^2}{\Gamma(1-r)\Gamma(1-r')\Gamma(r+r')}\,{}_{A+1}F_{B+1}\left(\begin{matrix}(a), r\ ;\\ (b), 1-r';\end{matrix}-x\right). \qquad (4.1.1.9)$$

If $x = -1$ and $r'=1-r$, we have

$$\int_C u^{r-1}(1-u)^{-r}\,{}_2F_1\left(\begin{matrix}a,d;\\ b\ ;\end{matrix}\frac{u}{1-u}\right)du = -\frac{(2\pi i)^2\Gamma(b)\Gamma(b-a-d)}{\Gamma(1-r)\Gamma(r)\Gamma(b-a)\Gamma(b-d)}, \qquad (4.1.1.10)$$

provided that $\mathrm{Re}(b-a-d) > 0$. Other results expressible in closed form similar to (4.1.1.4) and (4.1.1.10) may be obtained.

4.1.2 Double and Multiple Hypergeometric Functions

The results of the previous section may be generalised to cover the cases where double and multiple hypergeometric functions occur. We employ the generalised Kampé de Fériet function of Karlsson as the most convenient generalised hypergeometric function of several variables. The integrals (4.1.2.1) and (4.1.2.2) given below will be evaluated. These are

$$\int_C u^{r-1}(1-u)^{r'-1}\, F^{A:B}_{F:G}\left[\begin{matrix}(a):(b');.;(b^{(n)});\\(f):(g');.;(g^{(n)});\end{matrix} ux_1,.,ux_n\right] du \quad (4.1.2.1)$$

and

$$\int_C u^{r-1}(1-u)^{r'-1}\, F^{A:B}_{F:G}\left[\begin{matrix}(a):(b');.;(b^{(n)});\\(f):(g');.;(g^{(n)});\end{matrix} \frac{ux_1}{1-u},.,\frac{ux_n}{1-u}\right] du. \quad (4.1.2.2)$$

The multiple hypergeometric function of the integrand of (4.1.2.1) is expanded in series, and if it is possible to deform the contour of integration, C, so that this multiple series is uniformly convergent upon it, we may integrate term-by-term. The integral (4.1.2.1) may then be written in the form

$$\sum \frac{((a),m_1+.+m_n)((b'),m_1)..((b^{(n)}),m_n)x_1^{m_1}..x_n^{m_n}}{((f),m_1+.+m_n)((g'),m_1)..((g^{(n)}),m_n)m_1!..m_n!}$$

$$\times \int_C u^{r+m_1+.+m_n-1}(1-u)^{r'-1}du. \quad (4.1.2.3)$$

The inner integral may be evaluated by means of (4.1.4) as

$$\frac{(2\pi i)^2\ (-1)^{r+r'}\ (r,m_1+.+m_n)}{\Gamma(1-r)\Gamma(1-r')\Gamma(r+r')\ (r+r',m_1+.+m_n)}\ . \quad (4.1.2.4)$$

Hence, it follows that

$$\int_C u^{r-1}\ (1-u)^{r'-1}\, F^{A:B}_{F:G}\left[\begin{matrix}(a):(b');.;(b^{(n)});\\(f):(g');.;(g^{(n)});\end{matrix} ux_1,.,ux_n\right] du$$

$$= \frac{(2\pi i)^2\ (-1)^{r+r'}}{\Gamma(1-r)\Gamma(1-r')\Gamma(r+r')}\, F^{A+1:B}_{F+1:G}\left[\begin{matrix}(a),\ r\ \ :(b');.;(b^{(n)});\\(f),r+r':(g');.;(g^{(n)});\end{matrix} x_1,.,x_n\right].$$

(4.1.2.5)

Similarly,

$$\int_C u^{r-1}\ (1-u)^{r'-1}\, F^{A:B}_{F:G}\left[\begin{matrix}(a):(b');.;(b^{(n)});\\(f):(g');.;(g^{(n)});\end{matrix} \frac{ux_1}{1-u},.,\frac{ux_n}{1-u}\right] du$$

$$= \frac{(2\pi i)^2\ (-1)^{r+r'}}{\Gamma(1-r)\Gamma(1-r')\Gamma(r+r')}\, F^{A+1:B}_{F+1:G}\left[\begin{matrix}(a),\ r\ \ :(b');.;(b^{(n)});\\(f),1-r':(g');.;(g^{(n)});\end{matrix} -x_1,.,-x_n\right].$$

(4.1.2.6)

If $r=f_1$ and $r' = a_1-f_1$, (4.1.2.5) reduces to

$$\int_C u^{f_1-1}(1-u)^{f_1-a_1-1}\, F^{A:B}_{F:G}\left[\begin{matrix}(a):(b');.;(b^{(n)});\\(f):(g');.;(g^{(n)});\end{matrix}ux_1,.,ux_n\right]du$$

$$= \frac{(2\pi i)^2\ (-1)^{a_1}}{\Gamma(1-f_1)\Gamma(1-f_1+a_1)\Gamma(a_1)} F^{A-1:B}_{F-1:G}\left[\begin{matrix}a_2,.,a_A:(b');.;(b^{(n)});\\f_2,.,f_F:(f');.;(f^{(n)});\end{matrix}x_1,.,x_n\right], \tag{4.1.2.7}$$

and if $r'=1-r$, (4.1.2.6) may be written

$$\int_C u^{r-1}(1-u)^{-r}\, F^{A:B}_{F:G}\left[\begin{matrix}(a):(b');.;(b^{(n)});\\(f):(g');.;(g^{(n)});\end{matrix}\frac{ux_1}{1-u},.,\frac{ux_n}{1-u}\right]du$$

$$= \frac{(2\pi i)^2}{\Gamma(1-r)\Gamma(r)} F^{A:B}_{F:G}\left[\begin{matrix}(a):(b');.;(b^{(n)});\\(f):(g');.;(g^{(n)});\end{matrix}-x_1,.,-x_n\right]. \tag{4.1.2.8}$$

Other useful results may be deduced from (4.1.2.5) and (4.1.2.6), and many other similar formulae of a general nature exist.

4.1.2 Special Cases

This section is devoted to giving a few special cases of the double-loop integrals discussed above. Consider the integral

$$\int_C u^{r-1}(1-u)^{r'-1}\, P_n(1-2ux)du = \int_C u^{r-1}(1-u)^{r'-1}\, {}_2F_1\left(\begin{matrix}-n,n+1;\\1\ ;\end{matrix}ux\right)du. \tag{4.1.3.1}$$

This is now in the standard form (4.1.1.1), and so becomes

$$\frac{(-1)^{r+r'}\ (2\pi i)^2}{\Gamma(1-r)\Gamma(1-r')\Gamma(r+r')}\ {}_3F_2\left(\begin{matrix}r,-n,n+1;\\r+r',\ 1\ ;\end{matrix}x\right). \tag{4.1.3.2}$$

On letting $r+r'=n+1$ and $x=1$, (4.1.3.2) may be summed by Gauss's theorem, and we then have

$$\int_C u^{r-1}(1-u)^{n+1-r}\, P_n(1-2u)du = 4(-1)^{n+1}\pi\sin(\pi r)\left[\frac{(1-r,n)}{(r,n)}\right]^2. \tag{4.1.3.3}$$

The integral $$\int_C u^{r-1}(1-u)^{r'-1}\, L_n^c(ux)du \tag{4.1.3.4}$$

may be expressed in the form

$$\frac{(c+1,n)(-1)^{r+r'}\ (2\pi i)^2}{n!\Gamma(1-r)\Gamma(1-r')\Gamma(r+r')}\ {}_2F_2\left(\begin{matrix}r\ ,\ -n;\\r+r',c+1;\end{matrix}x\right) \tag{4.1.3.5}$$

by replacing the Laguerre polynomial by its hypergeometric

representation and again making use of (4.1.1.1). If, in addition, we take r = c+1, we have the interesting special case

$$\int_C u^c(1-u)^{r'-1}\, L_n^c(ux)du = \frac{(c+1,n)(-1)^{c+r'}\ (2\pi i)^2}{(c+r',n)\Gamma(1-c)\Gamma(1-r')\Gamma(a+r')} L_n^{c+r'-1}(x), \tag{4.1.3.6}$$

provided that c+r' > 0. Similarly, it follows that

$$\int_C u^{r-1}(1-u)^{r'-1}\, J_c(x\sqrt{u})du$$

$$= \frac{(x/2)^c(-1)^{r+r'+c/2}\ (2\pi i)^2}{\Gamma(c+1)\Gamma(1-r-c/2)\Gamma(1-r')\Gamma(r+r'+c/2)}\ {}_1F_2\left({r+c/2 \atop r+r'+c/2,c+1};-\frac{x^2}{4}\right) \tag{4.1.3.7}$$

and $$\int_C u^{r-1}(1-u)^{-1}\, J_c(x\sqrt{u})du = 4(-1)^{r+\frac{c}{2}+1}\ \pi\sin[\pi(r+\tfrac{c}{2})]J_c(x). \tag{4.1.3.8}$$

This last result is obtained from (4.1.3.7) by putting r'=0.

The expression (4.1.3.8) is a special case of the much more general result

$$\int_C u^{r-1}(1-u)^{-1}\ {}_AF_B((a);(b);ux)du = 4(-1)^{r+1}\ \pi\sin(\pi r)\ {}_AF_B\left({(a); \atop (b);}x\right) \tag{4.1.3.9}$$

which also follows from (4.1.1.1) by letting r'=0.

In considering special cases of integrals involving hypergeometric functions of several variables, we have, for example

$$\int_C u^{r-1}(1-u)^{r'-1}\ \Pi(cu,k\sqrt{u})du. \tag{4.1.3.10}$$

This is an integral of a complete elliptical integral of the third kind, which, in turn, may be written as an Appell function F_1. This integral may then be expressed as a special case of (4.1.2.5) which may then be used to evaluate it in the form

$$\frac{\pi\ (-1)^{r+r'}\ (2\pi i)^2}{2\Gamma(1-r)\Gamma(1-r')\Gamma(r+r')}\ F^{2:1;1}_{2:0;0}\left({r,1/2\ :1/2;1; \atop r+r',1:\ -\ ;-;}k^2,-c\right). \tag{4.1.3.11}$$

We now give an example of a double-loop integral involving an Appell function of the second kind:

$$\int_C u^{r-1}(1-u)^{r'-1}\ F_2(a,b,b';d,d';ux,[1-u]y)du$$

$$= \frac{(-1)^{r+r'}\ (2\pi i)^2}{\Gamma(1-r)\Gamma(1-r')\Gamma(r+r')}\ F^{1:2;2}_{1:1;1}\left({a\ \ :b,r;b',r'; \atop r+r':\ d\ ;\ \ d'\ ;}x,y\right), \tag{4.1.3.12}$$

by the use of (4.1.6) and (4.1.7). If $a=r+r'$, then the function on the right of the preceding expression splits up into the product of a pair of Gauss functions.

This section on examples of special double-loop integrals is concluded by a brief discussion of two integrals, involving respectively the Lauricella functions of the first and fourth kinds. These are

$$\int_C u^{r-1}(1-u)^{r'-1} F_A^{(n)}(a,b_1,..,b_n;d_1,..,d_n;ux_1,..,ux_n)du$$

$$= \frac{(2\pi i)^2 \; (-1)^{r+r'}}{\Gamma(1-r)\Gamma(1-r')\Gamma(r+r')} F^{2:1}_{1:1}\left[\begin{matrix} a,r :b_1;.;b_n; \\ r+r':d_1;.;d_n; \end{matrix} x_1,..,x_n\right] \qquad (4.1.3.13)$$

and

$$\int_C u^{r-1}(1-u)^{r'-1} F_D^{(n)}(a,b_1,..,b_n;d;\frac{ux_1}{1-u},..,\frac{ux_n}{1-u})du$$

$$= \frac{(2\pi i)^2 \; (-1)^{r+r'}}{\Gamma(1-r)\Gamma(1-r')\Gamma(r+r')} F^{2:1}_{2:0}\left[\begin{matrix} a, \; r :b_1;.;b_n; \\ d,1-r': \; - \; ; \end{matrix} x_1,..,x_n\right], \qquad (4.1.3.14)$$

and they follow from (4.1.2.5) and (4.1.2.6) respectively. The right-hand members of (4.1.3.13) and (4.1.3.14) reduce respectively to

$$4(-1)^{r+1} \pi\sin(\pi r)F_A^{(n)}(a,b_1,..,b_n;d_1,..,d_n;x_1,..,x_n) \qquad (4.1.3.15)$$

and

$$4\pi\sin(\pi r')F_D^{(n)}(a,b_1,..,b_n;d;x_1,..,x_n), \qquad (4.1.3.16)$$

where, in the former integral, we put $r'=0$, and in the latter, $r=1-r'$.

4.2 BARNES INTEGRALS

An important class of contour integrals is that where the contour of integration is the straight line, often indented, lying parallel to the imaginary axis, in the positive half-plane. Integrals with this type of contour, denoted by

$$\int_{c-i\infty}^{c+i\infty} f(u)\;du\;, \qquad (4.2.1)$$

were first introduced by Pincherle and then systematically studied by Barnes and Mellin. See Erdélyi et al. (1953) Vol.I page 49. It is assumed that c is real and positive.

Important cases of integrals of this type are

$$\frac{1}{2\pi i}\int_{c-i\infty}^{c+i\infty}\Gamma(a+u)\Gamma(b-u)\,du = \frac{\Gamma(a+b)}{2^{a+b}}, \tag{4.2.2}$$

where $\mathrm{Re}(a+b) > 0$,

and

$$\frac{1}{2\pi i}\int_{c-i\infty}^{c+i\infty}\Gamma(a+u)\Gamma(b+u)\Gamma(f-u)\Gamma(g-u)\,du = \frac{\Gamma(a+f)\Gamma(a+g)\Gamma(b+f)\Gamma(b+g)}{\Gamma(a+b+f+g)}, \tag{4.2.3}$$

where $\mathrm{Re}(a+b+f+g) < 1$.

See Titchmarsh (1948) page 194. The formula (4.2.3) is often referred to as Barnes's first lemma.

Also, by considering the Meijer G-function, see Section 5.2, we have

$$\Gamma\left[\begin{matrix}(a)\\(b)\end{matrix}\right]{}_AF_B\left(\begin{matrix}(a);\\(b);\end{matrix}x\right) = \frac{1}{2\pi i}\int_{c-i\infty}^{c+i\infty}\frac{\Gamma(a_1+u)..\Gamma(a_A+u)\Gamma(-u)}{\Gamma(b_1+u)..\Gamma(b_B+u)}(-x)^u\,du, \tag{4.2.4}$$

so that a number of other types of Barnes integrals expressible in closed form may be obtained by utilising the many summation theorems for the special cases of the generalised hypergeometric function of one variable.

We now discuss two integrals of Barnes type which involve a general function $F(u,z)$ which may be expanded as a series of the type

$$F(u,z) = \sum_{n=0}^{\infty}\Phi_n(u)E_n(z), \tag{4.2.5}$$

where $E_n(z)$ is an arbitrary coefficient independent of u. Suppose, first of all, that $\Phi_n(u)$ is of the form

$$\Phi_n(u) = (a+u,n)(b-u,n), \tag{4.2.6}$$

where $\mathrm{Re}(a+b) > 0$. Consider the integral

$$1/(2\pi i)\int_{c-i\infty}^{c+i\infty}\Gamma(a+u)\Gamma(b-u)F(u,z)\,du$$

$$= \sum_{n=0}^{\infty}E_n(z)\;/(2\pi i)\int_{c-i\infty}^{c+i\infty}\Gamma(a+n+u)\Gamma(b+n-u)\,du. \tag{4.2.7}$$

The inner integral on the right of (4.2.7) is of the same type as (4.2.2). The reversal of the operations of integration and summation is justified because the series in question is absolutely and uniformly convergent upon the contour of integration.

Hence,

$$\frac{1}{2\pi i}\int_{c-i\infty}^{c+i\infty}\Gamma(a+u)\Gamma(b-u)F(u,z)du = \frac{\Gamma(a+b)}{2^{a+b}}\sum_{n=0}^{\infty}E_n(z)(\tfrac{a+b}{2},n)(\tfrac{a+b+1}{2},n). \tag{4.2.8}$$

Similarly, if $\Phi_n(u) = (a+u,n)$, we have

$$\frac{1}{2\pi i}\int_{c-i\infty}^{c+i\infty}\Gamma(a+u)\Gamma(b-u)F(u,z)du = \frac{\Gamma(a+b)}{2^{a+b}}\sum_{n=0}^{\infty}E_n(z)(a+b,n)2^{-n}. \tag{4.2.9}$$

These two general results may be extended to cover double and multiple series.

4.2.1 Generalised Hypergeometric Function

We now suppose that the function $F(u,z)$ as given by (4.2.5) is of the form of a generalised hypergeometric function of one variable ${}_AF_B$. The formula (4.2.8) is now considered in the form

$$1/(2\pi i)\int_{c-i\infty}^{c+i\infty}\Gamma(a_1+u)\Gamma(a_2-u)\,{}_AF_B\left[\begin{matrix}a_1+u,a_2-u,a_3,..,a_A; & z\\ (b) & ;\end{matrix}\right]du$$

$$= \frac{\Gamma(a_1+a_2)}{2^{a_1+a_2}}\,{}_AF_B\left[\begin{matrix}[a_1+a_2]/2,[a_1+a_2+1]/2,a_3,..,a_A; & z\\ (b) & ;\end{matrix}\right], \tag{4.2.1.1}$$

$$\mathrm{Re}(a_1+a_2) > 0.$$

Also, (4.2.9) may be written

$$1/(2\pi i)\int_{c-i\infty}^{c+i\infty}\Gamma(a_1+u)\Gamma(d-u)\,{}_AF_B\left[\begin{matrix}a_1+u,a_2,..,a_A; & z\\ (b) & ;\end{matrix}\right]du$$

$$= \frac{\Gamma(a_1+d)}{2^{a_1+d}}\,{}_AF_B\left[\begin{matrix}a_1+d,a_2,..,a_A; & z/2\\ (b) & ;\end{matrix}\right], \quad \mathrm{Re}(a_1+d) > 0. \tag{4.2.1.2}$$

4.2.2 Double Hypergeometric Function

By means of similar processes, a number of integrals involving the Kampé de Fériet function may be evaluated. Consider the integral

$$\frac{1}{2\pi i}\int_{c-i\infty}^{c+i\infty}\Gamma(a_1+u)\Gamma(a_2-u)F^{A:B;B'}_{F:G;G'}\left[\begin{matrix}a\ \ +u,a_2-u,a_3,..,a_A:(b);(b'); & x,y\\ (f)\ \ :(g);(g'); & \end{matrix}\right]du,$$

$$\text{where } \mathrm{Re}(a_1+a_2) > 0. \tag{4.2.2.1}$$

A two-dimensional version of the formula (4.2.7) enables us to express this integral in the form

$$\frac{\Gamma(a_1+a_2)}{2^{a_1+a_2}} F^{A:B;B'}_{F:G;G'}\left[\begin{matrix}[a_1+a_2]/2,[a_1+a_2+1]/2,a_3,..,a_A:(b);(b');\\ (f)\qquad\qquad :(g);(g');\end{matrix} x,y\right].$$

(4.2.2.2)

Similarly, we may write

$$\frac{1}{2\pi i}\int_{c-i\infty}^{c+i\infty}\Gamma(a_1+u)\Gamma(d-u)F^{A:B;B'}_{F:G;G'}\left[\begin{matrix}a_1+u,a_2,..,a_A:(b);(b');\\ (f)\qquad :(g);(g');\end{matrix} x,y\right]du$$

$$=\frac{\Gamma(a_1+d)}{2^{a_1+d}}F^{A:B;B'}_{F:G;G'}\left[\begin{matrix}a_1+d,a_2,..,a_A:(b);(b');\\ (f)\qquad :(g);(g');\end{matrix} x/2,y/2\right], \tag{4.2.2.3}$$

$$\mathrm{Re}(a_1+d) > 0$$

and

$$\frac{1}{2\pi i}\int_{c-i\infty}^{c+i\infty}\Gamma(b_1+u)\Gamma(b_1'-u)F^{A:B;B'}_{F:G;G'}\left[\begin{matrix}(a):b_1+u,b_2,..,b_B;b_1'-u,b_2',..,b_B';\\ (f):\qquad (g)\quad ;\qquad (g')\quad ;\end{matrix} x,y\right]du$$

$$=\frac{\Gamma(b_1+b_1')}{2^{b_1+b_1'}}F^{A+1:B-1;B'-1}_{F\ :\ G\ ;\ G'}\left[\begin{matrix}(a),b_1+b_1':b_2,..,b_B;b_2',..,b_{B'}';\\ (f)\qquad :\ (g)\ ;\ (g')\ ;\end{matrix} x/2,y/2\right],$$

$$\mathrm{Re}(b_1+b_1') > 0. \tag{4.2.2.4}$$

4.2.3 Multiple Hypergeometric Function

As in many other instances, when considering integrals of Barnes type where hypergeometric functions of several variables are involved, the number of possibilities which arises increases rapidly with the number of variables of the functions concerned. The following example only will be given which is a generalisation of (4.2.2.1):-

$$\frac{1}{2\pi i}\int_{c-i\infty}^{c+i\infty}\Gamma(a_1+u)\Gamma(a_2-u)F^{A:B}_{F:G}\left[\begin{matrix}a_1+u,a_2-u,a_3,..,a_A:(b');.;(b^{(n)});\\ (f)\qquad :(g');.;(g^{(n)});\end{matrix} x_1,..,x_n\right]du$$

$$=\frac{\Gamma(a_1+a_2)}{2^{a_1+a_2}}F^{A:B}_{F:G}\left[\begin{matrix}[a_1+a_2]/2,[a_1+a_2+1]/2,a_3,..,a_A:(b');.;(b^{(n)});\\ (f)\qquad :(g');.;(g^{(n)});\end{matrix} x_1,..,x_n\right], \tag{4.2.3.1}$$

$$\mathrm{Re}(a_1+a_2) > 0.$$

4.2.4 Related Integrals

This section is devoted to an investigation of a number of integrals with respect to parameters of the hypergeometric function of each integrand and which are of a different type from those discussed above. While these results do not, strictly speaking, involve contour integrals, they are included here because they display a number of points of similarity with the integrals dealt with in the previous section.

We first of all note the formula given by Titchmarsh (1948) page 187:

$$\int_{-\infty}^{\infty} \frac{du}{\Gamma(a+u)\Gamma(b-u)} = \frac{2^{a+b-2}}{\Gamma(a+b-1)}, \qquad (4.2.4.1)$$

where Re(a+b) > 1, and which corresponds to (4.2.2) in the previous sections. This will be the basis of the discussion. If the hypergeometric function in the integrand of

$$\int_{-\infty}^{\infty} \frac{{}_AF_B((a);b_1+u,b_2-u,b_3,.,b_B;x)\,du}{\Gamma(b_1+u)\Gamma(b_2-u)} \qquad (4.2.4.2)$$

is expanded in series, then term-by-term integration is valid, and (4.2.4.2) may be written as

$$\sum_{m=0}^{\infty} \frac{(a_1,m)..(a_A,m)}{(b_3,m)..(b_B,m)} x^m \int_{-\infty}^{\infty} \frac{du}{\Gamma(b_1+m+u)\Gamma(b_2+m-u)}. \qquad (4.2.4.3)$$

The inner integral is evaluated by means of (4.2.4.1), so that (4.2.4.2) takes the form

$$\frac{2^{b_1+b_2-2}}{\Gamma(b_1+b_2-1)}\, {}_AF_B((a);\frac{b_1+b_2-1}{2},\frac{b_1+b_2}{2},b_3,.,b_B;x), \qquad (4.2.4.4)$$

provided that $\mathrm{Re}(b_1+b_2) > 1$. Similarly, we have a number of integrals involving hypergeometric functions of several variables, for example:

$$\int_{-\infty}^{\infty} F^{A:B}_{C:D}\left[\begin{matrix}(a) & :(b');.;(b^{(n)}); \\ c_1+u,c_2-u,c_3,.,c_C & :(d');.;(d^{(n)});\end{matrix} x_1,.,x_n\right] \frac{du}{\Gamma(c_1+u)\Gamma(c_2-u)}$$

$$= \frac{2^{c_1+c_2-2}}{\Gamma(c_1+c_2-1)} F^{A:B}_{C:D}\left[\begin{matrix}(a) & :(b');.;(b^{(n)}); \\ \frac{c_1+c_2-1}{2},\frac{c_1+c_2}{2},c_3,.,c_C & :(d');.;(d^{(n)});\end{matrix} x_1,.,x_n\right],$$

$$\mathrm{Re}(c_1+c_2) > 1 \qquad (4.2.4.5)$$

and

$$\int_{-\infty}^{\infty} F^{A:B}_{C:D}\left[\begin{matrix}(a) & :(b');.;(b^{(n)}); \\ c_1+u,c_2,.,c_C & :(d');.;(d^{(n)});\end{matrix} x_1,.,x_n\right] \frac{du}{\Gamma(c_1+u)\Gamma(f-u)}$$

$$= \frac{2^{c_1+f-2}}{\Gamma(c_1+f-1)} F^{A:B}_{C:D}\left[\begin{matrix}(a) & :(b');.;(b^{(n)}); \\ c_1+f-1,c_2,.,c_C & :(d');.;(d^{(n)});\end{matrix} 2x_1,.,2x_n\right],$$

$$\mathrm{Re}(c_1+f) > 1. \qquad (4.2.4.6)$$

These two results express integrals of generalised Kampé de Fériet functions with respect to their parameters in terms of other generalised Kampé de Fériet functions.

4.2.5 Special Cases

Very many integrals of Barnes type and related integrals involving special functions may be obtained from the formulae of Sections 4.2.1 to 4.2.4. A few examples are discussed here. Integrals which are of interest may be deduced by using the well-known result

$$\Gamma(a)\Gamma(1-a) = \pi\mathrm{cosec}(\pi a), \qquad (4.2.5.1)$$

see Erdélyi et al. (1953) Vol. I page 3. If we make use of this formula, we observe that (4.2.2.1) may be specialised to give the result

$$\int_{c-i\infty}^{c+i\infty} \mathrm{cosec}[\pi(a_1+u)]\,{}_AF_B(a_1+u,1-a_1-u,a_3,.,a_A;(b);z)du$$

$$= i\ {}_AF_B(1/2,1,a_3,.,a_A;(b);z) \qquad (4.2.5.2)$$

Similarly, from (4.2.1.2), we have

$$\int_{c-i\infty}^{c+i\infty} \mathrm{cosec}[\pi(a_1+u)]\,{}_AF_B(a_1+u,a_2,.,a_A;(b);z)du$$

$$= i\ {}_AF_B(1,a_2,.,a_A;(b);z/2). \qquad (4.2.5.3)$$

These two formulae may be generalised to give results where hypergeometric functions of more than one variable are involved. We note the following example:-

$$\int_{c-i\infty}^{c+i\infty} \mathrm{cosec}[\pi(b_1+u)]F^{A:B;B'}_{F:G;G'}\left[\begin{matrix}(a):b_1+u,b_2,.,b_B;1-b_1-u,b'_2,.,b'_{B'}; \\ (f):\quad (g)\quad ;\quad (g')\quad ;\end{matrix} x,y\right] \qquad (4.2.5.4)$$

$$\times du = \qquad \text{(continued)}$$

$$= i\, F^{A+1:B-1;B'-1}_{F\;:\;G\;;\;G'}\left[\begin{matrix}(a),1:b_2,..,b_B;b'_2,..,b'_B;\,x/2,y/2\\(f)\;:\;(g)\;;\;(g');\end{matrix}\right]. \quad (4.2.5.4)$$
(cont.)

A number of integrals involving simpler hypergeometric functions are now evaluated. The integral

$$1/(2\pi i)\int_{c-i\infty}^{c+i\infty}\Gamma(a+u)\Gamma(d-u)\,{}_1F_1(a+u;b;x)\,du \qquad (4.2.5.5)$$

is of the form (4.2.1.2), and, provided that $\mathrm{Re}(a+d) > 0$, this integral becomes

$$\frac{\Gamma(a+d)}{2^{a+d}}\,{}_1F_1(a+d;b;x/2). \qquad (4.2.5.6)$$

Further, if $d=b-a$ and $\mathrm{Re}(b) > 0$, we have the simpler form

$$1/(2\pi i)\int_{c-i\infty}^{c+i\infty}\Gamma(a+u)\Gamma(b-a-u)\,{}_1F_1(a+u;b;x)\,du = \frac{\Gamma(b)}{2^b}\,e^{x/2}. \qquad (4.2.5.7)$$

Now, two integrals involving the Gauss function ${}_2F_1$ are investigated. The first is obtained from (4.2.1.1), so that we have

$$\frac{1}{2\pi i}\int_{c-i\infty}^{c+i\infty}\Gamma(a+u)\Gamma(b-u)\,{}_2F_1\left(\begin{matrix}a+u,b-u;\\d\;;\end{matrix}x\right)du = \frac{\Gamma(a+b)}{2^{a+b}}\,{}_2F_1\left(\frac{a+b}{2},\frac{a+b+1}{2};d;x\right), \qquad (4.2.5.8)$$

which leads to the expression

$$\frac{1}{2\pi i}\int_{c-i\infty}^{c+i\infty}\Gamma(a+u)\Gamma(b-u)\,{}_2F_1\left(\begin{matrix}a+u,b-u;\\ [a+b]/2;\end{matrix}x\right)du = \frac{\Gamma(a+b)}{2^{a+b}}(1-x)^{-(a+b+1)/2}. \qquad (4.2.5.9)$$

From (4.2.1.2) it follows that

$$\frac{1}{2\pi i}\int_{c-i\infty}^{c+i\infty}\Gamma(a+u)\Gamma(d-u)\,{}_2F_1\left(\begin{matrix}a+u,b;\\f\;;\end{matrix}x\right)du = \frac{\Gamma(a+d)}{2^{a+d}}\,{}_2F_1(a+d,b;f;x/2) \qquad (4.2.5.10)$$

and so,

$$\frac{1}{2\pi i}\int_{c-i\infty}^{c+i\infty}\Gamma(a+u)\Gamma(d-u)\,{}_2F_1\left(\begin{matrix}a+u,b;\\a+d\;;\end{matrix}x\right)du = \frac{\Gamma(a+d)}{2^{a+d}}(1-x/2)^{-b}. \qquad (4.2.5.11)$$

In the equations (4.2.5.8) to (4.2.5.11), it is taken that $\mathrm{Re}(a+d) > 0$.

We now consider an integral involving a Legendre function $P_{c+u}(z)$, where the integration is carried out with respect to the parameter $c+u$.

$$\frac{1}{2\pi i}\int_{c-i\infty}^{c+i\infty} \Gamma(1+c+u)\Gamma(-c-u)P_{c+u}(z)du$$

$$= \frac{1}{2i}\int_{c-i\infty}^{c+i\infty} \operatorname{cosec}[\pi(1+c+u)]\,{}_2F_1(-c-u,1+c+u;1;[1-z]/2)du$$

$$= 1/2\ {}_2F_1(1.1/2;1;[1-z]/2) = (2z)^{-1/2}. \qquad (4.2.5.12)$$

The integrals

$$1/(2\pi i)\int_{c-i\infty}^{c+i\infty} \Gamma(b+u)\Gamma(b'-u)F_1(a,b+u,b'-u;d;x,y)du, \quad \mathrm{Re}(b+b') > 0, \qquad (4.2.5.13)$$

$$1/(2\pi i)\int_{c-i\infty}^{c+i\infty} \Gamma(a+u)\Gamma(b-u)F_4(a+u,b-u;d,d';x,y)du, \quad \mathrm{Re}(a+b) > 0 \qquad (4.2.5.14)$$

and

$$1/(2\pi i)\int_{c-i\infty}^{c+i\infty} \Gamma(a+u)\Gamma(a'-u)\,{}_2F_1(a+u,b;d;x)\,{}_2F_1(a'-u,b';d';y)du, \quad \mathrm{Re}(a+a') > 0, \qquad (4.2.5.15)$$

are special cases of formulae given in Section 4.2.2 where Kampé de Fériet functions are involved. When evaluated, the integrals (4.2.5.12) to (4.2.5.14) become, respectively

$$\frac{\Gamma(b+b')}{2^{b+b'}}\,{}_2F_1(a,b+b';c;[x+y]/2), \qquad (4.2.5.16)$$

$$\frac{\Gamma(a+b)}{2^{a+b}}\,F_4([a+b]/2,[a+b+1]/2;d,d';x,y) \qquad (4.2.5.17)$$

and

$$\frac{\Gamma(a+a')}{2^{a+a'}}F_2(a+a',b,b';d,d';x/2,y/2). \qquad (4.2.5.18)$$

If d=b+b', then (4.2.5.16) takes the simpler form

$$\frac{\Gamma(b+b')}{2^{b+b'}}\left(1-\frac{x+y}{2}\right)^{-a}. \qquad (4.2.5.18)$$

These results may easily be extended to cover integrals of Barnes type where hypergeometric functions of several variables occur in the integrands.

Chapter 5

Infinite Integrals

5.1 INTRODUCTION. A THEOREM ON THE INTEGRATION OF SERIES OVER AN INFINITE RANGE

Many infinite integrals involving hypergeometric functions, or functions reducible to hypergeometric form, such as, for example, Bessel functions, were evaluated by the classical authors by means of such techniques as contour integration or interchanging the order of integrations. An example of this type of approach is indicated in the discussion of Weber's second exponential integral in Watson (1944) Section 13.13. While it may be necessary in certain cases to employ such methods, the desired results can often be achieved by the much more easily applicable technique of term-by-term integration, provided that this process can be shown to be justified. This method has been exploited in previous chapters of this book where infinite ranges of integration do not arise. In view of the theoretical and practical importance of infinite integrals in general and infinite hypergeometric integrals in particular, it has been felt desirable that this chapter should be a good deal longer than its fellows.

First of all, it seems worthwhile to quote a theorem given by Bromwich (1931) page 500, to which the reader is referred for further details.

Theorem 1. If the series $\sum f_n(x)$ converges uniformly in any fixed interval $a \leq x \leq b$, where b is arbitrary, and if g(x) is continuous for all finite ranges of x, then

$$\int_a^\infty g(x)\left[\sum f_n(x)\right]dx = \sum \int_a^\infty g(x) f_n(x)\,dx, \qquad (5.1.1)$$

provided that either the integral $\int_a^\infty |g(x)|\left[\sum |f_n(x)|\right]dx$ or the series $\sum \int_a^\infty |g(x)| . |f_n(x)|\,dx$ is convergent. A large number of the integrals under consideration in this chapter may conveniently be evaluated by the application of this general result.

5.2 INTEGRALS OF LAPLACE TYPE

These integrals are, in many cases, readily evaluated by the use of *Theorem 1* of the previous section. Suppose that

$$\sum f_n(x) = \sum_{n=0}^{\infty} h_n x^n, \qquad (5.2.1)$$

where it is assumed that this series either terminates or is convergent for all finite values of the variable x. The type of integral to be considered is of the form

$$I = \int_0^{\infty} e^{-pt}\, t^{a-1} \sum_{n=0}^{\infty} h_n t^n \, dt, \qquad (5.2.2)$$

where, for convenience, Re(p), Re(a) > 0.

It is well-known that if a power series such as (5.2.1) converges, it converges uniformly, so that *Theorem 1* may be applied and we have

$$I = \sum_{n=0}^{\infty} [h_n \int_0^{\infty} e^{-pt}\, t^{a+n-1} \, dt]. \qquad (5.2.3)$$

The inner integral above may be evaluated in the form

$$\frac{\Gamma(a+n)}{p^{a+n}}, \qquad (5.2.4)$$

making use of the Euler integral for the gamma function. See Erdélyi et al. (1953) Vol. I page 1.

Hence,

$$I = (a)\, p^{-a} \sum_{n=0}^{\infty} h_n \,(a,n) p^{-n}. \qquad (5.2.5)$$

This result may easily be extended to multiple series, and if the coefficient h_n is suitably specialised, a large number of results involving hypergeometric series may be obtained. It is clear that integrals of the type (5.2.2) are closely related to the Laplace transform, and, in fact, the integral I is the Laplace transform of the function

$$t^{a-1} \sum^{\infty} h_n t^n, \qquad (5.2.6)$$

and so these integrals are of importance in applications. See Chapter 7.

Another general type of Laplace integral is also of importance. This occurs when the function to be integrated does not possess a series representation with a sufficiently large radius of convergence. Instead, it may happen that the function in question is capable of being represented as an integral such that the order of integration may be interchanged.

This last process is the subject of de la Vallée Poussin's Theorem, see Bromwich (1931) page 504, which states that the equation

$$\int_a^\infty dx \int_b^\infty f(x,y)dy = \int_b^\infty dy \int_a^\infty f(x,y)dx \tag{5.2.7}$$

holds, provided that both of the integrals

$$\int_a^\infty f(x,y)dx \qquad \text{and} \qquad \int_b^\infty f(x,y)dy \tag{5.2.8}$$

are convergent, and that either of the repeated integrals converges. This result also holds when applied to contour integrals.

5.2.1 Generalised Hypergeometric Function

The series may be taken to be a generalised hypergeometric function ${}_CF_D(xt^k)$, where k is a positive integer. When $C \leqq D-k$, the series representation of ${}_CF_D$ converges uniformly for all finite values of x, and if (5.2.2) is applied, we have the expression

$$\int_0^\infty e^{-st}\, t^{a-1}\, {}_CF_D((c);(d);xt^k)dt = \frac{\Gamma(a)}{s^a}\, {}_{C+k}F_D\left((c),\frac{a}{k},..,\frac{a+k-1}{k};\frac{xk^k}{s^k}\right),$$

$$\mathrm{Re}(s), \mathrm{Re}(a) > 0. \tag{5.2.1.1}$$

This result follows from the application of the gamma integral and the multiplication formula for the gamma function. See Erdélyi et al. (1953) Vol. I page 4.

The integral (5.2.1.1) also converges if $k > 0$ is not an integer, but the result obtained is not so conveniently expressed. If $C=D-k+1$, the hypergeometric series of the integrand of (5.2.1.1) has a radius of convergence of unity, so that (5.2.1.1) does not, in general, hold. In fact, if $C > D-k+1$, the series in question does not converge at all apart from the trivial case when $x=0$. The formula (5.2.1.1) holds, however, if one of the numerator parameters c_j is a negative integer which causes the series under discussion to terminate. Otherwise, we have a result which, at the best, may be regarded as an asymptotic representation of the integral on the right of (5.2.1.1).

A number of authors, notably MacRobert and Meijer have devised integral fromulae for the generalised hypergeometric function which can thus be given a definite meaning whatever the values of C and D. For further information on this topic, see Erdélyi et al. (1953) Vol. I pages 203 and 206.The Meijer G-function is defined by the contour integral (5.2.1.2) below. The contour of integration L runs from $-i\infty$ to $+i\infty$ so that all poles of $\Gamma(b_j-s)$, $j=1,..,m$, are on the right, and all poles of $\Gamma(1-a_k+s)$, $k=1,..,n$, are on the left, of L. For other paths of integration see Erdélyi et al. (1953) Vol. I page 207.

$$G^{m,n}_{p,q}\left(z\,\middle|\,\begin{matrix}a_j\\ b_j\end{matrix}\right) = \frac{1}{2\pi i}\int_L \frac{\prod_{j=1}^{m}\Gamma(b_j+s)\prod_{j=1}^{n}\Gamma(1-a_j-s)}{\prod_{j=m+1}^{q}\Gamma(1-b_j-s)\prod_{j=n+1}^{p}\Gamma(a_j+s)}\, z^{-s}\, ds. \qquad (5.2.1.2)$$

It seems pertinent to consider separately the cases when m=1 and n=1, since the G-function then takes simple forms in terms of the generalised hypergeometric function of one variable:

$$\frac{\prod_{j=2}^{q}\Gamma(1+b_1-b_j)\prod_{j=n+1}^{p}\Gamma(a_j-b_1)}{\prod_{j=1}^{n}\Gamma(1+b_1-a_j)\, z^{b_1}}\, G^{1,n}_{p,q}\left(z\,\middle|\,\begin{matrix}a_j\\ b_j\end{matrix}\right)$$

$$= {}_pF_{q-1}\left[\begin{matrix}1+b_1-a_1,\ldots,1+b_1-a_p;\\ 1+b_1-b_2,\ldots,1+b_1-b_q;\end{matrix}\,(-1)^{p-n-1}\, z\right], \qquad (5.2.1.3)$$

when $p < q$, or $p=q$ and $|z| < 1$. On the other hand, when $p > q$, or $p=q$ and $|z| > 1$, we have

$$\frac{\prod_{j=2}^{p}\Gamma(1+a_j-a_1)\prod_{j=m+1}^{q}\Gamma(a_1-b_j)}{\prod_{j=1}^{m}\Gamma(b_j-a_1-1)\, z^{a_1-1}}\, G^{m,1}_{p,q}\left(z\,\middle|\,\begin{matrix}a_j\\ b_j\end{matrix}\right)$$

$$= {}_qF_{p-1}\left[\begin{matrix}1+b_1-a_1,\ldots,1+b_q-a_1;\\ 1+a_2-a_1,\ldots,1+a_p-a_1;\end{matrix}\,(-1)^{q-m-1}/z\right]. \qquad (5.2.1.4)$$

We now consider the integral

$$\int_0^\infty e^{-st}\, t^{c-1}\, G^{m,n}_{p,q}\left(z/t\,\middle|\,\begin{matrix}a_j\\ b_j\end{matrix}\right) dt. \qquad (5.2.1.5)$$

Replace the G-function of the integrand by its contour integral representation (5.2.1.2). We may also reverse the order of the integrations by appealing to de la Vallée Poussin's theorem, see (5.2.7). We may then write, deforming the contour L if necessary,

$$\frac{1}{2\pi i}\int_L \frac{\prod_{j=1}^{m}\Gamma(b_j+u)\prod_{j=1}^{n}\Gamma(1-a_j-u)}{\prod_{j=m+1}^{q}\Gamma(1-b_j-u)\prod_{j=n+1}^{p}\Gamma(a_j+u)}\, z^{-u}\left[\int_0^\infty e^{-st}\, t^{c+u-1}\, dt\right] du. \qquad (5.2.1.6)$$

If the integral in t is evaluated as a gamma function, comparison with (5.2.1.2) enables us to write the integral (5.2.1.5) in the form

$$s^{-c}\, G^{m+1,n}_{p,q+1}\left(zs \,\middle|\, \begin{matrix} a_1,\ldots\ldots,a_p \\ b_1,..,b_m,c,b_{m+1},..,b_q \end{matrix}\right). \qquad (5.2.1.7)$$

An elementary change of the variable of integration in the defining integral of the G-function gives the formula

$$\int_0^\infty e^{-st}\, t^{c-1}\, G^{m,n}_{p,q}\left(zt \,\middle|\, \begin{matrix} a_j \\ b_j \end{matrix}\right) dt = s^{-c}\, G^{m,n+1}_{p+1,q}\left(z/s \,\middle|\, \begin{matrix} c,a_1,..,a_p \\ b_1,..,b_q \end{matrix}\right). \quad (5.2.1.8)$$

These results involving the Meijer G-function are of particular importance when evaluating Laplace integrals of the generalised hypergeometric function of one variable ${}_pF_q$ when p=q+1, since,in this case, the series representation of this function diverges when the modulus of its variable exceeds unity.

5.2.2 Double Hypergeometric Function

We may take the Laplace integral to be of the form

$$J = \int_0^\infty e^{-st}\, t^{a-1} \sum_{m,n=0}^{\infty} h_{m,n}\, t^{m+n}\, dt, \qquad (5.2.2.1)$$

where the single series of the integrand of (5.2.2) has been replaced by a double series. If this double series is uniformly convergent for all finite values of the variable of integration t, then term-by-term integration is possible by a straightforward extension of the reasoning which led to (5.2.3), and we have the following result which is analogous to (5.2.5):

$$J = \frac{\Gamma(a)}{s^a} \sum_{m,n=0}^{\infty} \frac{h_{m,n}(a,m+n)}{s^{m+n}}. \qquad (5.2.2.2)$$

The generalised double hypergeometric function of Kámpe de Fériet may be treated in this context. This function is given by the expression

$$F^{A:B;B'}_{C:D;D'}\left(\begin{matrix}(a):(b);(b');\\(c):(d);(d');\end{matrix}x,y\right) = \sum \frac{((a),m+n)((b),m)((b'),n)x^m y^n}{((c),m+n)((d),m)((d'),n)m!n!}. \qquad (5.2.2.3)$$

The Kampé de Fériet function converges for all finite values of its variables x and y if A+B+B' $\leq$ C+D+D'.

If at least one of the parameters $a_1,..,a_A$ is a non-positive integer or at least one of the parameters $b_1,..,b_B$ together with at least one of the parameters $b'_1,..,b'_{B'}$ are non-positive integers, then the Kampé de Fériet function terminates. We now specialise (5.2.2.2) using (5.2.2.3) as the double series. This yields the expression

$$\int_0^\infty e^{-st}\, t^{a-1} F^{C:D;D'}_{F:G;G'}\left({(c):(d);(d'); \atop (f):(g);(g');} xt, yt\right) dt$$

$$= \frac{\Gamma(a)}{s^a} F^{C+1:D;D'}_{F\ :G;G'}\left({(c),a:(d);(d'); \atop (f)\ :(g);(g');} x/s, y/s\right), \qquad (5.2.2.4)$$

$$\mathrm{Re}(a),\ \mathrm{Re}(s) > 0.$$

If we wish to consider integrals of functions of Kampé de Fériet type where (5.2.2.4) does not hold, due the convergence properties of these latter functions, it is necessary to employ double contour integral representations of such functions similar to the Meijer function.

5.2.3 Multiple Hypergeometric Function

The most convenient multiple hypergeometric function of general type is Karlsson's generalised Kampé de Fériet function. As indicated in Karlsson (1973), this function is reasonably compact in notation and is an adequate generalisation which includes most of the well-known hypergeometric functions and products of several of them. We now discuss the integral

$$K = \int_0^\infty e^{-st}\, t^{a-1} F^{C:D}_{E:F}\left[{(c):(d');.;(d^{(n)}); \atop (f):(g');.;(g^{(n)});} x_1 t^k,..,x_n t^k\right] dt, \qquad (5.2.3.1)$$

where k is a positive integer.

If

$$A+B+k \leq C+D, \qquad (5.2.3.2)$$

the inner multiple series converges for all finite values of t and the x_i, so that term-by-term integration may be carried out, and we have

$$K = \sum \frac{((c),m_1+.+m_n)((d'),m_1)..((d^{(n)}),m_n)x_1^{m_1}..x_n^{m_n}}{((f),m_1+.+m_n)((g'),m_1)..((g^{(n)}),m_n)m_1!..m_n!}$$

$$\times \int_0^\infty e^{-st}\, t^{a+km_1+.+km_n-1}\, dt. \qquad (5.2.3.3)$$

The preceding inner integral can be evaluated by using the integral of Laplace type for the gamma function and the multiplication formula for the gamma function. Hence, we may thus write

$$K = \frac{\Gamma(a)}{s^a} F^{C+k:D}_{F\ :G}\left[\begin{matrix}(c),\frac{a}{k},..,\frac{a+k-1}{k}:(d');.;(d^{(n)});\frac{k^k x_1}{s^k},..,\frac{k^k x_n}{s^k}\\ (f)\qquad :(g');.;(g^{(n)});\end{matrix}\right], \quad (5.2.3.4)$$

$$Re(a),\ Re(s) > 0.$$

If the condition (5.2.3.2) is not met, and if the multiple series concerned does not terminate, then we must employ a multiple contour integral representation of (5.2.3.1) similar to that of the G-function mentioned in Section 5.2.1. This contour integral formula may be written

$$F^{C:D}_{F:G}\left[\begin{matrix}(c):(d');.;(d^{(n)});\\ (f):(g');.;(g^{(n)});\end{matrix} x_1,..,x_n\right]$$

$$= L\int_{-i\infty}^{i\infty}.(n).\int_{-i\infty}^{i\infty}\Psi(t_1,..,t_n)\Gamma(-t_1)..\Gamma(-t_n)(-x_1)^{t_1}..(-x_n)^{t_n}\,dt_1..dt_n, \quad (5.2.3.5)$$

$$L = \frac{\prod_{j=1}^{F}\Gamma(f_j)\prod_{j=1}^{G}[\Gamma(g_j')..\Gamma(g_j^{(n)})]}{\prod_{j=1}^{C}\Gamma(c_j)\prod_{j=1}^{D}[\Gamma(d_j')..\Gamma(d_j^{(n)})](2\pi i)^n} \quad (5.2.3.6)$$

and $$\Psi(t_1,..,t_n) = \frac{\prod_{j=1}^{C}\Gamma(c_j+t_1+.+t_n)\prod_{j=1}^{D}[\Gamma(d_j'+t_1)..\Gamma(d_j^{(n)}+t_n)]}{\prod_{j=1}^{F}\Gamma(f_j+t_1+.+t_n)\prod_{j=1}^{G}[\Gamma(g_j'+t_1)..\Gamma(g_j^{(n)}+t_n)]}. \quad (5.2.3.7)$$

The contours of integration are indented if necessary in the usual manner, so as to separate the poles of the integrand at $t_j=0,1,2,..$ from the poles at $t_j = -c_1-N,.,-c_C-N$, $t_j=-d_1'-N,.,$ $-d_D'-N;.;-d_1^{(n)}-N,.,-d_D^{(n)}-N$ $(N=0,1,2,..)$ of the integrand. It is always possible to find such contours if the f's and the g's are not zero or negative integers.

Hence, if (5.2.3.2) does not hold and if the series representation of the generalised Kampé de Fériet function concerned does not terminate, we must write the integral K in the form (5.2.3.8) on the next page.

$$L\int_0^\infty e^{-st}\, t^{a-1}\left[\int_{-i\infty}^{i\infty}.(n).\int_{-i\infty}^{i\infty}\Psi(t_1,.,t_n)\Gamma(-t_1)..\Gamma(-t_n) \times (-t^k x_1)^{t_1}..(-t^k x_n)^{t_n}\, dt_1..dt_n\right]dt, \quad (5.2.3.8)$$

$$\mathrm{Re}(a),\ \mathrm{Re}(s) > 0.$$

If the order of integrations is interchanged, we may then evaluate the inner gamma integral in the usual way, and we have the result

$$K = Ls^{-a}\int_{-i\infty}^{i\infty}.(n).\int_{-i\infty}^{i\infty}\Psi(t_1,.,t_n)\Gamma(a+kt_1+.+kt_n)\Gamma(-t_1)..\Gamma(-t_n) \times \left(\frac{-x_1}{s^k}\right)^{t_1}..\left(\frac{-x_n}{s^k}\right)^{t_n} dt_1..dt_n. \quad (5.2.3.9)$$

This last integral is a special case of an n-fold generalisation of the Meijer G-function.

5.2.4 Special Cases

The vast majority of the special functions of applied mathematics may be expressed in one way or another as hypergeometric functions and only a few representative examples of Laplace integrals of functions of this type will be discussed here.

Consider the inregral (5.2.1.1) in which the integrand is taken to involve a confluent hypergeometric function. We thus have

$$\int_0^\infty e^{-st}\, t^{a-1}\, {}_1F_1(c;d;xt)dt = \Gamma(a)s^{-a}\, {}_2F_1(a,c;d;x/s) \quad (5.2.4.1)$$

by the use of (5.2.1.2); $\mathrm{Re}(a)$, $\mathrm{Re}(s) > 0$.

A number of further special cases of this result may be deduced. Suppose that $x=s$, then the Gauss function on the right of (5.2.4.1) may be summed by Gauss's summation theorem giving the formula

$$\int_0^\infty e^{-st}\, t^{a-1}\, {}_1F_1(c;d;st)dt = \frac{\Gamma(a)\Gamma(d)\Gamma(d-a-c)}{s^a\Gamma(d-c)\Gamma(d-a)}, \quad (5.2.4.2)$$

provided that, in addition to the preceding conditions, $\mathrm{Re}(d)$ exceeds $\mathrm{Re}(a+c)$, or that c is a non-positive integer.

Now let $x = -s$ and $d=1+c-a$. Kummer's theorem [Slater (1966) Appendix III] enables us to obtain the expression

$$\int_0^\infty e^{-st}\, t^{a-1}\, {}_1F_1(c;1+c-a;-st)dt = \frac{\Gamma(a)\Gamma(1+c-a)\Gamma(1+a/2)}{s^a\ \Gamma(1+c)\Gamma(1+c/2-a)}. \quad (5.2.4.3)$$

If $x=s/2$, Gauss's second theorem and Bailey's theorem give, respectively, the following:-

$$\int_0^\infty e^{-st}\, t^{a-1}\, {}_1F_1(c;\tfrac{1+c+a}{2};\tfrac{st}{2})dt = \frac{\Gamma(a)\Gamma(\frac{1}{2})\Gamma(\frac{1+c+a}{2})}{s^a\Gamma(\frac{1+c}{2})\Gamma(\frac{1+a}{2})} \tag{5.2.4.4}$$

and

$$\int_0^\infty e^{-st}\, t^{a-1}\, {}_1F_1(1-a;b;\tfrac{st}{2})dt = \frac{\Gamma(a)\Gamma(\frac{b}{2})\Gamma(\frac{1+b}{2})}{s^a\Gamma(\frac{a+b}{2})\Gamma(\frac{1+b-a}{2})}, \tag{5.2.4.5}$$

after appropriate specialisation of the parameters. On taking $d=a$, the hypergeometric function on the right of (5.2.4.1) reduces to a ${}_1F_0$ function, and this may be summed by the binomial theorem provided that $|x| < |s|$. Hence,

$$\int_0^\infty e^{-st}\, t^{a-1}\, {}_1F_1(c;a;xt)dt = \frac{\Gamma(a)}{s^a}(1-x/s)^{-c}. \tag{5.2.4.6}$$

Numerous special functions may be expressed as confluent hypergeometric functions, and we note the following two examples of integrals involving them:-

$$\int_0^\infty e^{-st}\, t^{a-1}\, L_n^c(st)dt = \frac{\Gamma(a)n!(c+1-a,n)}{s^a\,[(c+1,n)]^2} \tag{5.2.4.7}$$

and

$$\int_0^\infty e^{-st}\, t^{2c}\, I_c(xt)dt = \frac{(x/2)^c\Gamma(3c+1)}{(s+x)^{3c+1}\Gamma(c+1)}(1-2x/s)^{-c-1/2}. \tag{5.2.4.8}$$

The first integral involves a Laguerre polynomial and the second a modified Bessel function of the first kind.

We next suppose that the hypergeometric function of the integrand of (5.2.4.1) is of the form

$${}_0F_1(-;b;xt^k), \tag{5.2.4.9}$$

where $k=1$ or 2. This gives the expressions

$$\int_0^\infty e^{-st}\, t^{a-1}\, {}_0F_1(-;b;xt^k)dt$$

$$= \frac{\Gamma(a)}{s^a}\, {}_2F_1(a/2,\ [a+1]/2;b;4x/s^2) \qquad \text{if } k=2,$$

or

$$= \frac{\Gamma(a)}{s^a}\, {}_1F_1(a;b;x/s) \qquad \text{if } k=1. \tag{5.2.4.10}$$

If $b=a$ and $k=1$, we have

$$\int_0^\infty e^{-st}\, t^{a-1}\, {}_0F_1(-;a;xt)dt = s^{-a}\Gamma(a)\, e^{x/s}. \tag{5.2.4.11}$$

On putting k=2 and replacing the hypergeometric function of the integrand of (5.2.4.10) by its equivalent Bessel function, we have the well-known result [Erdélyi et al. (1953) Vol.II page 49]

$$\int_0^\infty e^{-st}\, t^{a-1}\, J_c(xt)dt = \frac{\Gamma(a+c)(x/2)^c}{\Gamma(c+1)s^{a+c}}\; {}_2F_1\left[\begin{matrix}\frac{a+c+1}{2}, \frac{a+c}{2}; \frac{-x^2}{} \\ c+1 \quad ; s^2\end{matrix}\right],$$

$$Re(s),\ Re(c+a) > 0. \tag{5.2.4.12}$$

Now let C=2 and D=k=1 in the integral (5.2.1.1). This leads to the formal result

$$\int_0^\infty e^{-st}\, t^{a-1}\, {}_2F_1(c_1,c_2;d;xt)dt = \frac{\Gamma(a)}{s^a}\; {}_3F_1(c_1,c_2,a;d;x/s), \tag{5.2.4.13}$$

which is valid if at least one of the quantities c_1 or c_2 is a negative integer. Otherwise, replace the Gauss function by its Barnes integral representation, when the left-hand member of (5.2.4.13) may be written as

$$\frac{\Gamma(d)}{\Gamma(c_1)\Gamma(c_2)} s^{-a}\; G^{1,3}_{3,2}\left(-\frac{x}{s}\middle|\begin{matrix}1-c_1, 1-c_2, 1-a \\ 0,\ 1-d\end{matrix}\right) \tag{5.2.4.14}$$

after a little re-arrangment.

The important identity of the G-function

$$G^{m,n}_{p,q}\left(x^{-1}\middle|\begin{matrix}a_j \\ b_j\end{matrix}\right) = G^{n,m}_{q,p}\left(x\middle|\begin{matrix}1-b_j \\ 1-a_j\end{matrix}\right) \tag{5.2.4.15}$$

may be obtained by an elementary change of the variable of integration in the integral defining the G-function, see (5.2.1.3). If this is applied to (5.2.4.14) we have the formula

$$\int_0^\infty e^{-st}\, t^{a-1}\, {}_2F_1(c_1,c_2;d;xt)dt = \frac{\Gamma(d)}{\Gamma(c_1)\Gamma(c_2)} s^{-a}\; G^{3,1}_{2,3}\left(-\frac{s}{x}\middle|\begin{matrix}1,\ d \\ c_1, c_2, a\end{matrix}\right). \tag{5.2.4.16}$$

The G-function may be expressed as a finite sum of hypergeometric functions where no two of the quantities c_1, c_2 or a differ by an integer. If this last condition does not hold, then logarithmic forms of the G-function occur due to the appearance of poles of higher order than unity in the integrand of (5.2.1.3). See Mathai and Saxena (1973) page 1.

If the Gauss function of (5.2.4.13) is replaced by the corresponding Legendre polynomial, we have

$$\int_0^\infty e^{-st}\, t^{a-1}\, P_n(1-2xt)dt = \frac{\Gamma(a)}{s^a}\, {}_3F_1(-n,n+1,a;1;x/s), \tag{5.2.4.17}$$

$$Re(a),\ Re(s) > 0.$$

The complete elliptical integral of the first kind may be expressed as a hypergeometric function by means fo the formula

$$K(k) = \frac{\pi}{2}\ {}_2F_1(1/2,1/2;1;k^2).$$

Hence, from (5.2.4.16) we have

$$\int_0^\infty e^{-st}\ t^{a-1}\ K(\sqrt{[kt]})dt = (2s^a)^{-1} G^{3,1}_{2,3}\left(-s/k\middle|\begin{matrix}1,\ 1\\ 1/2,1/2,a\end{matrix}\right). \tag{5.2.4.18}$$

Many other results of this type may be deduced.

Very many interesting formulae may be obtained from the general Laplace integral of the Kampé de Fériet function (5.2.2.3). Suppose that we consider this equation with the hypergeometric function on the left replaced by the Humbert functions Φ_2 and Ψ_2. We then obtain the results

$$\int_0^\infty e^{-st}\ t^{a-1}\Phi_2(b,b';c;xt,yt)dt = s^{-a}\Gamma(a)F_1(a,b,b';c;\tfrac{x}{s},\tfrac{y}{s}) \tag{5.2.4.19}$$

and

$$\int_0^\infty e^{-st}\ t^{a-1}\Psi_2(b;d,d';xt,yt)dt = s^{-a}\Gamma(a)F_4(a,b;d,d';\tfrac{x}{s},\tfrac{y}{s}), \tag{5.2.4.20}$$

respectively. A special case of (5.2.4.19) is readily deduced by letting c=a, giving

$$\int_0^\infty e^{-st}\ t^{a-1}\ \Phi_2(b,b';a;xt,yt)dt = s^{-a}\ \Gamma(a)\ (1-\tfrac{x}{s})^{-b}(1-\tfrac{y}{s})^{-b'}. \tag{5.2.4.21}$$

This formula generalises (5.2.4.6).

A number of Laplace integrals of products of hypergeometric functions may also be evaluated by means of (5.2.2.4). Thus,

$$\int_0^\infty e^{-st}\ t^{a-1}\ {}_1F_1(b_1;d_1;xt)\,{}_1F_1(b_2;d_2;yt)dt$$

$$= s^{-a}\ \Gamma(a)F_2(a,b_1,b_2;d_1,d_2;x/s,y/s) \tag{5.2.4.22}$$

and

$$\int_0^\infty e^{-st}\ t^{a-1}\ {}_0F_1(-;d_1;xt^2)\,{}_0F_1(-;d_2;yt^2)dt$$

$$= s^{-a}\ \Gamma(a)F_4(a/2,[a+1]/2;d_1,d_2;4x/s,4y/s). \tag{5.2.4.23}$$

From (5.2.4.22), we may obtain the following integral formula for the product of two incomplete gamma functions:-

$$\int_0^\infty e^{-st}\ t^{a-1}\ \gamma(b,xt)\ \gamma(c,yt)dt = \frac{\Gamma(a)x^{-b-c}}{bc\ s^{a-bc}}F_2(a,b,c;b+1,c+1;-\tfrac{x}{s},-\tfrac{y}{s}), \tag{5.2.4.24}$$

$$\mathrm{Re}(a-b-c),\ \mathrm{Re}(s) > 0.$$

When we interpret (5.2.4.23) as an integral of a product of two Bessel functions, we have

$$\int_0^\infty e^{-st}\, t^{a-1}\, J_c(xt) J_d(yt)\, dt =$$

$$\frac{(x/2)^c (y/2)^d\, \Gamma(a)}{\Gamma(c+1)\Gamma(d+1)}\, s^{-a}\, F_4\left(\frac{a+c+d}{2}, \frac{a+c+d+1}{2}; c, d; -\frac{x^2}{s^2}, -\frac{y^2}{s^2}\right),$$

$$\mathrm{Re}(a+c+d),\ \mathrm{Re}(s) > 0. \qquad (5.2.4.25)$$

If the hypergeometric function of the integrand of (5.2.2.3) is such that it doea not converge for all finite values of its variables, then we must employ a double contour integral of Barnes type instead of the series representation of the function under consideration. In this context, we make use of the double Barnes integral for the Appell function F_1 given by Appell and Kampé de Fériet (1926) page 40. This is

$$F_1(a,b,b';c;x,y) = \frac{\Gamma(c)}{(2\pi i)^2 \Gamma(a)\Gamma(b)\Gamma(b')}$$

$$\times \int_{-i\infty}^{i\infty} \int_{-i\infty}^{i\infty} \frac{\Gamma(a+u+v)\Gamma(b+u)\Gamma(b'+v)\Gamma(-u)\Gamma(-v)(-x)^u(-y)^v}{\Gamma(c+u+v)}\, du\, dv.$$

$$(5.2.4.26)$$

Some indication is now given as to the evaluation of the Laplace integral of the function F_1. Making use of (5.2.4.26), we see that, after interchanging the order of integration,

$$P = \frac{(2\pi i)^2 \Gamma(c)\Gamma(d)\Gamma(d')}{\Gamma(f)} \int_0^\infty e^{-st}\, t^{a-1}\, F_1(c,d,d';f;xt,yt)\, dt$$

$$= \int_{-i\infty}^{i\infty} \int_{-i\infty}^{i\infty} \frac{\Gamma(c+u+v)\Gamma(d+u)\Gamma(d'+v)\Gamma(-u)\Gamma(-v)(-x)^u\, (-y)^v}{\Gamma(f+u+v)}$$

$$\times \left[\int_0^\infty e^{-st}\, t^{a+u+v-1} dt\right] du\, dv. \qquad (5.2.4.27)$$

The inner integral may be evaluated as a gamma function, and so we have

$$P = \frac{\Gamma(a)}{s^a} \int_{-i\infty}^{i\infty} \int_{-i\infty}^{i\infty} \frac{\Gamma(c+u+v)\Gamma(a+u+v)\Gamma(d+u)\Gamma(d'+v)\Gamma(-u)\Gamma(-v)}{\Gamma(f+u+v)}$$

$$\times (-x/s)^u\, (-y/s)^v\, du\, dv. \qquad (5.2.4.28)$$

In order to obtain a representation of this last result in terms of convergent series, the above integral may be written as an integral of Barnes type of a G-function of one variable. Some rather lengthy manipulation eventually leads to the sum of six double hypergeometric series of higher order with arguments s/x and s/y.

If the function F_1 under consideration terminates, then the situation is much more straightforward, and only one terminating Kampé de Fériet function results;

$$\int_0^\infty e^{-st}\, t^{a-1}\, F_1(-N,d,d';f;xt,yt)\,dt$$

$$= \sum \frac{(-N,m+n)(d,m)(d',n)}{(f,m+n)} \frac{x^m y^n}{m!n!} \int_0^\infty e^{-st}\, t^{a+m+n-1}\, dt \qquad (5.2.4.29)$$

$$= \frac{\Gamma(a)}{s^a} F^{2:1}_{1:0}\left({-N,a:d;d'; \atop f\ :-;-\ ;} x/s, y/s\right), \qquad (5.2.4.30)$$

$$\mathrm{Re}(s),\ \mathrm{Re}(a) > 0.$$

When special cases of the generalised Kampé de Fériet function of several variables are considered, we obtain the following results which give integrals of functions which are convergent for all finite values of their variables:-

$$\int_0^\infty e^{-st}\, t^{a-1}\, \Phi_2^{(n)}(b_1,..,b_n;c;x_1 t,..,x_n t)\,dt$$

$$= \frac{\Gamma(a)}{s^a} F_D^{(n)}(a,b_1,..,b_n;c;x_1/s,..,x_n/s), \qquad (5.2.4.31)$$

$$\int_0^\infty e^{-st}\, t^{a-1}\, \Psi_2^{(n)}(b;d_1,..,d_n;x_1 t,..,x_n t)\,dt$$

$$= \frac{\Gamma(a)}{s^a} F_C^{(n)}(a,b;d_1,..,d_n;x_1/s,..,x_n/s) \qquad (5.2.4.32)$$

and

$$\int_0^\infty e^{-st}\, t^{a-1}\, \Phi_3^{(n)}(b_1,..,b_{n-1},-;c;x_1 t,..,x_n t)$$

$$= \frac{\Gamma(a)}{s^a} F_{D1}^{(n)}(a,b_1,..,b_{n-1},-;c;x_1/t,..,x_n/s), \qquad (5.2.4.33)$$

$$\mathrm{Re}(a),\ \mathrm{Re}(s) > 0.$$

As in the case of the Appell functions, when the Laplace integrals of terminating Lauricella functions are discussed, we have such formulae as

$$\int_0^\infty e^{-st}\, t^{a-1}\, F_C^{(n)}(-N,b;c_1,..,c_n;x_1 t,..,x_n t)\,dt$$

$$= \frac{\Gamma(a)}{s^a} F^{3:0}_{0:1}\left({-N,a,b:\ \text{—}\ ; \atop \text{—}\ :c_1;.;c_n;} x_1/s,..,x_n/s\right), \qquad (5.2.4.34)$$

$\mathrm{Re}(a),\ \mathrm{Re}(s) > 0$; see (5.2.3.1).

If it is desired to evaluate a Laplace integral of a non-terminating Lauricella function, the situation is much more complicated and we must tackle the problem in a way similar to that outined in dealing with the function F_1 in the expression (5.2.4.27).

Of considerable practical importance are the Laplace integrals of certain cases of the generalised Kampé de Fériet function which consist of products of several single hypergeometric functions. For example,

$$\int_0^\infty e^{-st}\, t^{a-1}\, F^{0:1}_{0:1}\left[\begin{matrix}-:b_1;.;b_n;\\ -:d_1;.;d_n;\end{matrix}\; x_1t,.,x_nt\right] dt$$

$$= \int_0^\infty e^{-st}\, t^{a-1}\, {}_1F_1(b_1;d_1;x_1t)\,.\,.\,{}_1F_1(b_n;d_n;x_nt)dt$$

$$= \frac{\Gamma(a)}{s^a} F_A^{(n)}(a,b_1,.,b_n;d_1,.,d_n;x_1/s,.,x_n/s) \qquad (5.2.4.35)$$

and

$$\int_0^\infty e^{-st}\, t^{a-1}\, {}_0F_1(-;d_1;x_1t^2)\,.\,.\,{}_0F_1(-;d_n;x_nt^2)dt$$

$$= \frac{\Gamma(a)}{s^a} F_C^{(n)}(a/2,[a+1]/2;d_1,.,d_n;4x_1/s,.,4x_n/s). \qquad (5.2.4.36)$$

In the expressions (5.2.4.35) and (5.2.4.36), Re(a), Re(s) > 0. Since the hyperbolic functions may be expressed as confluent hypergeometric functions, such as

$$\sinh z = ze^{-z}\, {}_1F_1(1;2;2z), \qquad (5.2.4.37)$$

we have

$$\int_0^\infty e^{-st}\, t^{a-1}\, \sinh(x_1t)\,.\,.\,\sinh(x_nt)dt$$

$$= \frac{2^n\, x_1\,.\,.\,x_n\Gamma(a+n)}{(s+x_1+.+x_n)^{a+n}} F_A^{(n)}(a+n,1,1,.,1;2,2,.,2;$$
$$2x_1/[s+x_1+.+x_n],.,2x_n/[s+x_1+.+x_n]),$$

where $\mathrm{Re}(s+x_1+.+x_n),\mathrm{Re}(a+n) > 0$. (5.2.4.38)

Similarly,

$$\int_0^\infty e^{-st}\, t^{a-1}\, J_{c_1}(x_1t)\,.\,.\,J_{c_n}(x_nt)dt = \frac{(x_1/2)^{c_1}\,.\,.\,(x_n/2)^{c_n}\Gamma(a+c_1+.+c_n)}{\Gamma(c_1+1)\,.\,.\,\Gamma(c_n+1)s^{a+c_1+.+c_n}}$$

$$\times F_C^{(n)}([a+c_1+.+c_n]/2,[a+c_1+.+c_n+1]/2;1+c_1,.,1+c_n;-x_1^2/s^2,.,-x_n^2/s^2),$$

$$\mathrm{Re}(a+c_1+.+c_n),\ \mathrm{Re}(s) > 0. \qquad (5.2.4.39)$$

Now, the Chebyshev polynomial of the first kind may be expressed as a Gauss function by means of the formula

$$T_m(z) = {}_2F_1(-m,m;1/2;\tfrac{1-z}{2}). \qquad (5.2.4.40)$$

Thus, (5.2.3.4) gives the formula

$$\int_0^\infty e^{-st}\, t^{a-1}\, T_{m_1}(1-2x_1t)..T_{m_n}(1-2x_nt)dt$$

$$= \frac{\Gamma(a)}{s^a} F^{1:2}_{0:1}\left[\begin{matrix} a:-m_1,m_1;.;-m_n,m_n; \\ -:\ 1/2\ \ ;.;1/2\ \ ; \end{matrix} x_1/s,.,x_n/s\right], \qquad (5.2.4.41)$$

$Re(s), Re(a) > 0.$

5.3 INFINITE INTEGRALS ASSOCIATED WITH CONFLUENT HYPERGEOMETRIC FUNCTIONS AND BESSEL FUNCTIONS

An important class of infinite integrals includes those whose integrands involve a confluent hypergeometric function along with another hypergeometric function. Since the exponential function and the Bessel function are particular cases of the confluent hypergeometric function, Laplace transforms, Hankel transforms and Fourier transforms may be included under this heading. Certain devices for the evaluation of such integrals are given by Watson (1944) page 381, and these include

(i) Expanding the hypergeometric function(s) of the integrand in powers of their arguments and integrating term-by-term.

(ii) Replacing the hypergeometric function(s) by their Barnes integral representations and reversing the order of integration.

Watson (1944) Chapter 13 is a frequently-used source of information on infinite integrals involving Bessel functions, and such integrals are of importance in many branches of applied mathematics as well as being of interest to pure mathematicians.

The formula

$$\int_0^\infty e^{-st}\, t^{a-1}\, {}_1F_1(c;d;xt)dt = \frac{\Gamma(a)}{s^a}\, {}_2F_1(a,c;d;x/s), \qquad (5.3.1)$$

$Re(a), Re(s) > 0,$

has already been given, see (5.2.4.1), and its special case

$$\int_0^\infty e^{-st}\, t^{a-1}\, {}_1F_1(c;d;st)dt = \frac{\Gamma(a)\Gamma(d)\Gamma(d-c-a)}{s^a\ \Gamma(d-c)\Gamma(d-a)}, \qquad (5.3.2)$$

$Re(a), Re(s) > 0, Re(d) > Re(c+a)$, see (5.2.4.2).

If $c = -N$, where N is a non-negative integer, then the third condition of convergence of (5.3.2) may be dispensed with since the series involved terminate, and we have

$$\int_0^\infty e^{-st}\, t^{a-1}\, {}_1F_1(-N;d;st)dt = \frac{\Gamma(a)(d-a,N)}{s^a\ \ (d,N)}\,. \qquad (5.3.3)$$

An interesting integral involving a modified Bessel function of the second kind

$$\int_0^\infty t^a\, K_c(t)dt = 2^{a-1}\,\Gamma\left(\frac{a+c+1}{2}\right)\Gamma\left(\frac{a-c+1}{2}\right),\ \mathrm{Re}(a\pm c) > -1, \quad (5.3.4)$$

has been given by Luke. See Abramowitz and Stegun (1965) page 486. This may be obtained by replacing the function $K_c(t)$ by its Barnes integral in the form of a G-function, and reversing the order of integration. The results (5.3.1) to (5.3.4) will be used in what follows.

5.3.1 Generalised Hypergeometric Function

The integrals to be considered now are mostly of the form

$$\int_0^\infty e^{-st}\, t^{a-1}\, {}_1F_1(c;d;xt)\, {}_FF_G((f);(g);yt^k)dt,\ \mathrm{Re}(a),\mathrm{Re}(s) > 0. \quad (5.3.1.1)$$

If $F \leqq G+k$, where k is a positive integer, the generalised hypergeometric function of the integrand converges uniformly for all finite values of the variable. If this function is expanded in series, we may then integrate term-by-term. The application of (5.3.1) enables us to evaluate the inner integral which occurs in the form

$$\frac{\Gamma(a+km)}{s^a}\ {}_2F_1(c,a+km;d;x/s) \qquad (5.3.1.2)$$

provided that $|x| < |s|$. Hence, (5.3.1.1) may be written as

$$\frac{\Gamma(a)}{s^a}\sum\frac{(a,km+n)((f),m)}{((g),m)\ m!n!}(ys^{-k})^m\ (xs^{-1})^n. \qquad (5.3.1.3)$$

If k=1 and $F \leqq G+1$, a Kampé de Fériet function results:

$$\frac{\Gamma(a)}{s^a}\, F^{1:1;F}_{0:1;G}\left(\begin{matrix}a:c;(f);\\ -:d;(g);\end{matrix}x/s,y/s\right), \qquad \mathrm{Re}(a),\mathrm{Re}(s) > 0. \qquad (5.3.1.4)$$

If c = -N, the condition $F \leqq G+1$ may be relaxed. Suppose, further, that x=s, when the inner integral mentioned above may be given in closed form by means of (5.3.3). The integral (5.3.1.1) then becomes

$$\frac{\Gamma(a)(d-a,N)}{s^a\ \ (d,N)}\ {}_{F+2k}F_{G+k}\left[\begin{matrix}(f),\frac{a}{k},..,\frac{a+k-1}{k}, & \frac{1+a-d}{k},..,\frac{1+a-d+k-1}{k}; & \frac{k^k\ y}{s^k}\\ (g),\frac{1+a-d-N}{k},...,\frac{1+a-d-N+k-1}{k} & ; & \end{matrix}\right].$$

(5.3.1.5)

Many interesting special cases of (5.3.1.5) may be deduced, see Section 5.3.4 for a few examples.

A formula similar to (5.3.1.3) may be deduced for the evaluation of the integral

$$\int_0^\infty e^{-st}\, t^{a-1}\, J_c(xt)\, {}_FF_G((f);(g);yt^k)\,dt, \tag{5.3.1.6}$$

by employing the formula (5.2.4.12). The expression (5.3.1.6) may be written in the form

$$\frac{\Gamma(a+c)(x/2)^c}{s^{a+c}\Gamma(c+1)} \sum \frac{(a+c,km+2n)((f),m)(y/s^k)^m(-x^2/[4s^2])^n}{((g),m)(c+1,n)\ m!\ \ n!}, \tag{5.3.1.7}$$

where $F+k+1 \leqq G$, unless the function ${}_FF_G$ is terminating. In addition, Re(s), Re(a+c) > 0. Furthermore, if we let k=2, the result (5.3.1.7) may be written in the form of the Kampé de Fériet function

$$\frac{\Gamma(a+c)(x/2)^c}{\Gamma(c+1)\ s^{a+c}}\, F^{2:F;0}_{0:G;1}\left[\begin{matrix}\frac{a+c}{2},\frac{a+c+1}{2}:(f);\ -\ ; \\ \text{—}\ :(g);c+1;\end{matrix}\ y,-\frac{x^2}{s^2}\right]. \tag{5.3.1.8}$$

This section is concluded by considering the integral

$$A = \int_0^\infty t^a\, K_c(t)\, {}_FF_G((f);(g);xt^{2k})\,dt, \tag{5.3.1.9}$$

where Re(a±c) > -1 and k is a positive integer. If $F+k \leqq G$, term-by-term integration using the formula (5.3.4) gives the following expression after a little reduction:-

$$A = 2^{a-1}\Gamma(\tfrac{a+c+1}{2})\Gamma(\tfrac{a-c+1}{2})$$

$$\times\ {}_{F+2k}F_G\left[\begin{matrix}(f),\frac{a+c+1}{2k},..,\frac{a+c+k}{2k},\frac{a-c+1}{2k},..,\frac{a-c+k}{2k}; \\ (g)\qquad ;\end{matrix}\ k^k x\right]. \tag{5.3.1.10}$$

If the ${}_FF_G$ function of the preceding integrand is terminating, then the restriction $F+k \leqq G$ may be lifted.

5.3.3 Multiple Hypergeometric Function

If the integral under consideration takes the form

$$H = \int_0^\infty e^{-st}\, t^{a-1}\, {}_1F_1(-N;d;st) \sum_{m,n=0}^\infty h_{m,n} t^{m+n}\,dt, \tag{5.3.2.1}$$

where the double series of the integrand is either terminating or uniformly convergent for all finite values of the variable of integration t, the term-by-term integration is justified and we have

$$H = \frac{(d-a,N)\Gamma(a)}{(d,N)s^a}\sum h_{m,n}\frac{(a,m+n)(1+a-b,m+n)}{(1+a-b-N,m+n)}, \qquad (5.3.2.2)$$

by the application of (5.3.3). It is now assumed that the double series is a Kampé de Fériet function. This now gives the expression

$$\int_0^\infty e^{-st}\, t^{a-1}\, {}_1F_1(-N;d;st)F^{B:C;C'}_{F:G;G'}\left({(b):(c);(c'); \atop (f):(g);(g');}xt,yt\right)dt$$

$$= \frac{(d-a,N)\Gamma(a)}{(d,N)\quad s^a}F^{B+2:C;C'}_{F+1:G;G'}\left({(b),a,1+a-d:(c);(c'); \atop (f),1+a-d-N:(g);(g');}\frac{x}{s},\frac{y}{s}\right), \qquad (5.3.2.3)$$

where $\mathrm{Re}(a)$, $\mathrm{Re}(s) > 0$ and $B+C \leqq F+G$ and $B+C' \leqq F+G'$, unless the Kampé de Fériet function on the left of (5.3.2.3) terminates.

The formulae (5.2.4.12) and (5.3.4) may be used respectively to obtain the two results now given.

$$\int_0^\infty e^{-st}\, t^{a-1}\, J_c(xt)F^{B:D;D'}_{F:G;G'}\left({(b):(d);(d'); \atop (f):(g);(g');}yt^2,zt^2\right)dt$$

$$= \frac{\Gamma(a+c)(x/2)^c}{\Gamma(c+1)\ s^{a+c}}\ F^{(3)}\left[{\frac{a+c}{2},\frac{a+c+1}{2}::(b):-:-:(d);(d');\ -\ ; \atop \text{---}\quad ::(f):-:-;(g);(g');c+1;}\frac{4y^2}{s^2},\frac{4z^2}{s^2},-\frac{x^2}{s^2}\right]$$

$$\mathrm{Re}(s),\ \mathrm{Re}(a+c) > 0 \qquad (5.3.2.4)$$

and

$$\int_0^\infty t^a\ K_c(t)\ F^{B:D;D'}_{F:G;G'}\left({(b):(d);(d'); \atop (f):(g);(g');}xt^2,yt^2\right)dt$$

$$= 2^{a-1}\Gamma\left(\frac{a+c+1}{2}\right)\Gamma\left(\frac{a-c+1}{2}\right)F^{B+2:D;D'}_{F\ :G;G'}\left[{(b),\frac{a+c+1}{2},\frac{a-c+1}{2}:(d);(d'); \atop (f)\qquad :(g);(g');}4x,4y\right]$$

$$\mathrm{Re}(a\pm c) > -1. \qquad (5.3.2.5)$$

5.3.3 Multiple Hypergeometric Function

The methods of the previous section may be generalised fairly easily so as to apply to integrals involving hypergeometric functions of several variables. We thus obtain the formula

$$\int_0^\infty e^{-st}\, t^{a-1}\, {}_1F_1(-N;d;st)F^{B:C}_{F:G}\left[{(b):(c');.;(c^{(n)}); \atop (f):(g');.;(g^{(n)});}x_1t,..,x_nt\right]dt$$

$$= \frac{\Gamma(a)(d-a,N)}{s^a\ (d,N)}\ F^{B+2:C}_{F+1:G}\left[{(b),a,1+a-d:(c');.;(c^{(n)}); \atop (f),1+a-d-N:(g');.;(g^{(n)});}\frac{x_1}{s},..,\frac{x_n}{s}\right], \qquad (5.3.3.1)$$

$\mathrm{Re}(s)$, $\mathrm{Re}(a) > 0$, $B+C \leqq F+G$.

The extension of (5.3.2.4) to multiple series does not yield an elegent expression, but before proceeding to the discussion of special cases of integrals associated with Bessel functions or confluent hypergeometric functions, we note the following result which is an extension of (5.3.2.5);-

$$\int_0^\infty t^a K_c(t)\, F^{B:D}_{F:G}\left[\begin{matrix}(b):(d');.;(d^{(n)});\\(f):(g');.;(g^{(n)});\end{matrix} x_1t^2,.,x_nt^2\right] dt$$

$$= 2^{a-1}\,\Gamma(\tfrac{a+c+1}{2})\Gamma(\tfrac{a-c+1}{2})$$

$$\times F^{B+2:D}_{F\ :G}\left[\begin{matrix}(b),\frac{a+c+1}{2},\frac{a-c+1}{2}:(d');.;(d^{(n)});\\(f)\qquad :(g');.;(g^{(n)});\end{matrix} 4x_1,.,4x_n\right],$$

$$\mathrm{Re}(a\pm c) > -1,\ B+D+2 \leqq F+G. \qquad (5.3.3.2)$$

5.3.4 Special cases

A number of special cases of (5.3.1.4) which are of some interest may be deduced. First of all, if we let F=G=k=1, an integral of a pair of confluent hypergeometric functions may be evaluated:

$$\int_0^\infty e^{-st}\, t^{a-1}\, {}_1F_1(c;d;xt)\, {}_1F_1(f;g;yt)dt = \frac{\Gamma(a)}{s^a}F_2(a,c,f;d,g;\tfrac{x}{s},\tfrac{y}{s}),$$

$$\mathrm{Re}(a),\ \mathrm{Re}(s) > 0. \qquad (5.3.4.1)$$

When the parameters and variables are further specialised, as appropriate, we have several more results. A few examples are now given.

$$\int_0^\infty e^{-st}\, t^{a-1}\, {}_1F_1(c;d;xt)\, {}_1F_1(f;a;yt)dt$$

$$= \frac{\Gamma(a)}{s^a}\left(\frac{y}{y-s}\right)^f F_1(c,a-f,f;d;x/s,y/[s-y]). \qquad (5.3.4.2)$$

We next consider an integral involving the product of a pair of Laguerre polynomials. This is

$$\int_0^\infty e^{-st}\, t^{a-1}\, L_N^c(st)L_M^d(st)dt$$

$$= \frac{\Gamma(a)\,(c+1,N)\,(d+1,M)}{s^a\quad N!\qquad M!}\sum\frac{(a,m)(-N,m)}{(c+1,m)m!}\ {}_2F_1(a+m,-M;d+1;1). \qquad (5.3.4.3)$$

The inner Gauss function may be summed by Vandermonde's theorem [Slater (1966) Appendix III], so that the right-hand member of (5.3.4.3) becomes

$$\frac{\Gamma(a)\,(c+1,N)\,(d+1-a,M)}{s^a\quad N!\qquad M!}\ {}_3F_2(a,a-d,-N;c+1,a-d-M;1). \qquad (5.3.4.4)$$

If a=c+N-M, the preceding Clausen function may be summed by Saalschütz's theorem and so we have

$$\int_0^\infty e^{-st}\, t^{c+N-M-1}\, L_N^c(st) L_M^d(st)\,dt$$

$$= \frac{(c-M)(-M,N)(c-d-2M,N)(d+1-a,M)}{(c+N-M)N!(-d-M,N)M!([c-d]/2,N)([c-d+1]/2-M,N)}\, 4^{-N}, \quad (5.3.4.5)$$

$$\mathrm{Re}(s) > 0,\ \mathrm{Re}(c) \geq M-N.$$

If F=0, G=1 and k=2, then the ${}_FF_G$ function in the integrand of (5.3.1.1) may be replaced by a Bessel function. Thus

$$\int_0^\infty e^{-st}\, t^{a-1}\, {}_1F_1(c;d;xt) J_g(yt)\,dt$$

$$= \frac{(y/2)^g \Gamma(a+g)}{s^{a+g}\Gamma(g+1)} H_4(a+g,c;g+1,d;-\frac{y^2}{4s^2},\frac{x}{s}), \quad (5.3.4.6)$$

$$\mathrm{Re}(s),\ \mathrm{Re}(a+g) > 0.$$

H_4 is a Horn function, and its series representation is given by Erdélyi et al. (1953) Vol. I page 225 as

$$H_4(a,b;c,d;x,y) = \sum \frac{(a,2m+n)(b,n)}{(c,m)(d,n)} \frac{x^m y^n}{m!n!}\ . \quad (5.3.4.7)$$

An example of an integral involving a Laguerre polynomial and a Chebyshev polynomial is now given.

$$\int_0^\infty e^{-st}\, t^{a-1}\, L_n^c(st) T_m(1-2yt)\,dt = \frac{\Gamma(a)(c+1-a,n)}{s^a\ n!}\, {}_3F_2\left[\begin{matrix}-m,m,a\ ; \\ \frac{1}{2},\frac{a-c-n}{2};\end{matrix}\ y\right]. \quad (5.3.4.8)$$

If we consider (5.3.1.6) with F=2 and G=k=1 such that the Gauss function in the integrand is terminating, an integral involving a Bessel function and a Gegenbauer function (for example) may be investigated.

$$\int_0^\infty e^{-st}\, t^{a-1} J_c(xt) C_n^d(1-2yt)\,dt$$

$$= \frac{(2d,n)\Gamma(a+c)(x/2)^c}{s^{a+c}\ n!\ \Gamma(c+1)} \sum \frac{(a+c,m+2p)(-n,m)(n+2d,m)y^m}{(d+1/2,m)(c+1,p)\ m!p!} \left(-\frac{x^2}{4s^2}\right)^p, \quad (5.3.4.9)$$

$$\mathrm{Re}(s),\ \mathrm{Re}(a+c) > 0.$$

A function related to H_4, but of higher order, appears. If d=-1/2-n, this result simplifies as follows:-

$$\int_0^\infty e^{-st}\, t^{a-1} J_c(xt) C_n^{-1/2-n}(1-2yt)\,dt = \quad (5.3.4.10)$$

(continued)

(continued)

$$= \frac{(-1-2n,n)\Gamma(a+c)(x/2)^c}{s^{a+c}\quad n!\;\Gamma(c+1)} \sum \frac{(a+c,m+2p)(-1-n,m)y^m}{(c+1,p)\;m!p!}\left(-\frac{x^2}{4s^2}\right)^p. \quad (5.3.4.10)$$

The series on the right of (5.3.4.10) terminates in y and converges for $|x| < 2|s|$.

A further example may be furnished by the integral of a Bessel function and a Legendre polynomial,

$$\int_0^\infty e^{-st}\, t^{a-1} J_c(xt) P_n(1-2y^2t^2)\,dt. \quad (5.3.4.11)$$

This may be evaluated in the form

$$\frac{\Gamma(a+c)(x/2)^c}{s^{a+c}\Gamma(c+1)}\, F^{2:2;0}_{0:1;0}\left[\begin{matrix}\frac{a+c}{2},\frac{a+c+1}{2}: -n,1+n; & - & ; y,-\frac{x^2}{s^2}\\ \text{—} \quad : \quad 1 \quad ; & 1+c; & \end{matrix}\right]. \quad (5.3.4.12)$$

$Re(a+c)$, $Re(s) > 0$.

The Kampé de Fériet function above converges if $|x| < |s|$.

Before proceeding to the discussion of special cases of integrals of Bessel functions and double hypergeometric functions, we investigate integrals involving a modified Bessel function of the second kind and firstly, a Bessel function of the first kind, and secondly, a Jacobi polynomial.

$$\int_0^\infty t^a K_c(t) J_d(xt)\,dt = \frac{(x/2)^d}{\Gamma(d+1)}\int_0^\infty t^{a+d} K_c(t)\,{}_0F_1(-;d+1;-\frac{x^2t^2}{4})\,dt. \quad (5.3.4.13)$$

The integral on the right of (5.3.1.13) may be evaluated by means of (5.3.1.10) and we obtain the well-known result

$$\int_0^\infty t^a K_c(t) J_d(xt)\,dt = 2^{a-1}\Gamma\left(\frac{a+d+c+1}{2}\right)\Gamma\left(\frac{a+d-c+1}{2}\right)$$

$$\times\; {}_2F_1\left(\frac{a+d+c+1}{2},\frac{a+d-c+1}{2};d+1;-x^2\right). \quad (5.3.4.14)$$

See Watson (1944) page 410.

The Jacobi polynomial below may be written in hypergeometric form and so, (5.3.1.16) may be used to evaluate the integral

$$\int_0^\infty t^a K_c(t) P_n^{f,g}(1-2xt)\,dt, \quad (5.3.4.15)$$

noting that the associated Gauss function is terminating. It thus follows that the above integral may be written as

$$2^{a-1}\Gamma\left(\frac{a+c+1}{2}\right)\Gamma\left(\frac{a-c+1}{2}\right)$$

$$\times\,{}_4F_1(-n,f+1+g+n,[a+c+1]/2,[a-c+1]/2;f+1;x),\quad Re(a\pm c) > -1. \quad (5.3.4.16)$$

Integrals involving confluent hypergeometric functions and the confluent Appell functions may be evaluated as special cases of (5.3.2.3). For example,

$$\int_0^\infty e^{-st}\, t^{a-1}\, {}_1F_1(-N;b;st)\Psi_2(c;d,d';xt,yt)dt$$

$$= \frac{(b-a,N)\Gamma(a)}{(b,N)\quad s^a} F^{3:0;0}_{1:1;1}\left(\begin{matrix}c,a,1+a-b:-;-\ ;\\ 1+a-b-N\ \ :d;d';\end{matrix}x,y\right),\ \mathrm{Re}(s),\mathrm{Re}(a) > 0. \qquad (5.3.4.17)$$

If $c=1+a-b-N$, then this expression simplifies so that the right-hand member now involves an Appell function of the fourth kind. This is

$$\frac{(b-a,N)\Gamma(a)}{(b,N)\quad s^a} F_4(a,1+a-b;d,d';x,y). \qquad (5.3.4.18)$$

We also have

$$\int_0^\infty e^{-st}\, t^{a-1}\, {}_1F_1(-N;b;st)\Phi_2(c,c';d;xt,yt)dt$$

$$= \frac{(b-a,N)\Gamma(a)}{(b,N)\quad s^a} F^{2:1;1}_{2:0;0}\left(\begin{matrix}a,1+a-b\ \ :c;c';\\ d,1+a-b-N:-;-\ ;\end{matrix}x,y\right),\mathrm{Re}(a),\mathrm{Re}(s)>0, \qquad (5.3.4.19)$$

and its special case

$$\int_0^\infty e^{-st}\, t^{d-1}\, {}_1F_1(-N;b;st)\Phi_2(c,c';d;xt,yt)dt$$

$$= \frac{(b-d,N)\Gamma(d)}{(b,N)\quad s^d} F_1(1+d-b,c,c';1+d-b-N;x,y), \qquad (5.3.4.20)$$

$$\mathrm{Re}(s),\ \mathrm{Re}(d) > 0.$$

Furthermore, the Kampé de Fériet function on the left of (5.3.2.3) may be so specialised as to give integrals involving products of three confluent hypergeometric functions such as

$$\int_0^\infty e^{-st}\, t^{a-1}\, {}_1F_1(-N;b;st)\,{}_1F_1(c;d;xt)\,{}_1F_1(c';d';yt)dt$$

$$= \frac{(b-a,N)\Gamma(a)}{(b,N)\quad s^a} F^{2:1;1}_{1:1;1}\left(\begin{matrix}a,1+a-b:c;c';\\ 1+a-b-N:d;d';\end{matrix}x,y\right),\ \mathrm{Re}(s),\mathrm{Re}(a)>0. \qquad (5.3.4.21)$$

The integral (5.3.2.3) may also be specialised in several other ways, most of which yield rather cumbersome forms of the general triple hypergeometric function of Srivastava. However, if we consider the integral of three Bessel functions of the first kind,an elegent result follows. This expression has already been given, see (5.2.5.39) when $n=3$.

An intersting integral involving the product of two Bessel functions of the first kind and a modified Bessel function of the second kind may be evaluated as a special case of (5.3.2.5):-

$$\int_0^\infty t^a K_c(t)J_f(xt)J_g(yt)dt = \frac{x^f y^g 2^{a-1}}{\Gamma(f+1)\Gamma(g+1)} \Gamma(\frac{a+f+g+c+1}{2})\Gamma(\frac{a+f+g-c+1}{2})$$

$$\times F_4(\frac{a+f+g+c+1}{2},\frac{a+f+g-c+1}{2};f+1,g+1;-x^2,-y^2), \qquad (5.3.4.22)$$

$$Re(a+f+g\pm c) > -1.$$

Similarly, we may evaluate the following integral involving the product of a pair of Hermite polynomials:-

$$\int_0^\infty t^a K_c(t)He_{2n}(xt)He_{2m}(yt)dt = \frac{(2n)!}{n!}\frac{(2m)!}{m!}(-2)^{-n-m} 2^{a-1}$$

$$\times \Gamma(\frac{a+c+1}{2})\Gamma(\frac{a-c+1}{2})F^{2:1;1}_{0:1;1}\left[\begin{matrix}\frac{a+c+1}{2},\frac{a-c+1}{2}:-n;-m; \\ \text{---} \quad :\frac{1}{2};\frac{1}{2};2x,2y\end{matrix}\right], \qquad (5.3.4.23)$$

$$Re(a\pm c) > -1.$$

We now consider the integral

$$\int_0^\infty e^{-st} t^{a-1} {}_1F_1(-N;b;st)\Psi_2^{(n)}(c;d_1,..,d_n;x_1t,..,x_nt)dt, \qquad (5.3.4.24)$$

which may be tackled by using (5.3.3.1), when it will be seen to be equal to

$$\frac{\Gamma(a)(b-a,N)}{s^a \quad (b,N)} F^{3:0}_{1:1}\left(\begin{matrix}c,a,1+a-b: & ; \\ 1+a-b-N & ;d_1;.;d_n;\end{matrix} x_1/s,..,x_n/s\right), \qquad (5.3.4.25)$$

$$Re(a),Re(s) > 0.$$

It thus follows that

$$\int_0^\infty e^{-st} t^{a-1} {}_1F_1(-N;b;st)\Psi_2^{(n)}(1+a-b-N;d_1,..,d_n;x_1t,..,x_nt)dt$$

$$= \frac{\Gamma(a)(b-a,N)}{s^a \quad (b,N)} F_C^{(n)}(a,1+a-b;d_1,..,d_n;x_1/s,..,x_n/s), \qquad (5.3.4.26)$$

where the Function $F_C^{(n)}$ was first defined by Lauricella (1893).

Similarly, we have an integral involving the function Φ_2:-

$$\int_0^\infty e^{-st} t^{a-1} {}_1F_1(-N;b;st)\Phi_2^{(n)}(c_1,..,c_n;a;x_1t,..,x_nt)dt$$

$$= \frac{\Gamma(a)(b-a,N)}{s^a \quad (b,N)} F_D^{(n)}(1+a-b,c_1,..,c_n;1+a-b-N;x_1/s,..,x_n/s), \qquad (5.3.4.27)$$

see Lauricella (1893).

In addition, we have an integral involving the product of several confluent hypergeometric functions:-

$$\int_0^\infty e^{-st}\, t^{a-1}\, {}_1F_1(-N;b;st)\,{}_1F_1(c_1;d_1;x_1t)\,..\,{}_1F_1(c_n;d_n;x_nt)\,dt$$

$$= \frac{\Gamma(a)\,(b-a,N)}{s^a\;(b,N)} F^{2:1}_{1:1}\left[\begin{matrix} a,1+a-b:c_1;.;c_n;x_1/s,.,x_n/s \\ 1+a-b-N:d_1;.;d_n; \end{matrix}\right]. \qquad (5.3.4.28)$$

This section is concluded by giving an integral involving several Bessel functions of the first kind and a modified Bessel function of the second kind. This integral may be evaluated by specialising the formula (5.3.3.3).

$$\int_0^\infty t^a\, K_c(t) J_{d_1}(x_1t)\,..\,J_{d_n}(x_nt)\,dt = \frac{x_1^{d_1}..x_n^{d_n}\, 2^{a-1}}{\Gamma(d_1+1)\,..\,\Gamma(d_n+1)}$$

$$\times\; \Gamma\left(\frac{a+d_1+.+d_n+c+1}{2}\right)\Gamma\left(\frac{a+d_1+.+d_n-c+1}{2}\right)$$

$$\times\; F_C^{(n)}\left[\frac{a+d_1+.+d_n+c+1}{2},\frac{a+d_1+.+d_n-c+1}{2};d_1+1,.,d_n+1;-x_1^2,.,-x_n^2\right].$$

$$(5.3.4.29)$$

5.4 INFINITE INTEGRALS AS INVERSES OF BARNES INTEGRALS

A convenient and practical way of deducing certain types of infinite integrals involving hypergeometric functions is to make use of the Mellin transform theorem, see Slater (1966) page 148:

$$\text{If}\quad f(x) = \frac{1}{2\pi i}\int_{c-i\infty}^{c+i\infty} x^{-s}\, g(s)\,ds, \qquad (5.4.1)$$

$$\text{then}\quad g(s) = \int_0^\infty x^{s-1}\, f(x)\,dx, \qquad (5.4.2)$$

provided that g(s) exists in the Lebesque sense over the range zero to infinity. By a simple exponential change of the variable x in both integrals, the Laplace and Fourier transforms may be deduced.

The integral (5.4.1) is a Barnes contour integral, and many cases have been discussed in the literature. Probably the most important general class of integrals of this type is the Meijer G-function. This function has been discussed in Section 5.2.1, and its definition in terms of an integral of Barnes type is given as equation (5.2.1.2).

We now quote an important theorem due to Slater (1966) page 143.

Slater's Mixed Integral Theorem If

$$I(z) = \frac{1}{2\pi i}\int_{-i\infty}^{i\infty} \Gamma\left[{(a)+s,(b)-s,(g)+s,(h)-s \atop (c)+s,(d)-s,(j)+s,(k)-s}\right]$$

$$\times {}_{A+B+E}F_{C+D+F}\left({(a)+s,(b)-s,(e); \atop (c)+s,(d)-s,(f);}x\right) z^s\, ds, \qquad (5.4.3)$$

where $\Gamma\left[{(a) \atop (b)}\right] = \dfrac{\Gamma(a_1)..\Gamma(a_A)}{\Gamma(b_1)..\Gamma(b_B)}$,

$$\sum_{A,\infty}(z) = \sum_{\mu=1}^{A}\Gamma\left[\begin{matrix}(a)'-a_\mu,(b)+a_\mu,(g)-a_\mu,(h)+a_\mu \\ (c)\ -a_\mu,(d)+a_\mu,(j)-a_\mu,(k)+a_\mu\end{matrix}\right]$$

$$\times \sum_{m,n=0}^{\infty}\frac{((b)+a_\mu,2m+n)((h)+a_\mu,m+n)((e),m)(1+a_\mu-(c),n)}{(1+a_\mu-(a)',n)(1+a_\mu-(g),m+n)((f),m)((d)+a_\mu,2m+n)}$$

$$\times \frac{(1+a_\mu-(j),m+n)x^m z^{-a_\mu-m-n}}{((k)+a_\mu,m+n)}\frac{(-1)^{m(A+G-C-J)}}{m!n!}$$

$$+ \sum_{\mu=1}^{G}\Gamma\left[\begin{matrix}(a)-g_\mu,(b)+g_\mu,(g)'-g_\mu,(h)+g_\mu \\ (c)-g_\mu,(d)+g_\mu,(j)\ -g_\mu,(k)+g_\mu\end{matrix}\right]$$

$$\times \sum_{m,n=0}^{\infty}\frac{((a)-g_\mu,m-n)((b)+g_\mu,m+n)(1+g_\mu-(j),n)((h)+g_\mu,n)}{((c)-g_\mu,m-n)((d)+g_\mu,m+n)(1+g_\mu-(g)',n)((k)+g_\mu,n)}$$

$$\times \frac{((e),m)\ x^m\ z^{-g_\mu-n}}{((f),m)}\frac{(-1)^{n(G-J)}}{m!n!} \qquad (5.4.4)$$

and

$$\sum_{B,\infty}(z) = \sum_{\nu=1}^{B}\Gamma\left[\begin{matrix}(a)+b_\nu,(b)'-b_\nu,(g)+b_\nu,(h)-b_\nu \\ (c)+b_\nu,(d)\ -b_\nu,(j)+b_\nu,(k)-b_\nu\end{matrix}\right]$$

$$\times \sum_{m,n=0}^{\infty}\frac{((a)+b_\nu,2m+n)((g)+b_\nu,m+n)(1+b_\nu-(d),n)(1+b_\nu-(k),m+n)}{(1+b_\nu-(b)',n)(1+b_\nu-(k),m+n)((c+b_\nu,2m+n)((j)+b_\nu,m+n)}$$

$$\times\ ((e),m)x^m z^{b_\nu+m+n}(-1)^{n(B+H-D-K)}/[((f),m)m!n!]+ \qquad (5.4.5)$$

(cont.)

(continued)

$$+\sum_{\nu=1}^{H}\Gamma\left[\begin{matrix}(a)+h_\nu,(b)-h_\nu,(g)+h_\nu,(h)'-h_\nu\\(c)+h_\nu,(d)-h_\nu,(j)+h_\nu,(k)\ -h_\nu\end{matrix}\right]$$

$$\times\sum_{m,n=0}^{\infty}\frac{((a)+h_\nu,m+n)((b)-h_\nu,m-n)((g)+h_\nu,n)((e),m)}{(1+h_\nu-(h)',n)((c)+h_\nu,m+n)((d)-h_\nu,m-n)((f),m)}$$

$$\times\frac{(1+h_\nu-(k),n)\,x^m z^{h_\nu+n}}{((j)+h_\nu,n)}\frac{(-1)^{n(H-K)}}{m!n!}, \qquad (5.4.5)$$

where the series ${}_{A+B+E}F_{C+D+F}(x)$ is absolutely and uniformly convergent in x, then, provided that

$$\tfrac{\pi}{2}(A+G+B+H-C-D-J-K) > |\arg z| ,$$

we have

$$\text{(i)} \qquad I(z) = \sum_{A,\infty}(z) \sim \sum_{B,\infty}(z) \qquad (5.4.6)$$

either (a) when $A+G+D+K > B+H+C+J$
or (b) when $A+G+D+K = B+H+C+J$ and $|z| > 1$,

$$\text{and (ii)} \qquad I(z) = \sum_{B,\infty}(z) \sim \sum_{A,\infty}(z) \qquad (5.4.7)$$

either (a) when $A+G+D+K < B+H+C+J$
or (b) when $A+G+D+K = B+H+C+J$ and $|z| < 1$.

Also, provided that z=1 and

$$\mathrm{Re}[\textstyle\sum(c+d+j+k-a-b-g-h)] > 0,$$

$$\text{(iii)} \qquad I(1) = \sum_{A,\infty}(1) = \sum_{B,\infty}(1) \qquad (5.4.8)$$

when $\qquad A+G-C-J = B+H-D-K \geq 0.$

This theorem has been established in detail by Slater (1966) page 143 by the use of contour integration. All the results of this section may be extended to multiple integrals and multiple series by appealing to de la Vallée Poussin's theorem, see (5.2.7) and (5.2.8).

5.4.1 Generalised Hypergeometric Function

In order to deduce Mellin integrals of the type (5.4.2) from Slater's Mixed Integral Theorem, the number of parameters and general conditions must be so adjusted that either $\sum_{A,\infty}(z)$ or $\sum_{B,\infty}(z)$ involves only one generalised hypergeometric function of one variable. It is also possible to apply the Mellin inversion theorem to the definition of the Meijer G-function, (5.2.1.3), so that, by a trivial change of variable, we have

$$\int_0^\infty x^{s-1} G^{m,n}_{p,q}\left(x\left|\begin{matrix}a_j\\b_j\end{matrix}\right.\right)dx = \frac{\prod_{j=1}^{m}\Gamma(b_j+s)\prod_{j=1}^{n}\Gamma(1-a_j-s)}{\prod_{j=m+1}^{q}\Gamma(1-b_j-s)\prod_{j=n+1}^{p}\Gamma(a_j+s)}, \qquad (5.4.1.1)$$

where, for convergence,

$$-\min_{1\leq j\leq m}\mathrm{Re}(b_j) < \mathrm{Re}(s) < 1-\max_{1\leq j\leq n}\mathrm{Re}(a_j) . \qquad (5.4.1.2)$$

We note that the special case of the G-function, (5.2.1.3), after a straightforward change of variable, enables us to write

$$\int_0^\infty x^{s-1} {}_AF_C((a);(c);kx)dx = k^{-s}\frac{\Gamma(s)\prod_{j=1}^{C}\Gamma(c_j)\prod_{j=1}^{A}\Gamma(a_j-s)}{\prod_{j=1}^{A}\Gamma(a_j)\prod_{j=1}^{C}\Gamma(b_j-s)}, \qquad (5.4.1.3)$$

where $0 < \mathrm{Re}(s) < \max \mathrm{Re}(a_j)$.

We now consider more general types of Mellin integrals which are consequences of Slater's Mixed Integral Theorem. Take A=1 and G=B=H=J=D=K=C=0 in this theorem when we have

$$\frac{1}{2\pi i}\int_{-i\infty}^{i\infty}\Gamma(a-s)\,{}_{E+1}F_F\left(\begin{matrix}a-s,(e);\\(f)\;\;\;\;;\end{matrix}x\right)z^{-s}\,ds = z^{-a}\exp\left(\frac{1}{z}\right){}_EF_F\left(\begin{matrix}(e);\\(f);\end{matrix}-\frac{x}{z}\right) \qquad (5.4.1.4)$$

which on inversion gives the result

$$\int_0^\infty z^{s-a-1}\exp\left(\frac{1}{z}\right){}_EF_F\left(\begin{matrix}(e);\\(f);\end{matrix}-\frac{x}{z}\right)dz = \Gamma(a-s)\,{}_{E+1}F_F\left(\begin{matrix}a-s,(e);\\(f)\;\;\;\;;\end{matrix}x\right), \qquad (5.4.1.5)$$

$$\mathrm{Re}(a-s) > 0, \qquad \frac{\pi}{2} > |\arg z| .$$

Similarly, if we let G=1 and A=B=C=D=E=F=x=0 and G=1 and A=H=J=0 E=C=K=F=0, it follows respectively that

$$\int_0^\infty z^{s-g-1}\,{}_{H+J}F_K\left(\begin{matrix}1+g-(j),(h)+g;\\(k)+g\qquad ;\end{matrix}(-1)^{(1-J)}z^{-1}\right)dz$$

$$= \Gamma\left[\begin{matrix}g-s,(h)+s,(j)-g,(k)+g\\(j)-s,(k)+s,(h)+g\end{matrix}\right], \qquad (5.4.1.6)$$

where $K > H+J-1$, $\mathrm{Re}(g-s)$, $\mathrm{Re}((h)+s) > 0$ and $\frac{\pi}{2}(1+H-J-K)>|\arg z|$

and

$$\int_0^\infty z^{s-g-1}\,{}_BF_D\left(\begin{matrix}(b)+g;\\(d)+g;\end{matrix}x-\frac{1}{z}\right)dz = \Gamma\left[\begin{matrix}(d)+g,(b)+s,g-s\\(b)+g,\ (g)+s\end{matrix}\right]{}_BF_D\left(\begin{matrix}(b)+s;\\(d)+s;\end{matrix}x\right), \quad (5.4.1.7)$$

where $D > B-1$, $\mathrm{Re}((b)+s),\mathrm{Re}(g-s) > 0$ and $\frac{\pi}{2}(1+B-D) > |\arg z|$.

5.4.2 Double Hypergeometric Function

Within its region of convergence, the Kampé de Fériet function of two variables may be expanded as a series of generalised hypergeometric functions of one variable:

$$F^{A:B;B'}_{C:D;D'}\left(\begin{matrix}(a):(b);(b');\\(c):(d);(d');\end{matrix}x,y\right)$$

$$= \sum_{m=0}^{\infty}\frac{((a),m)((b),m)x^m}{((c),m)((d),m)m!}\,{}_{A+B'}F_{C+D'}\left(\begin{matrix}(a)+m,(b');\\(c)+m,(d');\end{matrix}y\right). \qquad (5.4.2.1)$$

If the inner hypergeometric function is replaced by its Barnes integral, the processes of integration and summation may be interchanged, so that the right-hand member of (5.4.2.1) may be written

$$\sum_{m=0}^{\infty}\frac{((a),m)((b),m)x^m}{((c),m)((d),m)m!}\,\Gamma\left[\begin{matrix}(c)+m,(d')\\(a)+m,(b')\end{matrix}\right]\frac{1}{2\pi i}\int_{-i\infty}^{i\infty}\Gamma\left[\begin{matrix}(a)+m+s,(b')+s,-s\\(c)+m+s,(d')+s\end{matrix}\right]y^s\,ds. \qquad (5.4.2.2)$$

Hence,

$$\Gamma\left[\begin{matrix}(c),(d')\\(a),(b')\end{matrix}\right]\frac{1}{2\pi i}\int_{-i\infty}^{i\infty}\Gamma\left[\begin{matrix}(a)+s,(b')+s,-s\\(c)+s,(d')+s\end{matrix}\right]y^s\,{}_{A+B}F_{C+D}\left(\begin{matrix}(a)+s,(b);\\(c)+s,(d);\end{matrix}x\right)ds$$

$$= F^{A:B:B'}_{C:D:D'}\left(\begin{matrix}(a):(b);(b');\\(c):(d);(d');\end{matrix}x,y\right), \qquad (5.4.2.3)$$

where $A+B+B' \leqq C+D+D'+1$ and $\frac{\pi}{2}|A+B'+1-C-D'| > |\arg y|$. $\quad$ (5.4.2.4)

On inversion, we have

$$\int_0^\infty y^{s-1}F^{A:B;B'}_{C:D;D'}\left(\begin{matrix}(a):(b);(b');\\(c):(d);(d');\end{matrix}x,y\right)dy = \Gamma\left[\begin{matrix}(a)-s,(b')-s,s,(c),(d')\\(c)-s,(d')-s\qquad (a),(b')\end{matrix}\right]$$

$$\times\ {}_{A+B}F_{C+D}\left(\begin{matrix}(a)-s,(b);\\(c)-s,(d);\end{matrix}x\right), \qquad (5.4.2.5)$$

with the additional restriction

$$\mathrm{Re}((a)-s),\mathrm{Re}((b')-s),\mathrm{Re}(s) > 0.$$

Barnes integrals leading to Kampé de Fériet functions may also be deduced from Slater's Mixed Integral Theorem (5.4.3) and (5.4.4). If we invert by the Mellin transform theorem, we obtain the expressions (5.4.2.6) and (5.4.2.7) by taking A=1, G=B=D=0 and G=1, A=C=0, respectively.

$$\int_0^\infty z^{s-a-1} F^{H+J:E;C}_{K\ :F;0}\left(\begin{matrix}(h)+a,1+a-(j):(e);1+a-(c);\\(k)+a\qquad :(f);\ \ —\ \ ;\end{matrix}(-1)^{(1-C-J)}x/z,1/z\right)dz$$

$$= \Gamma\left[\begin{matrix}a-s,(h)+s,(e)-a,(j)-a,(k)+a\\(e)-s,(j)-s,(k)+s,(h)+a\end{matrix}\right] \qquad (5.4.2.6)$$

where $1+K>H+C+J$, $\mathrm{Re}(a-s),\mathrm{Re}((h)+s)>0$ and $\frac{\pi}{2}|1+H-C-J-K|>|\arg z|$.

$$\int_0^\infty z^{s-g-1} F^{B:E;H+J}_{D:F;\ K}\left(\begin{matrix}(b)+g:(e);(h)+g,1+g-(j);\\(d)+g:(f);\quad (k)+g\qquad ;\end{matrix}x,(-1)^{1-J}/z\right)dz$$

$$= \Gamma\left[\begin{matrix}(b)+s,g-s,(h)+s,(d)+g,(j)-g,(k)+g\\(d)+s,\ (j)-s,\ (k)+s,\ (b)+g,\ (h)+g\end{matrix}\right] {}_{B+E}F_{D+F}\left(\begin{matrix}(b)+s,(e);\\(d)+s,(f);\end{matrix}x\right), \qquad (5.4.2.7)$$

where $D+K+1>B+H+F$, $\mathrm{Re}((b)+s),\mathrm{Re}(g-s),\mathrm{Re}((h)+s) > 0$ and $\frac{\pi}{2}|1+B+H-D-J-K| > |\arg z|$.

5.4.3 Multiple Hypergeometric Function

First of all, we give a single Barnes integral for the generalised Kampé de Fériet function of several variables. The function

$$F^{A:B}_{C:D}\left[\begin{matrix}(a):(b');.;(b^{(n)});\\(c):(d');.;(d^{(n)});\end{matrix}x_1,.,x_n\right] \qquad (5.4.3.1)$$

may be expanded, within its region of convergence, into the (n-1) - fold series of generalised hypergeometric functions of one variable which follows:-

$$\sum\frac{((a),m_1+.+m_{n-1})((b'),m_1)..((b^{(n-1)}),m_{n-1})x_1^{m_1}..x_{n-1}^{m_{n-1}}}{((c),m_1+.+m_{n-1})((d'),m_1)..((d^{(n-1)}),m_{n-1})m_1!..m_{n-1}!}$$

$$\times\ {}_{A+B}F_{C+D}\left[\begin{matrix}(a)+m_1+.+m_{n-1},((b^{(n)});\\(C)+m_1+.+m_{n-1},((d^{(n)});\end{matrix}x_n\right]. \qquad (5.4.3.2)$$

The hypergeometric function of (5.4.3.2) has the following Barnes integral representation:-

$$\Gamma\begin{bmatrix}(c)+m_1+.+m_{n-1},(d^{(n)})\\(a)+m_1+.+m_{n-1},(b^{(n)})\end{bmatrix}\frac{1}{2\pi i}\int_{-i\infty}^{i\infty}\Gamma\begin{bmatrix}(a)+m_1+.+m_{n-1}+s,(b^{(n)})+s,-s\\(c)+m_1+.+m_{n-1}+s,(d^{(n)})+s\end{bmatrix}x_n^s ds. \tag{5.4.3.3}$$

We may thus write (5.4.3.1) in the form

$$\Gamma\begin{bmatrix}(c),(d^{(n)})\\(a),(b^{(n)})\end{bmatrix}\frac{1}{2\pi i}\int_{-i\infty}^{i\infty}\Gamma\begin{bmatrix}(a)+s,(b^{(n)})+s,-s\\(c)+s,(d^{(n)})+s\end{bmatrix}x_n^s$$

$$\times\ F^{A:B}_{C:D}\begin{bmatrix}(a)+s:(b');.;(b^{(n-1)});\\(c)+s:(d');.;(d^{(n-1)});\end{bmatrix}x_1,.,x_{n-1}\Bigg]ds, \tag{5.4.3.4}$$

provided that $A+B \leqq C+D+1$ and $\frac{\pi}{2}|A+B+1-C-D| > |\arg x_n|$.

The Mellin transform theorem, see (5.4.1) and (5.4.2), may now be applied and we have

$$\int_0^\infty x_n^{s-1}\ F^{A:B}_{C:D}\begin{bmatrix}(a):(b');.;(b^{(n)});\\(c):(d');.;(d^{(n)});\end{bmatrix}x_1,.,x_n\Bigg]dx_n$$

$$=\Gamma\begin{bmatrix}(a)-s,(b^{(n)})-s,s,(c),(d^{(n)})\\(c)-s,(d^{(n)})-s,\ (a),\ (b^{(n)})\end{bmatrix}$$

$$\times\ F^{A:B}_{C:D}\begin{bmatrix}(a)-s:(b');.;(b^{(n-1)});\\(c)-s:(d');.;(d^{(n-1)});\end{bmatrix}x_1,.,x_{n-1}\Bigg], \tag{5.4.3.5}$$

provided also that $\mathrm{Re}((a)-s),\mathrm{Re}((b^{(n)})-s),\mathrm{Re}(s) > 0$.

Many other Mellin integral relations involving hypergeometric functions of several variables exist. These results have not, so far, been worked out.

5.4.4 Special Cases

The generalised hypergeometric function of (5.4.1.5) may be so specialised as to enable us to deduce a large number of Mellin integrals of the special functions. Hence, we have

$$\int_0^\infty x^{s-1}\ I_c(p\sqrt{x})dx = \frac{(p/2)^c}{\Gamma(c+1)}\int_0^\infty x^{s+c/2-1}\ {}_0F_1(-;c+1;p^2x/4)dx$$

$$= \frac{\Gamma(s+c/2)\ p^{-2s}}{2^c\Gamma(c/2+1-s)},\quad \mathrm{Re}(s+c/2) > 0. \tag{5.4.4.1}$$

This gives a Mellin integral of a modified Bessel function.

Another example is

$$\int_0^\infty x^{s-1}\ \gamma(a,px)dx = -p^{-s-a}\ s\Gamma(s+a), \quad \text{Re}(s+a),\text{Re}(s) > 0, \quad (5.4.4.2)$$

whereby an integral of an incomplete gamma function is evaluated. The integral

$$\int_0^\infty z^{s-1}\ \exp(1/z)\ \underline{H}_c(p\sqrt{z})dz, \qquad (5.4.4.3)$$

involving a Struve function, may be evaluated as a special case of (5.4.1.7) by expressing the integrand in terms of a ${}_1F_2$ series. Hence, (5.4.4.3) takes the form

$$\frac{p^{c/2+1}\Gamma(1-s-c/2)}{\sqrt{\pi}2^c\Gamma(c+3/2)}\ {}_2F_2\left({1-s-c/2,1;\atop 3/2,3/2+c;}-p^2/4\right), \qquad (5.4.4.4)$$

$$\pi/2 > |\arg z|,\ \text{Re}(1-s-c/2) > 0,$$

and this, in turn, yields the two results

$$\int_0^\infty z^{s-1}\ \exp(1/z)\ \underline{H}_{-1/2}(p\sqrt{z})dz = \frac{\sqrt{2}\ p^{3/4}\ \Gamma(5/4-s)}{\sqrt{\pi}}\ {}_1F_1\left({5/4-s;\atop 3/2\ ;}-p^2/4\right),$$

$$\pi/2 > |\arg z|,\ \text{Re}(5/4-s) > 0 \qquad (5.4.4.5)$$

and

$$\int_0^\infty z^{-5/4}\ \exp(1/z)\ \underline{H}_{-1/4}(p\sqrt{z})dz = \sqrt{2}\ p^{3/4}\ \exp(-p^2/4), \qquad (5.4.4.6)$$

$$\pi/2 > |\arg z|\ .$$

We now give two integrals each involving an Appell function, by specialising (5.4.2.5).

$$\int_0^\infty y^{s-1}\ F_1(a,b,b';c;x,y)dy = \Gamma\left[{a-s,b'-s,s,c\atop c-s,\ a,\ b'}\right]{}_2F_1\left({a-s,b;\atop c-s\ ;}x\right) \quad (5.4.4.7)$$

and

$$\int_0^\infty y^{s-1}\ F_2(a,b,b';c,c';x,y)dy = \Gamma\left[{a-s,b'-s,s,d'\atop d'-s,a,\ b'}\right]{}_2F_1\left({a-s,b;\atop d\ ;}x\right).$$

$$(5.4.4.8)$$

The functions F_1 and F_2 have recently arisen in a number of applications. See Exton (1976) Chapters 7 and 8 for example.

Equation (5.4.3.5) yields Mellin integrals of the four Lauricella functions $F_A^{(n)}$, $F_B^{(n)}$, $F_C^{(n)}$ and $F_D^{(n)}$, see Lauricella (1893). These results are now given.

$$\int_0^\infty x_n^{s-1} F_A^{(n)}(a,b_1,..,b_n;d_1,..,d_n;x_1,..,x_n)dx_n$$

$$= \Gamma\left[{a-s,b_n-s,s,d_n \atop d_n-s,\ a,\ b_n}\right]F_A^{(n-1)}(a-s,b_1,..,b_{n-1};d_1,..,d_{n-1};x_1,..,x_{n-1}),$$

$$\mathrm{Re}(a-s),\ \mathrm{Re}(b_n-s),\ \mathrm{Re}(s) > 0, \qquad (5.4.4.9)$$

$$\int_0^\infty x_n^{s-1} F_B^{(n)}(a_1,..,a_n,b_1,..,b_n;c;x_1,..,x_n)dx_n$$

$$= \Gamma\left[{a_n-s,b_n-s,s,c \atop c-s,\ a_n,b_n}\right]F_B^{(n-1)}(a_1,..,a_{n-1},b_1,..,b_{n-1};c-s;x_1,..,x_{n-1}),$$

$$\mathrm{Re}(a_n-s),\ \mathrm{Re}(b_n-s),\ \mathrm{Re}(s) > 0, \qquad (5.4.4.10)$$

$$\int_0^\infty x_n^{s-1} F_C^{(n)}(a,b;d_1,..,d_n;x_1,..,x_n)dx_n$$

$$= \Gamma\left[{a-s,b-s,s,d_n \atop d_n-s,\ a,b}\right]F_C^{(n-1)}(a-s,b-s;d_1,..,d_{n-1};x_1,..,x_{n-1}), \qquad (5.4.4.11)$$

$$\mathrm{Re}(a-s),\ \mathrm{Re}(b-s),\ \mathrm{Re}(s) > 0$$

and

$$\int_0^\infty x_n^{s-1} F_D^{(n)}(a,b_1,..,b_n;c;x_1,..,x_n)dx_n$$

$$= \Gamma\left[{a-s,b_n-s,s,c \atop c-s,\ \ a,b_n}\right]F_D^{(n-1)}(a-s,b_1,..,b_{n-1};c-s;x_1,..,x_{n-1}), \qquad (5.4.4.12)$$

$$\mathrm{Re}(a-s),\ \mathrm{Re}(b_n-s),\ \mathrm{Re}(s) > 0.$$

In the expressions (5.4.4.9) to (5.4.4.12), we have the additional restriction that $\pi > |\arg x_n|$.

This concluded the discussion of infinite integrals.

Chapter 6

Multiple Integrals

6.1 MULTIPLE EULER INTEGRALS

Integrals of this type are generalisations of the single Euler integrals discussed in Chapter 1. We consider the two general forms which follow:-

$$\int_0^1 .(n). \int_0^1 u_1^{a_1-1}(1-u_1)^{b_1-a_1-1} .. u_n^{a_n-1}(1-u_n)^{b_n-a_n-1} f(u_1,..,u_n)du_1..du_n \tag{6.1.1}$$

and

$$\int .(n). \int_R u_1^{a_1-1} .. u_n^{a_n-1}(1-u_1-.-u_n)^{c-a_1-.-a_n-1} f(u_1,..,u_n)du_1..du_n, \tag{6.1.2}$$

where R is the region $0 \leq u_1,..,0 \leq u_n$, $1 \geq u_1+.+u_n$. It is supposed that the function $f(u_1,..,u_n)$ takes the form

$$f(u_1,..,u_n) = \sum_{m_1,..,m_n=0}^{\infty} A_{m_1,..,m_n} u_1^{m_1}..u_n^{m_n}, \tag{6.1.3}$$

$A_{m_1,..,m_n}$ being an arbitrary coefficient independent of the u_i.

Many other types of multiple Euler integral exist, but they are not considered explicitly here as their theory may be developed in a similar manner to that of (6.1.1) and (6.1.2).

If we substitute (6.1.3) into (6.1.1), assuming that this series converges uniformly over the range of integration in question, term-by term integration is permissable and (6.1.1) becomes

$$\sum A_{m_1,..,m_n} \int_0^1 u_1^{a_1+m_1-1}(1-u_1)^{b_1-a_1-1}du_1 .. \int_0^1 u_n^{a_n+m_n-1}(1-u_n)^{b_n-a_n-1}du_n. \tag{6.1.4}$$

Each of the integrals of (6.1.4) is of beta type and may be written

$$B(a_i+m_i,b_i-a_i) = \frac{\Gamma(a_i)\Gamma(b_i-a_i)}{\Gamma(b_i)} \frac{(a_i,m_i)}{(b_i,m_i)}, \tag{6.1.5}$$

$$1 \leq i \leq n.$$

Hence, the integral (6.1.1) becomes

$$\frac{\Gamma(a_1)\Gamma(b_1-a_1)}{\Gamma(b_1)} \cdots \frac{\Gamma(a_n)\Gamma(b_n-a_n)}{\Gamma(b_n)} \sum \frac{(a_1,m_1)\cdots(a_n,m_n)}{(b_1,m_1)\cdots(b_n,m_n)} A_{m_1,\dots,m_n},$$

provided that $\mathrm{Re}(a_i), \mathrm{Re}(b_i-a_i) > 0$. (6.1.6)

If (6.1.2) is treated in a similar way, it takes the form

$$\frac{\Gamma(a_1)\cdots\Gamma(a_n)\Gamma(c-a_1-\cdots-a_n)}{\Gamma(c)} \sum \frac{(a_1,m_1)\cdots(a_n,m_n)}{(c,m_1+\dots+m_n)} A_{m_1,\dots,m_n}, \quad (6.1.7)$$

provided that $\mathrm{Re}(a_i)$, $\mathrm{Re}(c-a_1-\cdots-a_n) > 0$. The formulae (6.1.6) and (6.1.7) are used as a basis for deducing a number of results involving various types of hypergeometric functions. See Whittaker and Watson (1952) page 258.

6.1.1 Generalised Hypergeometric Function

Suppose that $f(u_1,\dots,u_n)$ is a generalised hypergeometric function of one variable. The expression (6.1.6) gives the formula

$$\int_0^1 \cdot (n) \cdot \int_0^1 u^{a_1-1}(1-u_1)^{b_1-a_1-1} \cdots u_n^{a_n-1}(1-u_n)^{b_n-a_n-1}$$
$$\times \; {}_CF_D((c);(d);x_1u_1+\dots+x_nu_n)\,du_1\cdots du_n$$
$$= \frac{\Gamma(a_1)\Gamma(b_1-a_1)}{\Gamma(b_1)} \cdots \frac{\Gamma(a_n)\Gamma(b_n-a_n)}{\Gamma(b_n)} F^{C:1}_{D:1}\left[\begin{matrix}(c):a_1;\dots;a_n;\\(d):b_1;\dots;b_n;\end{matrix}\, x_1,\dots,x_n\right], \quad (6.1.1.1)$$

and (6.1.7) enables us to write down the result

$$\int \cdot (n) \cdot \int_R u_1^{a_1-1} \cdots u_n^{a_n-1}(1-u_1-\cdots-u_n)^{b-a_1-\cdots-a_n-1}$$
$$\times \; {}_CF_D((c);(d);x_1u_1+\dots+x_nu_n)\,du_1\cdots du_n$$
$$= \frac{\Gamma(a_1)\cdots\Gamma(a_n)\Gamma(b-a_1-\cdots-a_n)}{\Gamma(b)} F^{C\;:1}_{D+1:0}\left[\begin{matrix}(c)\;:a_1;\dots;a_n;\\(d),b:\quad - \quad;\end{matrix}\, x_1,\dots,x_n\right]. \quad (6.1.1.2)$$

6.1.2 Hypergeometric Functions of Several Variables

Two examples of multiple Euler integrals involving multiple hypergeometric functions will now be evaluated by letting $f(u_1,\dots,u_n)$ of Section 6.1 be a generalised Kampé de Fériet function. From (6.1.6) and (6.1.7) we have, respectively,

$$\int_0^1 .(n). \int_0^1 u_1^{a_1-1}(1-u_1)^{b_1-a_1-1}..u_n^{a_n-1}(1-u_n)^{b_n-a_n-1}$$

$$\times F^{C:D}_{F:G}\left[\begin{matrix}(c):(d');.;(d^{(n)});\\(f):(g');.;(g^{(n)});\end{matrix} u_1x_1,.,u_nx_n\right] du_1..du_n$$

$$= \frac{\Gamma(a_1)\Gamma(b_1-a_1)}{\Gamma(b_1)}..\frac{\Gamma(a_n)\Gamma(b_n-a_n)}{\Gamma(b_n)}$$

$$\times F^{C:D+1}_{F:G+1}\left[\begin{matrix}(c):(d'),a_1;.;(d^{(n)}),a_n;\\(f):(g'),b_1;.;(g^{(n)}),b_n;\end{matrix} x_1,.,x_n\right] \quad (6.1.2.1)$$

and

$$\int_R .(n). \int u_1^{a_1-1}..u_n^{a_n-1}(1-u_1-.-u_n)^{b-a_1-.-a_n-1}$$

$$\times F^{C:D}_{F:G}\left[\begin{matrix}(c):(d');.;(d^{(n)});\\(f):(g');.;(g^{(n)});\end{matrix} u_1x_1,.,u_nx_n\right] du_1..du_n$$

$$= \frac{\Gamma(a_1)..\Gamma(a_n)\Gamma(b-a_1-.-a_n)}{\Gamma(b)}$$

$$\times F^{C\ :D+1}_{F+1:\ G}\left[\begin{matrix}(c)\ :(d'),a_1;.;(d^{(n)}),a_n;\\(f),b:\ (g')\ ;.;\ (g^{(n)})\ ;\end{matrix} x_1,.,x_n\right]. \quad (6.1.2.2)$$

6.1.3 Special Cases

A number of particular cases of the results of Sections 6.1.1 and 6.1.2 are now given. These are intended merely to be examples of how these cases arise. Using the formula (6.1.1.1), it is possible to write down multiple Euler integrals of the Bessel function and the Laguerre polynomial which are, respectively,

$$\int_0^1 .(n). \int_0^1 u_1^{a_1-1}(1-u_1)^{b_1-a_1-1}..u_n^{a_n-1}(1-u_n)^{b_n-a_n-1}$$
$$\times (u_1x_1+.+u_nx_n)^{1/2}\ J_c(\surd[u_1x_1+.+u_nx_n])du_1..du_n$$

$$= \frac{\Gamma(a_1)\Gamma(b_1-a_1)}{\Gamma(b_1)}..\frac{\Gamma(a_n)\Gamma(b_n-a_n)}{\Gamma(b_n)} \cdot \frac{2^{-c}}{\Gamma(c+1)}$$

$$\times F^{0:1}_{1:1}\left[\begin{matrix}:a_1;.;a_n;\\c+1:b_1;.;b_n;\end{matrix} -\frac{x_1}{4},.,-\frac{x_n}{4}\right] \quad (6.1.3.1)$$

and

$$\int_0^1 .(n). \int_0^1 u_1^{a_1-1}(1-u_1)^{b_1-a_1-1}..u_n^{a_n-1}(1-u_n)^{b_n-a_n-1}$$
$$\times\ L_m^c(x_1u_1+.+x_nu_n)du_1..du_n$$

$$= \frac{\Gamma(a_1)\Gamma(b_1-a_1)}{\Gamma(b_1)}..\frac{\Gamma(a_n)\Gamma(b_n-a_n)}{\Gamma(b_n)}\cdot\frac{(c+1,m)}{m!}$$
$$\times\ F^{1:1}_{1:1}\left[\begin{matrix}-m:a_1;.;a_n;\\c+1:b_1;.;b_n;\end{matrix}x_1,.,x_n\right]. \qquad (6.1.3.2)$$

An integral involving a Gegenbauer polynomial is now evaluated by means of (6.1.1.2):

$$\int_0 .(n). \int_0 u_1^{a_1-1}..u_n^{a_n-1}(1-u_1-.-u_n)^{b-a_1-.-a_n-1}$$
$$\times\ C_m^d(1-2x_1u_1-.-2x_nu_n)du_1..du_n$$

$$= \frac{(2d,m)}{m!}\cdot\frac{\Gamma(a_1)..\Gamma(a_n)\Gamma(b-a_1-.-a_n)}{\Gamma(b)}$$
$$\times\ F^{2:1}_{2:0}\left[\begin{matrix}-m,m+2d:a_1;.;a_n;\\c+1/2,d:\ -\ ;\end{matrix}x_1,.,x_n\right]. \qquad (6.1.3.3)$$

Similarly, we may write down the following two integrals involving Lauricella functions using the results (6.1.2.1) and (6.1.2.2), respectively:-

$$\int_0^1 .(n). \int_0^1 u_1^{a_1-1}(1-u_1)^{b_1-a_1-1}..u_n^{a_n-1}(1-u_n)^{b_n-a_n-1}$$
$$\times\ F_A^{(n)}(c,d_1,.,d_n;f_1,.,f_n;u_1x_1,.,u_nx_n)du_1..du_n$$

$$= \frac{\Gamma(a_1)\Gamma(b_1-a_1)}{\Gamma(b_1)}..\frac{\Gamma(a_n)\Gamma(b_n-a_n)}{\Gamma(b_n)}$$
$$\times\ F^{1:2}_{0:2}\left[\begin{matrix}c:d_1,a_1;.;d_n,a_n;\\-:f_1,b_1;.;f_n,b_n;\end{matrix}x_1,.,x_n\right] \qquad (6.1.3.4)$$

and

$$\int .(n). \int_R u_1^{a_1-1}..u_n^{a_n-1}\ (1-u_1-.-u_n)^{b-a_1-.-a_n-1}$$
$$\times\ F_D^{(n)}(b,d_1,.,d_n;f;u_1x_1,.,u_nx_n)du_1..du_n = \qquad (6.1.3.5)$$

(continued)

(continued)

$$= \frac{\Gamma(a_1)..\Gamma(a_n)\Gamma(b-a_1-.-a_n)}{\Gamma(b)} F_B^{(n)}(d_1,.,d_n,a_1,.,a_n;f;x_1,.,x_n). \tag{6.1.3.5}$$

6.2 MULTIPLE LAPLACE AND HANKEL INTEGRALS

The repeated application of the integral formula of Laplace type for the gamma function leads to the development of multiple integrals of Laplace type. The function $f(u_1,.,u_n)$ as given by (6.1.3) above is substituted into the multiple Laplace integral

$$\int_0^\infty .(n). \int_0^\infty e^{-u_1-.-u_n} u_1^{a_1-1}..u_n^{a_n-1} f(u_1,.,u_n)du_1..du_n. \tag{6.2.1}$$

If the series representation of $f(u_1,.,u_n)$, that is (6.1.3), converges for all finite values of its variables $u_1,.,u_n$, then the multiple integration and multiple summation may be interchanged, and the use of the gamma integral enables us to write (6.2.1) in the form

$$\Gamma(a_1)..\Gamma(a_n)\sum A_{m_1,.,m_n}(a_1,m_1)..(a_n,m_n), \tag{6.2.2}$$

$$\mathrm{Re}(a_1),.,\mathrm{Re}(a_n) > 0.$$

Similarly, if we employ Hankel's formula for the gamma function, we have the result

$$\int_{-\infty}^{(0+)} .(n). \int_{-\infty}^{(0+)} e^{u_1+.+u_n} u_1^{-b_1}..u_n^{-b_n} f(u_1^{-1},.,u_n^{-1})du_1..du_n$$

$$= 1/[\Gamma(b_1)..\Gamma(b_n)]\sum A_{m_1,.,m_n}/[(b_1,m_1)..(b_n,m_n)]. \tag{6.2.3}$$

A number of variants of (6.2.2) and (6.2.3) are obtainable by the same methods.

6.2.1 Multiple Hypergeometric Function

The integrals (6.2.1) and (6.2.3) may be specialised in a number of ways so that multiple Laplace and Hankel integrals of hypergeometric functions of one or more variables may be evaluated. We give two examples.

$$\int_0^\infty .(n). \int_0^\infty e^{-u_1-.-u_n} u_1^{a_1-1}..u_n^{a_n-1} \tag{6.2.1.1}$$

$$\times F_{F:G}^{C:D}\left[\begin{matrix}(c):(d');.;(d^{(n)}); \\ (f):(g');.;(g^{(n)});\end{matrix} x_1u_1,.,x_nu_n\right] du_1..du_n =$$

(cont.)

(continued)

$$= \Gamma(a_1)..\Gamma(a_n)F^{C:D+1}_{F:\ G}\left[\begin{matrix}(c):(d'),a_1;.;(d^{(n)}),a_n;\\(f):\ (g')\ \ ;.;\ (g^{(n)})\ \ ;\end{matrix}x_1,.,x_n\right],$$

(6.2.1.1)

$$\mathrm{Re}(a_1),.,\mathrm{Re}(a_n) > 0$$

and

$$\int_{-\infty}^{(o+)}.(n).\int_{-\infty}^{(o+)} e^{u_1+.+u_n}\, u_1^{-b_1}..u_n^{-b_n}$$

$$\times F^{C:D}_{F:G}\left[\begin{matrix}(c):(d');.;(d^{(n)});\\(f):(g');.;(g^{(n)});\end{matrix}\frac{x_1}{u_1},.,\frac{x_n}{u_n}\right]du_1..du_n$$

$$= 1/[\Gamma(b_1)..\Gamma(b_n)]F^{C:\ D}_{F:G+1}\left[\begin{matrix}(c):\ (d')\ \ ;.;\ (d^{(n)})\ \ ;\\(f):(g'),b_1;.;(g^{(n)}),b_n;\end{matrix}x_1,.,x_n\right].$$

(6.2.1.2)

An integral formula may readily be obtained by combining (6.2.1.1) and (6.2.1.2).

6.2.2 Special Cases

The results of the previous two sections may be further specialised, and of the very many possibilities which may arise, we give the expressions below.

$$\int_0^\infty.(n).\int_0^\infty e^{-u_1-.-u_n}\, u_1^{a_1-1}..u_n^{a_n-1}\,{}_1F_1\left(\begin{matrix}b;\\c;\end{matrix}u_1x_1+.+u_nx_n\right)du_1..du_n$$

$$= \Gamma(a_1)..\Gamma(a_n)F_D^{(n)}(b,a_1,.,a_n;c;x_1,.,x_n),\ \mathrm{Re}(a_i) > 0 \qquad (6.2.2.1)$$

and

$$\int_{-\infty}^{(o+)}.(n).\int_{-\infty}^{(o+)} e^{u_1+.+u_n}\, u_1^{-b_1}..u_n^{-b_n}\,{}_2F_0(-N,d;-;\frac{x_1}{u_1}+.+\frac{x_n}{u_n})du_1..du_n$$

$$= 1/[\Gamma(b_1)..\Gamma(b_n)]F_C^{(n)}(-N,d;b_1,.,b_n;x_1,.,x_n). \qquad (6.2.2.2)$$

The first of these two formulae may be expressed in terms of the Laguerre polynomial, and the second in terms of the Hermite polynomial. If we consider examples of integrals of multiple hypergeometric functions, then we have

$$\int_0^\infty.(n).\int_0^\infty e^{-u_1-.-u_n}\, u_1^{a_1-1}..u_n^{a_n-1}\Psi_2^{(n)}(b;c_1,.,c_n;x_1u_1,.,x_nu_n)du_1..du_n$$

$$= \Gamma(a_1)..\Gamma(a_n)F_A^{(n)}(b,a_1,.,a_n;c_1,.,c_n;x_1,.,x_n),\ \mathrm{Re}(a_i) > 0$$

(6.2.2.3)

and

$$\int_{-\infty}^{(o+)}.(n).\int_{-\infty}^{(o+)} e^{u_1+.+u_n} u_1^{-a_1}..u_n^{-a_n} \Phi_2^{(n)}(a_1,.,a_n;c;\frac{x_1}{u_1},.,\frac{x_n}{u_n}) \times du_1..du_n$$

$$= 1/[\Gamma(a_1)..\Gamma(a_n)]_0F_1(-;c;x_1+.+x_n). \qquad (6.2.2.4)$$

6.3 MULTIPLE BARNES INTEGRALS

The integrals under consideration here are of the two types

$$\frac{1}{(2\pi i)^n}\int_{c'-i\infty}^{c'+i\infty}.(n).\int_{c^{(n)}-i\infty}^{c^{(n)}+i\infty} \Gamma(a_1+u_1)\Gamma(b_1-u_1)..\Gamma(a_n+u_n)\Gamma(b_n-u_n) \times f(u_1,.,u_n)du_1..du_n \qquad (6.3.1)$$

and

$$\frac{1}{(2\pi i)^n}\int_{c'-i\infty}^{c'+i\infty}.(n).\int_{c^{(n)}-i\infty}^{c^{(n)}+i\infty} \frac{\Gamma((a)+u_1+.+u_n)\Gamma((b')+u_1)..\Gamma((b^{(n)})+u_n)}{\Gamma((f)+u_1+.+u_n)\Gamma((g')+u_1)..\Gamma((g^{(n)})+u_n)}$$

$$\times \Gamma(-u_1)..\Gamma(-u_n)(-x_1)^{u_1}..(-x_n)^{u_n} f(u_1,.,u_n)du_1..du_n, \qquad (6.3.2)$$

where the quantities $c',.,c^{(n)}$ are real and non-negative. The function $f(u_1,.,u_n)$ is expanded as the series

$$f(u_1,.,u_n) = \sum_{m=0}^{\infty} A_m(u_1,.,u_n), \qquad (6.3.3)$$

where the coefficient $A_m(u_1,.,u_n)$ is to be specified, depending upon the type of integral under consideration.

Suppose that $A_m(u_1,.,u_n)$ is given by

$$A_m(u_1,.,u_n) = (a_1+u_1,m)(b_1-u_1,m)..(a_n+u_n,m)(b_n-u_n,m) \times z^m B_m, \qquad (6.3.4)$$

where B_m is an arbitrary coefficient independent of $u_1,.,u_n$ and z. The integral (6.3.1) then takes the form

$$\sum_{m=0}^{\infty} B_m z^m \frac{1}{2\pi i}\int_{c'-i\infty}^{c'+i\infty} \Gamma(a_1+m+u_1)\Gamma(b_1+m-u_1)du_1...$$

$$\times \frac{1}{2\pi i}\int_{c^{(n)}-i\infty}^{c^{(n)}+i\infty} \Gamma(a_n+m+u_n)\Gamma(b_n+m-u_n)du_n. \qquad (6.3.5)$$

If $\text{Re}(a_i+b_i) > 0$, $1 \leqq i \leqq n$, then each of the integrals of (6.3.5) may be evaluated by means of Titchmarsh's formula (4.2.2), so that the integral (6.3.1) may be written as

$$\sum_{m=0}^{\infty} B_m \,([a_1+b_1]/2,m)([a_1+b_1+1]/2,m) \times ([a_n+b_n]/2,m)([a_n+b_n+1]/2,m)\,z^m. \quad (6.3.6)$$

If we now take $A_m(u_1,.,u_n)$ to be of the form

$$\frac{((a)+u_1+.+u_n,m)((b')+u_1,m)..((b^{(n)})+u_n,m)}{((f)+u_1+.+u_n,m)((g')+u_1,m)..((g^{(n)})+u_n,m)} B_m \, z^m \quad (6.3.7)$$

and make use of the multiple Barnes integral representation of the generalised Kampé de Fériet function (5.2.3.6), we may show that the integral (6.3.2) is equal to the expression

$$\sum_{m=0}^{\infty} B_m \frac{((a),m)((b'),m)..((b^{(n)}),m)\,z^m}{((f),m)((g'),m)..((g^{(n)}),m)} F^{A:B}_{F:G}\left[\begin{matrix}(a):(b');.;(b^{(n)});\\(f):(g');.;(g^{(n)});\end{matrix}\, x_1,.,x_n\right] \quad (6.3.8).$$

A number of results of interest where multiple Barnes type integrals of numerous types of hypergeometric functions may be deduced form the expressions (6.3.6) and (6.3.8) if the function $f(u_i)$ is suitably specialised. It must be stressed that other multiple Barnes integral formulae similar to the above expressions may be obtained by similar means, for example, if (6.3.3) is presented as a multiple series. Finally, we point out that certain multiple integrals related to those discussed above may be deduced using the formula (4.2.4.1) in place of (4.2.2) above.

Chapter 7

Applications

7.1 GENERALISED UNIVARIATE STATISTICAL DISTRIBUTION

Many distributions of one variate may be expressed as combinations of hypergeometric functions, and the density functions, characteristic functions and cumulative distribution functions of mathematical statistics then involve integrals of hypergeometric functions. In this section, we examine a few of the properties of two distributions which generalise, respectively, the well-known beta and gamma distributions associated with many classical problems in statistics.

7.1.1 A Generalised Finite Distribution

We consider the family of distributions which have the probability density function

$$f(x) = Kx^{d-1}\,{}_{A'}F_{B'}\left[\begin{matrix}(a');\\(b');\end{matrix}h_1x\right]\cdots{}_{A^{(n)}}F_{B^{(n)}}\left[\begin{matrix}(a^{(n)});\\(b^{(n)});\end{matrix}h_nx\right], \quad 0 \leqq 1 \leqq x,$$

$$f(x) = 0 \text{ elsewhere}. \tag{7.1.1.1}$$

$A'+1 \leqq B', .., A^{(n)}+1 \leqq B^{(n)}$; d is real and positive and the moduli of the quantities $h_1, .., h_n$ are all less than or equal to unity. The constant K is to be determined so that

$$\int_{-\infty}^{\infty} f(x)\,dx = \int_0^1 f(x)\,dx = 1. \tag{7.1.1.2}$$

It will be seen that if $A'=1, B'=0$ and $n=1$, then the density function under consideration is that of the beta distribution. Hence,

$$K^{-1} = \int_0^1 x^{d-1}\,{}_{A'}F_{B'}\left[\begin{matrix}(a');\\(b');\end{matrix}h_1x\right]\cdots{}_{A^{(n)}}F_{B^{(n)}}\left[\begin{matrix}(a^{(n)});\\(b^{(n)});\end{matrix}h_nx\right]dx$$

$$= d^{-1}\,F^{1:A';.;A^{(n)}}_{1:B';.;B^{(n)}}\left[\begin{matrix}d:(a');.;(a^{(n)});\\d+1:(b');.;(b^{(n)});\end{matrix}h_1,..,h_n\right]. \tag{7.1.1.3}$$

The characteristic function, denoted by $\phi(t)$, may be represented as $\langle e^{itx}\rangle$, where the angle brackets denote a mathematical expectation. We may thus write

$$\phi(t) = \int_{-\infty}^{\infty} e^{itx} f(x)\, dx = \int_{0}^{1} e^{itx} f(x)\, dx \qquad (7.1.1.4)$$

in the case presently under consideration and we have

$$\phi(t) = K\, F^{1:0;A';.;A^{(n)}}_{1:0;B';.;B^{(n)}}\left[\begin{matrix} d :-;(a');.;(a^{(n)}); \\ d+1:-;(b');.;(b^{(n)}); \end{matrix} it,h_1,.,h_n\right] . \qquad (7.1.1.5)$$

The cumulative distribution F(x) is given by

$$F(x) = \int_{-\infty}^{x} f(t)\, dt = \int_{0}^{x} f(t)\, dt$$

$$= K\,\frac{x^d}{d}\, F^{1:A';.;A^{(n)}}_{1:B';.;B^{(n)}}\left[\begin{matrix} d :(a');.;(a^{(n)}); \\ d+1:(b');.;(b^{(n)}); \end{matrix} h_1x,.,h_nx\right] . \qquad (7.1.1.6)$$

If we put x equal to unity, then F(x) = 1, as would be expected from statistical considerations. The largest order statistic and the various moments of the distribution characterised by (7.1.1.1) may readily be deduced from the above results.

7.1.2 A Generalised Infinite Distribution

Similar expressions may be obtained for the generalised gamma distribution whose density is

$$f(x) = Ke^{-px}\, x^{d-1}\, {}_{A'}F_{B'}\left[\begin{matrix}(a'); \\ (b');\end{matrix} h_1x\right] .. {}_{A^{(n)}}F_{B^{(n)}}\left[\begin{matrix}(a^{(n)}); \\ (b^{(n)});\end{matrix} h_nx\right], \qquad (7.1.2.1)$$

where $A' \leqq B', ., A^{(n)} \leqq B^{(n)}$ and d and p are real and positive. We give the following results without proof:

$$K^{-1} = \frac{\Gamma(d)}{p^d} F^{1:A';.;A^{(n)}}_{0:B';.;B^{(n)}}\left[\begin{matrix} d:(a');.;(a^{(n)}); \\ -:(b');.;(b^{(n)}); \end{matrix} h_1/p,.,h_n/p\right], \qquad (7.1.2.2)$$

$$\phi(t) = \frac{K\ \Gamma(d)}{(p-it)^d} F^{1:A';.;A^{(n)}}_{0:B';.;B^{(n)}}\left[\begin{matrix} d:(a');.;(a^{(n)}); \\ -:(b');.;(b^{(n)}); \end{matrix} \frac{h_1}{p-it},.,\frac{h_n}{p-it}\right], \qquad (7.1.2.3)$$

and

$$F(x) = \frac{\Gamma(d)}{(px)^d} K\, F^{1:0;A';.;A^{(n)}}_{1:0;B';.;B^{(n)}}\left[\begin{matrix} d :-;(a');.;(a^{(n)}); \\ d+1:-;(b');.;(b^{(n)}); \end{matrix} -px,\frac{h_1}{p},.,\frac{h_n}{p}\right] . \qquad (7.1.2.4)$$

See Karlsson (1973) for the generalised Kampé de Fériet function.

7.2 GENERALISED MULTIVARIATE STATISTICAL DISTRIBUTION

Two families of distributions in several variates are now considered. They may be regarded as generalising the beta and gamma distributions in a manner different from that in the preceding section.

7.2.1 A Generalised Finite Distribution in Several Variates

The distribution now discussed has density given by

$$f(x_1,..,x_n) = K\, x_1^{d_1-1}..x_n^{d_n-1}\ {}_AF_B\left({(a);\atop (b);} h_1x_1+.+h_nx_n\right) \qquad (7.2.1.1)$$

for $0 \leqq x_1 \leqq 1,..,0 \leqq x_n \leqq 1$, and $f(x_1,..,x_n) = 0$ otherwise.
It is assumed that $d_1,h_1,..,d_n,h_n$ are real and positive and that $A+1 \leqq B$. As in the previous cases, the constant K is so determined that

$$\int_{-\infty}^{\infty}.(n).\int_{-\infty}^{\infty} f(x_1,..,x_n)dx_1..dx_n = \int_0^1.(n).\int_0^1 f(x_1,..,x_n)dx_1..dx_n = 1. \qquad (7.2.1.2)$$

Hence,

$$K^{-1} = (d_1..d_n)^{-1}\ F^{A:1;.;1}_{B:1;.;1}\left[\begin{matrix}(a): d_1\ ;.;\ d_n\ ; \\ (b):d_1+1;.;d_n+1;\end{matrix} h_1,.,h_n\right]. \qquad (7.2.1.3)$$

The multivariate characteristic function is given by

$$\phi(t_1,.,t_n) = K\int_0^1.(n).\int_0^1 \exp(it_1x_1+,+it_nx_n)x_1^{d_1-1}..x_n^{d_n-1}$$

$$_AF_B\left({(a);\atop (b);} h_1x_1+.+h_nx_n\right)dx_1..dx_n$$

$$= \frac{K}{d_1..d_n}\sum\frac{((a),m_1+.+m_n)(d_1,m_1)..(d_n,m_n)h_1^{m_1}..h_n^{m_n}}{((b),m_1+.+m_n)(d_1+1,m_1)..(d_n+1,m_n)m_1!..m_n!}$$

$$\times\ {}_1F_1(d_1+m_1;d_1+m_1+1;it_1)..{}_1F_1(d_n+m_n;d_n+m_n+1;it_n). \qquad (7.2.1.4)$$

The cumulative distribution function may be written

$$F(x_1,.,x_n) = \frac{x_1^{d_1}}{d_1}..\frac{x_n^{d_n}}{d_n}\ K\ F^{A:1;.;1}_{B:1;.;1}\left[\begin{matrix}(a): d_1\ ;.;\ d_n\ ; \\ (b):d_1+1;.;d_n+1;\end{matrix} h_1x_1,.,h_nx_n\right]. \qquad (7.2.1.5)$$

7.2.2 A Generalised Infinite Distribution of Several Variates

The density function to be studied here is given by

$$f(x_1,.,x_n) = K\exp(-p_1x_1-.-p_nx_n)x_1^{d_1-1}..x_n^{d_n-1}\ {}_AF_B\left({(a);\atop (b);} x_1h_1+.+x_nh_n\right)$$

for $0 \leqq x_1 \leqq \infty,.,0 \leqq x_n \leqq \infty$ and zero otherwise. $\qquad$ (7.2.2.1)

The numbers $p_1,d_1,..,p_n,d_n$ are real and positive, and $A \leq B$. As in Section 7.1.2, we write down the following results directly:-

$$K^{-1} = \frac{\Gamma(d_1)}{p_1^{d_1}}..\frac{\Gamma(d_n)}{p_n^{d_n}} F^{A:1;.;1}_{B:0;.;0}\left[\begin{matrix}(a):d_1;.;d_n;h_1 & & h_n\\ (b):\text{———} & ;\overline{p_1},.., & \overline{p_n}\end{matrix}\right], \tag{7.2.2.2}$$

$$\phi(t_1,..,t_n) = \frac{K\Gamma(d_1)..\Gamma(d_n)}{(p_1-it_1)^{d_1}..(p_n-it_n)^{d_n}}$$

$$\times F^{A:1;.;1}_{B:0;.;0}\left[\begin{matrix}(a):d_1;.;d_n; & h_1 & & h_n\\ (b):\text{———} & ;\overline{p_1-it_1},.., & & \overline{p_n-it_n}\end{matrix}\right] \tag{7.2.2.3}$$

and

$$F(x_1,..,x_n) = x_1^{d_1}..x_n^{d_n}\, K \sum\frac{((a),m_1+.+m_n)(d_1,m_1)..(d_n,m_n)}{((b),m_1+.+m_n)(d_1+1,m_1)..(d_n+1,m_n)}$$

$$\times \frac{(x_1h_1)^{m_1}..(x_nh_n)^{m_n}}{m_1!.....m_n!}\;{}_1F_1\left[\begin{matrix}d_1+m_1;\\ d_1+m_1+1;\end{matrix} -p_1x_1\right]..\;{}_1F_1\left[\begin{matrix}d_n+m_n;\\ d_n+m_n+1;\end{matrix} -p_nx_n\right]. \tag{7.2.2.4}$$

7.3 THE STUDY OF THE ANGULAR DISPLACEMENT OF A SHAFT

In this section, we shall consider the problem of determining the angular displacement or twist $T(x,t)$ in a shaft of circular cross-section with its axis along the x-axis. If the ends $x=0$ and $x=L$ are free, the displacement $T(x,t)$ due to initial twist must satisfy the boundary-value problem

$$\frac{\partial^2T}{\partial t^2} = k^2\frac{\partial^2T}{\partial x^2},\quad \frac{\partial T(0,t)}{\partial x} = \frac{\partial T(L,t)}{\partial t} = \frac{\partial T(x,0)}{\partial t} = 0 \text{ and } T(x,0)=f(x), \tag{7.3.1}$$

where k is a constant. See Bajpai (1971).

We write $f(x)$ in the form

$$f(x) = (\sin\frac{\pi x}{2L})^{2r-b-1}(\cos\frac{\pi x}{2L})^{b-1}\,G^{m,n}_{p,q}\left(z[\tan\frac{x}{2L}]^{2d}\middle|\begin{matrix}(a)\\(c)\end{matrix}\right), \tag{7.3.2}$$

where d is a positive integer. We require to evaluate the integral

$$I = \int_0^L \cos(\frac{\pi r\phi}{L})(\sin\frac{\pi\phi}{2L})^{2r-b-1}(\cos\frac{\pi\phi}{2L})^{b-1}\,G^{m,n}_{p,q}\left(z[\tan\frac{\pi\phi}{2L})^{2d}\middle|\begin{matrix}(a)\\(c)\end{matrix}\right)d\phi. \tag{7.3.3}$$

The G-function is expressed in the form of an integral of Barnes type, and the order of the integrations is interchanged. We may then write

$$I = \int_C \frac{\prod_{j=1}^{m}\Gamma(c_j-s)\prod_{j=1}^{n}\Gamma(1-a_j+s)z^s}{\prod_{j=m+1}^{q}\Gamma(1-c_j+s)\prod_{j=n+1}^{p}\Gamma(a_j-s)} \int_0^L \cos(\frac{\pi r\phi}{L})(\sin\frac{\pi\phi}{2L})^{2r-b+2sd-1} \times (\cos\frac{\pi\phi}{2L})^{b-2sd-1}\frac{d\phi ds}{2\pi i}. \qquad (7.3.4)$$

The reversal of the order of the two integrations is justified because both the integrals concerned are absolutely convergent.

Now, the inner integral of (7.3.4) may be evaluated using the formula

$$\int_0^L \cos(\frac{\pi r\phi}{L})(\sin\frac{\pi\phi}{2L})^{2r-b-1}(\cos\frac{\pi\phi}{2L})^{b-1}d\phi = \frac{L2^{2r-b}\Gamma(\frac{2r-b}{2})\Gamma(b)}{\sqrt{\pi}\Gamma(\frac{1-2r+b}{2})\Gamma(2r)}, \qquad (7.3.5)$$

$$r > b > 0.$$

If we now employ the multiplication formula for the gamma function, we have

$$I = \frac{L(2d)^{2r-1}2^{1-d}}{\pi^d\Gamma(2-r)(2\pi i)}$$

$$\times \int_C \frac{\prod_{j=1}^{m}\Gamma(c_j-s)\prod_{j=1}^{n}\Gamma(1-a_j+s)\prod_{j=0}^{d-1}\Gamma(\frac{r-b/2+j}{2}+s)\prod_{j=0}^{2d-1}\Gamma(\frac{b+j}{2d}-s)z^s}{\prod_{j=m+1}^{q}\Gamma(1-c_j+s)\prod_{j=n+1}^{p}\Gamma(a_j-s)\prod_{j=0}^{d-1}\Gamma(\frac{1/2+b/2-r+j}{d}-s)}ds. \qquad (7.3.6)$$

On interpreting the integral on the right of (7.3.6) as a G-function, we have the expression

$$I = \frac{L(2d)^{2r-1}2^{1-d}}{\pi^d\Gamma(2r)} G^{m+2d,\,n+d}_{p+2d,q+2d}\left(z\,\middle|\,\begin{matrix}\Delta(d,1-r+b/2),(a),\Delta(d,1/2-r+b/2)\\ \Delta(2d,b),\ (c)\end{matrix}\right), \qquad (7.3.7)$$

where $\Delta(d,a)$ represents the set of parameters

$$a/d,\ (a+1)/d,..,(a+d-1)/d.$$

The solution of the problem under consideration can be written as the Fourier series

$$F(x,t) = a_0/2 + \sum_{s=1}^{\infty} a_s \cos(\frac{s\pi x}{2})\cos(\frac{s\pi at}{L}), \qquad (7.3.8)$$

where $a_1, a_2, ..$ are the coefficients in the Fourier cosine series for f(x) in the interval (0,L). If t=0, then by virtue of (7.3.2), we have

$$(\sin\frac{\pi x}{2L})^{2r-b-1}(\cos\frac{\pi x}{2L})^{b-1}\, G^{m,n}_{p,q}\left(z[\tan\frac{\pi x}{2L}]^{2d}\,\middle|\,\begin{matrix}(a)\\(c)\end{matrix}\right)=a_0/2+\sum_{s=0}^{\infty} a_s\cos\frac{s\pi x}{L}\,. \tag{7.3.9}$$

Multiplying both sides of (7.3.9) by $\cos(\frac{r\pi x}{L})$ and integrating with respect to x from 0 to L, we obtain the expression

$$\int_0^L \cos\frac{\pi rx}{L}(\sin\frac{\pi x}{2L})^{2r-b-1}(\cos\frac{\pi x}{2L})^{b-1}\, G^{m,n}_{p,q}\left(z[\tan\frac{\pi x}{2L}]^{2d}\,\middle|\,\begin{matrix}(a)\\(c)\end{matrix}\right)dx$$

$$= \frac{a_0}{2}\int_0^L \cos\frac{r\pi x}{L}\,dx + \sum_{s=1}^{\infty} a_s\int_0^L \cos\frac{s\pi x}{L}\cos\frac{r\pi x}{L}\,dx. \tag{7.3.10}$$

Hence, the result

$$a_r = \frac{4d^{2r-1}}{(2\pi)^d\Gamma(2r)}\, G^{m+2d,\ n+d}_{p+2d,q+2d}\left(z\,\middle|\,\begin{matrix}\Delta(d,1-r+b/2),(a),\Delta(d,1/2-r+b/2)\\ \Delta(2d,b),\quad (c)\end{matrix}\right) \tag{7.3.11}$$

follows from the orthogonality property of the cosine function, and we obtain the solution of the problem in the form

$$T(x,t) = \frac{4}{(2\pi)^d}\sum_{s=1}^{\infty}\frac{(2d)^{2s-1}}{\Gamma(2s)}\, G^{m+2d,\ n+d}_{p+2d,q+2d}\left[z\,\middle|\,\begin{matrix}\Delta(d,1-s+\frac{b}{2}),(a),(d,1/2-s+b/2)\\ \Delta(2d,b),\quad (c)\end{matrix}\right]. \tag{7.3.12}$$

The G-function of Meijer may be specialised so as to be of the form of very many of the special functions of mathematical physics and chemistry, for example Bessel or Legendre functions.

7.4 THE VIBRATION OF A THIN ELASTIC PLATE

The governing equation of small transverse vibrations of a thin elastic plate bounded by the two parallel planes z = ±h, but otherwise of unlimited extent may be written

$$b^2(\frac{\partial^2}{\partial r^2} + 1/r\frac{\partial}{\partial r})^2 w + \frac{\partial^2 w}{\partial t^2} = \frac{Z(r,t)}{2\rho h}, \tag{7.4.1}$$

where the vibrations are symmetrical about the z-axis, and b is a real constant. We consider the solution of the problem of free symmetrical vibrations of a very large plate, when the forcing function Z(r,t) is then identically zero. It is supposed that the vibrations result from the initial conditions

$$w = f(r), \quad \frac{\partial w}{\partial t} = 0 \qquad (7.4.2)$$

at time t=0.

The equation (7.4.1) is multiplied throughout by $rJ_0(\xi r)$ and we then integrate with respect to r from zero to infinity and write

$$\int_0^\infty r\, w(r,t) J_0(\xi r)\, dr = \bar{w}(\xi,t). \qquad (7.4.3)$$

If F is any function of r such that $r\partial F/\partial r$ tends to zero at r=0 and r=∞, we have

$$\int_0^\infty \left(\frac{\partial^2}{\partial r^2} + \frac{1}{r}\frac{\partial}{\partial r}\right) F(r) J_0(\xi r) dr = -\xi^2\, \bar{F}(\xi), \qquad (7.4.4)$$

where $$\bar{F}(\xi) = \int_0^\infty rF(r)J_0(\xi r)dr. \qquad (7.4.5)$$

Now put $$F(r) = \left(\frac{\partial^2}{\partial r^2} + \frac{1}{r}\frac{\partial}{\partial r}\right)w, \qquad (7.4.6)$$

when we see that

$$\int_0^\infty r\left(\frac{\partial^2}{\partial r^2} + \frac{1}{r}\frac{\partial}{\partial r}\right)w\; J_0(\xi r)dr = \xi^4\, \bar{w}(\xi,t). \qquad (7.4.7)$$

Thus, (7.4.1) now becomes

$$\frac{1}{b^2}\frac{d^2\bar{w}}{dt^2} + \xi^4\bar{w} = 0 \qquad (7.4.8)$$

and we may then write

$$\bar{w}(\xi,t) = A(\xi)\cos(b\xi^2 t) + B(\xi)\sin(b\xi^2 t). \qquad (7.4.9)$$

The initial conditions are now employed to give the results

$$A(\xi) = \int_0^\infty uf(u)J_0(\xi u)\, du \text{ and } B(\xi) = 0, \qquad (7.4.10)$$

and so, $$\bar{w}(\xi,t) = \cos(b\xi^2 t)\int_0^\infty uf(u)J_0(\xi u)du. \qquad (7.4.11)$$

The Hankel inversion formula now enables us to write

$$\begin{aligned} w(r,t) &= \int_0^\infty \xi J_0(\xi r)\cos(b\xi^2 t)d\xi \int_0^\infty uf(u)J_0(u\xi)du \\ &= \int_0^\infty uf(u)du\int_0^\infty \xi J_0(\xi r)J_0(\xi u)\cos(b\xi^2 t)dt. \end{aligned} \qquad (7.4.12)$$

Now, by Weber's second exponential integral

$$\int_0^\infty \xi J_0(\xi r)J_0(\xi r)e^{-p\xi^2}\,d\xi = \frac{1}{2p}\exp[-(u^2+r^2)/4p]I_0(\frac{ur}{2p}), \quad (7.4.13)$$

where $I_0(z) = J_0(iz)$, so that, substituting $p = -ibt$ and equating real parts, we obtain

$$\int_0^\infty \xi J_0(\xi u)J_0(\xi r)\cos(b\xi^2 t)\,d\xi = \frac{1}{2bt}J_0(\frac{ur}{2bt})\sin(\frac{u^2+r^2}{4bt}), \quad (7.4.14)$$

giving finally

$$w(r,t) = \frac{1}{2bt}\int_0^\infty uf(u)J_0(\frac{ur}{2bt})\sin(\frac{u^2+r^2}{4bt})\,du. \quad (7.4.15)$$

See Sneddon (1951) page 134.

7.5 HEAT PRODUCTION IN A CYLINDER

The differential equation

$$\frac{\partial\theta}{\partial t} = \kappa(\frac{\partial^2\theta}{\partial r^2} + \frac{1}{r}\frac{\partial\theta}{\partial r}) + T(r,t) \quad (7.5.1)$$

occurs when the problem of diffusion of heat in a cylinder of radius a when any sources of heat within it lead to an axially symmetrical temperature distribution. It is also assumed that the rate of generation of heat is independent of the temperature and that the cylinder is of infinite length so that the problem is independent of the longitudinal coordinate. Suppose that the surface $r = a$ is maintained at zero temperature. We choose ξ_i to be a zero of the Bessel function of order zero $J_0(\xi_i a)$, when

$$\int_0^a r(\frac{\partial^2\theta}{\partial r^2} + \frac{1}{r}\frac{\partial\theta}{\partial r})J_0(\xi_i r)\,dr = -\xi_i^2\bar{\theta}_J, \quad (7.5.2)$$

where $\bar{\theta}_J(\xi_i)$ is the finite Hankel transform of order zero of the function $\theta(r)$.

If we multiply equation (7.5.1) throughout by $rJ_0(\xi_i r)$ and integrate with respect to r from 0 to a, we have

$$\frac{d\,\bar{\theta}_J}{dt} + \kappa\xi_i^2\bar{\theta}_J = \bar{T}_J(\xi_i,t) \quad (7.5.3)$$

together with the initial condition $\bar{\theta}_J(\xi_i,0) = 0$.

Hence,

$$\bar{\theta}_J = \int_0^t \bar{T}_J(\xi_i,\tau)e^{-\kappa\xi_i^2(t-\tau)}\,d\tau, \quad (7.5.4)$$

and if we apply the inversion theorem for the Hankel transform of order zero, we have

$$\theta = 2a^{-2}\sum_{i=1}^\infty J_0(r\xi_i)[J_1(a\xi_i)]^{-2}\int_0^t \bar{T}_J(\xi_i,\tau)e^{-\kappa\xi_i^2(t-\tau)}\,d\tau. \quad (7.5.5)$$

An important special case arises if the source function T(r,t) is separable, that is, if it can be written in the form

$$T(r,t) = \kappa/k\ f(r)\ g(t), \qquad (7.5.6)$$

so that, finally, we have the expression

$$\theta = 2\kappa/(ka^2) \sum_{i=1}^{\infty} J_0(r\xi_i)[J_1(a\xi_i)]^{-2}\ \bar{f}_J(\xi_i)\int_0^t g(\tau)e^{-\kappa\xi_i^2(t-\tau)}d\tau. \qquad (7.5.7)$$

See Sneddon (1951) page 202.

7.6 MOTION OF A VISCOUS FLUID UNDER A SURFACE LOAD

Consider a semi-infinite, incompressible, viscous fluid under the action of a radially symmetrical pressure applied to the free surface. If we are dealing with small accelerations and high viscosity, the initial terms in the equations of motion may be neglected. With these restrictions, the equations of motion of a fluid in a gravitational field may be written in vector notation as

$$\mu\nabla^2\underline{v} = \text{grad } p - (0,0,\rho g) \qquad (7.6.1)$$

provided that the positive z axis is taken as pointing downwards. If cylindrical symmetry is assumed, we may write

$$\frac{1}{r}\frac{\partial}{\partial r}\left(r\frac{\partial v_r}{\partial r}\right) - \frac{v_r}{r^2} + \frac{\partial^2 v_r}{\partial z^2} = \frac{1}{\mu}\frac{\partial \bar{p}}{\partial r} \qquad (7.6.2)$$

and

$$\frac{1}{r}\frac{\partial}{\partial r}\left(r\frac{\partial v_z}{\partial r}\right) + \frac{\partial^2 v_z}{\partial z^2} = \frac{1}{\mu}\frac{\partial \bar{p}}{\partial z}, \qquad (7.6.3)$$

where $\bar{p} = p - g\rho z$.

Since the velocity $\underline{v}$ must satisfy the equation of continuity, we also have

$$\frac{1}{r}\frac{\partial}{\partial r}(rv_r) + \frac{\partial v_z}{\partial z} = 0, \qquad (7.6.4)$$

where cylindrical polar coordinates are used throughout. The components of stress in the z-direction are

$$\sigma_z = -p + 2\mu\frac{\partial v_z}{\partial z} \qquad (7.6.5)$$

and

$$\tau_{rz} = \partial\left(\frac{\partial v_r}{\partial z} + \frac{\partial v_z}{\partial r}\right). \qquad (7.6.6)$$

At a free surface, the shear stress is zero and the normal component of stress is equal to the applied pressure. It is assumed that the stress and components of velocity vanish at infinity and that ζ is small compared with other distances which enter into the problem, where the equation of the free surface is $z = \zeta(r,t)$, $z=0$ being the equation of the undisturbed free surface. If we replace all quantities by their values at $z=0$ except $g\rho z = g\rho\zeta(r,t)$, then $\sigma_z = -\sigma(r,t)$ when $z=\zeta(r,t)$, where $\sigma(r,t)$ is the applied pressure.

Hence, we may write

$$\bar{p} + g\rho\zeta(r,t) - 2\mu\frac{\partial v_z}{\partial z} = \sigma(r,t) \text{ on } z=0 \qquad (7.6.7)$$

and

$$\frac{\partial \zeta}{\partial t} = (v_z)_{z=0}. \qquad (7.6.8)$$

Also, the shear stress on the free surface is zero, so that we now have

$$(\tau_{rz})_{z=0} = 0. \qquad (7.6.9)$$

We now take the Hankel transform of order unity with respect to r of equation (7.6.2) and so we have

$$\left(\frac{d^2}{dz^2} - \xi^2\right)R = -\frac{\xi}{\mu}P, \qquad (7.6.10)$$

where

$$R = \int_0^\infty r v_r J_1(\xi r)\,dr \qquad \text{and} \qquad P = \int_0^\infty r\bar{p}J_0(\xi r)\,dr. \qquad (7.6.11)$$

Similarly, equation (7.6.3) takes the form

$$\left(\frac{d^2}{dz^2} - \xi^2\right)Z = \frac{1}{\mu}\frac{dP}{dz}, \qquad (7.6.12)$$

where

$$Z = \int_0^\infty r v_z J_0(\xi r)\,dr. \qquad (7.6.13)$$

The equation of continuity becomes

$$\xi R + \frac{dZ}{dz} = 0. \qquad (7.6.14)$$

If P and R are eliminated from the three equations (7.6.10), (7.6.12) and (7.6.14), we find that Z satisfies the fourth order differential equation

$$\left(\frac{d^2}{dz^2} - \xi^2\right)^2 Z = 0 \qquad (7.6.15)$$

whose solution which tends to zero as $z \to \infty$ is

$$Z = [A(\xi) + B(\xi)\xi z]e^{-\xi z}. \qquad (7.6.16)$$

Hence, from (7.6.13), we have

$$R = [A(\xi) - B(\xi) + B(\xi)\xi z]e^{-\xi z}. \qquad (7.6.17)$$

If we multiply equation (7.6.6) throughout by $rJ_1(\xi r)$ and integrate over r, we have, as a result of integrating by parts, that

$$\frac{1}{\mu}\int_0^\infty r\tau_{rz}J_1(\xi r)dr = \frac{dR}{dz} - \xi Z = -2\xi(A - B + B\xi z)e^{-\xi z}. \qquad (7.6.18)$$

If we put z=0, the boundary condition (7.6.8) gives $A=B=\alpha(\xi)/\xi$, say, and so we find that

$$Z = [\alpha(\xi)/\xi](1+\xi z)e^{-\xi z} \quad \text{and } R = z\alpha(\xi)e^{-\xi z}. \qquad (7.6.19)$$

By the Hankel inversion theorem, we have

$$v_r = z\int_0^\infty \xi\alpha(\xi)e^{-\xi z}J_1(\xi r)d\xi, \qquad (7.6.20)$$

$$v_z = \int_0^\infty (1+\xi z)\alpha(\xi)e^{-\xi z}J_0(\xi r)d\xi \qquad (7.6.21)$$

and

$$\bar{p} = 2\mu\int_0^\infty \alpha(\xi)e^{-\xi z}J_0(\xi r)d\xi. \qquad (7.6.22)$$

We now substitute these equations into (7.6.7) and (7.6.8), and after differentiating with respect to t, we have

$$2\mu\int_0^\infty \xi\frac{\partial\alpha}{\partial t}J_0(\xi r)d\xi + g\rho\frac{\partial\zeta}{\partial t} = \frac{\partial\sigma}{\partial t} \qquad (7.6.23)$$

and

$$\frac{\partial\zeta}{\partial t} = \int_0^\infty \alpha(\xi)J_0(\xi r)d\xi, \qquad (7.6.24)$$

from which we have the equation

$$\int_0^\infty \xi\left(2\mu\frac{\partial\alpha}{\partial t} + \frac{g\rho\alpha}{\xi}\right)J_0(\xi r)d\xi = \frac{\partial\sigma}{\partial t}. \qquad (7.6.25)$$

Inverting this equation by means of the Hankel transform theorem, we have

$$2\mu\frac{\partial\alpha}{\partial t} + \frac{g\rho}{\xi}\alpha = \int_0^\infty r\frac{\partial\sigma(r,t)}{\partial t}J_0(\xi r)dr. \qquad (7.6.26)$$

The first order differential equation (7.6.26) has the solution

$$\alpha(\xi) = K(\xi)e^{-g\rho t/(2\mu\xi)} + \frac{1}{2\mu}\int_0^t e^{g\rho(\tau-t)/(2\mu\xi)}\int_0^\infty r\frac{\partial\sigma(r,t)}{\partial t}J_0(\xi r)drd\tau, \qquad (7.6.27)$$

where $K(\xi)$ is determined from the initial conditions.

If $\sigma(r,0)$ is denoted by σ_0, integration by parts gives the formula

$$\int_0^t \frac{\partial\sigma}{\partial t} e^{g\rho t/(2\mu\xi)} dt = \sigma e^{\rho g t/(2\mu\xi)} - \sigma_0 - \frac{g\rho}{2\mu\xi}\int_0^t e^{g\rho t/(2\mu\xi)} dt, \quad (7.6.28)$$

and we have finally

$$\alpha(\xi) = K(\xi)e^{-g\rho t/(2\mu\xi)} + \frac{1}{2\mu}\int_0^\infty r\sigma J_0(\xi r)dr - \frac{1}{2\mu}e^{-g\rho t/(2\mu\xi)}\int_0^\infty r\sigma_0 J_0(\xi r)dr$$

$$- \frac{g\rho}{4\mu^2\xi} e^{-g\rho t/(2\mu\xi)}\int_0^t\int_0^\infty \sigma e^{g\rho t/(2\mu\xi)} J_0(\xi r) r dr\, dt \quad (7.6.29)$$

which, together with (7.6.20), (7.6.21) and (7.6.22) gives the final solution of the problem. The above analysis was originally motivated by a discussion of the plastic recoil of the earth after the disappearance of the Pleistocene ice sheets.
See Sneddon (1951) page 307.

7.7 THE TIME-DOMAIN SYNTHESIS OF SIGNALS

The classical problem known as the 'time-domain synthesis problem' occurs in electrical network theory and one version of this problem may be stated as follows: Given an electrical signal described by a real-valued conventional function $f(t)$ on $0 < t < \infty$, construct an electrical network consisting of only a finite number of components R, C and I which are all fixed, linear and positive, such that its output $f_N(t)$, resulting from a delta-function input $\delta(t)$, approximates $f(t)$ on $0 < t < \infty$ in some sense.

In order to tackle this problem, we expand the function $f(t)$ into a convergent series

$$f(t) = \sum_{n=0}^{\infty} g_n(t) \quad (7.7.1)$$

of real-valued functions $g_n(t)$ such that every partial sum

$$f_N(t) = \sum_{n=0}^{N} g_n(t), \quad N=0,1,2,.. \quad (7.7.2)$$

possesses the two properties:

(i) $f_N(t) \equiv 0$ for $-\infty < t < 0$.

(ii) The Laplace transform $F_N(s)$ of $f_N(t)$ is a rational function having a zero at $s=\infty$, and all its poles in the left-half s-plane (Re $s < 0$) except possibly for a simple pole at the origin.

After choosing N in (7.7.2) sufficiently large to satisfy whatever approximation criterion is being used, an orthonormal series expansion may now be employed. The Laguerre transformation yields an immediate solution in the form

$$f_N(t) = 1_+(t)\sum_{n=0}^{N}(f,\psi_n)\psi_n(t), \qquad (7.7.3)$$

where

$$\begin{aligned} 1_+(t) &= 1 \text{ for } t > 0 \\ &= 1/2 \text{ for } t = 0 \\ &= 0 \text{ for } t < 0, \end{aligned} \qquad (7.7.4)$$

and

$$(f,\psi_n) = \int_0^\infty f(t)\psi_n(t)dt. \qquad (7.7.5)$$

$\psi_n(t)$ is the generalised orthonormal Laguerre function given by

$$\psi_n(t) = \sqrt{[\Gamma(n+1)/\Gamma(\alpha+n+1)]}x^{\alpha/2}\, e^{-x/2}\, L_n^\alpha(x),$$
$$n=0,1,2,\ldots\,, \qquad (7.7.6)$$

and

$$L_n^\alpha(x) = (\alpha+1,n)/n!\,{}_1F_1(-n;\alpha+1;x),\ \alpha > -1. \qquad (7.7.7)$$

This case is an example of the use of a hypergeometric integral in an orthonormal series expansion. See Zemanian (1968) page 275.

7.8 A DIRICHLET PROBLEM FOR THE INTERIOR OF THE UNIT SPHERE

Consider the following Dirichlet problem for the interior of a unit sphere: In spherical polar coordinates (r,θ,ϕ) such that $0 \leqq r \leqq 1$, $0 \leqq \theta \leqq \pi$ and $0 \leqq \phi \leqq 2\pi$, we assume that the potential v does not depend upon the azimuthal coordinate ϕ, that is, $v \equiv v(r,\theta)$. Upon putting $\mu = \cos\theta$, we may write Laplace's equation as

$$r\frac{\partial^2(rv)}{\partial r^2} + \frac{\partial}{\partial\mu}(1-\mu^2)\frac{\partial v}{\partial\mu} = 0, \qquad (7.8.1)$$

where $v = v(r, \cos^{-1}\mu)$, $0 < r < 1$ and $-1 < \mu < 1$. Furthermore, we require that $v(r,\theta)$ remains bounded near $r=0$.

Let us impose the boundary condition

$$v(r, \cos^{-1}\mu) \to f(u) \text{ as } r \to 1- . \qquad (7.8.2)$$

We now apply the Legendre transformation with respect to μ to the differential equation (7.8.1), so that the transform $\bar{v}$ satisfies

$$r\frac{d^2(r\bar{v})}{dr^2} - n(n+1)\bar{v} = 0. \qquad (7.8.3)$$

The solution of (7.8.3) is

$$\bar{v}(r,n) = A(n)r^n + B(n)r^{-n-1}, \qquad (7.8.4)$$

where $A(n)$ and $B(n)$ are arbitrary functions of n. Since $v(r,\theta)$

near r=0, we choose B(n)=0, and the A(n) are determined by matching the transform of the boundary condition (7.8.2).

Hence,
$$\bar{v}(r,n) = [f(\mu),\psi_n(\mu)]r^n, \qquad (7.8.5)$$

where
$$\psi_n(\mu) = \sqrt{(n+1/2)}P_n(\mu) = \sqrt{(n+1/2)}\,{}_2F_1(-n,n+1;1;\tfrac{1-\mu}{2}). \qquad (7.8.6)$$

The equation (7.8.5) may be inverted, and we obtain the required solution
$$v(r,\theta) = \sum_{n=0}^{\infty} (f,\psi_n)r^n\psi_n(\cos\theta), \qquad (7.8.7)$$

where
$$(f,\psi_n) = \sqrt{(n+1/2)} \int_{-1}^{1} {}_2F_1(-n,n+1;1;\tfrac{1-\mu}{2})f(\mu)d\mu. \qquad (7.8.8)$$

See Sneddon (1966b) page 215.

7.9 DUAL INTEGRAL EQUATIONS WITH LEGENDRE FUNCTION KERNELS

In various physical problems, dual integral equations are of frequent occurrence. The kernels of such pairs of equations may consist of virtually any special function, and here we consider briefly the pair of equations
$$\int_0^{\infty} f(\tau)P_{-1/2+i\tau}(\cosh\alpha)d\tau = g(\alpha),\quad 0 \leq \alpha \leq \alpha_1 \qquad (7.9.1)$$
$$\int_0^{\infty} \tau\tanh(\pi\tau)f(\tau)P_{-1/2+i\tau}(\cosh\alpha)d\tau = h(\alpha), \alpha>\alpha_1. \qquad (7.9.2)$$

Multiply equation (7.9.1) by $\sinh\alpha(\cosh x - \cosh\alpha)^{-1/2}$ and integrate with respect to α from 0 to x and then differentiate with respect to x, so that, if $0 \leq x \leq \alpha_1$, we have
$$\pi^{-1/2}\int_0^{\infty} f(\tau)d\tau \frac{d}{dx}\int_0^{x} \frac{\sinh\alpha}{\sqrt{(\cosh x - \cosh\alpha)}}P_{-1/2+i\tau}(\cosh\alpha)d\alpha = F_1(x), \qquad (7.9.3)$$

where $F_1(x)$ is defined in terms of the prescribed function $g(\alpha)$ by means of the equation
$$F_1(x) = \pi^{-1/2}\frac{d}{dx}\int_0^{x} g(\alpha)\sinh\alpha(\cosh x - \cosh\alpha)^{-1/2}d\alpha. \qquad (7.9.4)$$

If we now multiply (7.9.2) by $\sinh\alpha(\cosh\alpha - \cosh x)^{-1/2}$ and integrate with respect to α from x to ∞, we find that
$$\sqrt{(2/\pi)}\int_x^{\infty} f(\tau)\cos(\tau x)d\tau = F_2(x),\quad x\geq\alpha, \qquad (7.9.5)$$

where $$F_2(x) = \pi^{-1/2} \int_x^\infty h(\alpha)\sinh\alpha(\cosh\alpha - \cosh x)^{-1/2}d\alpha, \quad x \geqq \alpha_1. \tag{7.9.6}$$

The function $F(x)$ is now defined by the equations

$$F(x) = \begin{cases} F_1(x) & 0 \leqq x \leqq \alpha_1 \\ F_2(x) & x > \alpha_1, \end{cases} \tag{7.9.7}$$

whence $F(x)$ is the Fourier cosine transform of $f(\tau)$ and so $f(\tau)$ is the Fourier cosine transform of $F(x)$. This gives the solution of the problem in question together with (7.9.7), (7.9.4) and (7.9.5). Integral equations of this type occur when problems leading to the separation of Laplace's equation in toroidal coordinates are considered. See Sneddon (1972) page 407.

7.10 THE LINEAR FLOW OF HEAT IN AN ANISOTROPIC SOLID MOVING IN A CONDUCTING MEDIUM

The problem of conduction of heat in anisotropic materials has gained much interest recently. Problems of this type occur mainly in wood technology, soil mechanics and the mechanics of solids of fibrous structure.

In this section, we shall consider the linear heat flow in a finite solid with conductivity $K=K_0(1-x^2)$, and moving in a conducting medium with constant velocity. Saxena and Nagara (1974) have recently discussed this problem with the help of integrals involving Jacobi polynomials. Suppose that the solid rod $-1 \leqq x \leqq 1$ is moving along the direction of its length. The flux vector is given by

$$\underline{f} = -K \text{ grad } u, \tag{7.10.1}$$

where u is the temperature and is a function of position and time. For one-dimensional flow, the single component of the flux vector along any plane at a distance x from the origin is given by

$$f = -K\frac{\partial u}{\partial x} + pcu\nu\ . \tag{7.10.2}$$

Here, we have assumed that the solid is moving with a constant velocity ν along the direction of the x-axis. Also, p and c are the density and the specific heat of the solid respectively. Both these quantities are assumed to be constant in this study. Hence, by an appeal to the law of continuity and the fundamental laws of heat transfer, we arrive at the following differential equation of heat conduction:

$$\frac{\partial}{\partial x}[(1-x^2)\frac{\partial u}{\partial x}] - pc\nu/\lambda\ \frac{\partial u}{\partial x} + Q(x)/\lambda = pc/\lambda\ \frac{\partial u}{\partial t}, \tag{7.10.3}$$

with the law of conductivity $K = \lambda(1-x^2)$. $Q(x)$ is the intensity of a continuous source of heat situated inside the solid.

Let the initial temperature of the rod be given by

$$u(x,0) = g(x). \tag{7.10.4}$$

Equation (7.10.3) is easily comparable with the Jacobi equation

$$(1-x^2)y'' + [(\beta-\alpha)-(\alpha+\beta+2)x]y' + n(n+\alpha+\beta+1)y = 0, \tag{7.10.5}$$

which has the Jacobi polynomials

$$P_n^{(\alpha,\beta)}(x) = \frac{(\alpha+1,n)}{n!}\,{}_2F_1(-n,\alpha+1+\beta+n;\alpha+1;\frac{1-x}{2}) \tag{7.10.6}$$

as its solution. If we take

$$\nu = \frac{\alpha-\beta}{q},\ Q(x) = -(\alpha+\beta)\lambda x\frac{\partial u}{\partial x}, \quad q = \frac{c\rho}{\lambda},$$

then the solution of (7.10.3) can be written in the form

$$u = \sum_{n=0}^{\infty} A_n e^{-B_n t} P_n^{(\alpha,\beta)}(x). \tag{7.10.7}$$

Substituting this into equation (7.10.3), we have

$$B_n = \frac{n}{q}(n+\alpha+\beta+1).$$

In order to find the value of A_n, we make use of the initial condition (7.10.4). This gives

$$g(x) = \sum_{n=0}^{\infty} A_n P_n^{(\alpha,\beta)}(x). \tag{7.10.8}$$

Multiplying both sides of (7.10.8) by $(1-x)^\alpha(1+x)^\beta P_n^{(\alpha,\beta)}(x)$, and integrating with respect to x from -1 to 1, we obtain the result

$$A_n = G_n/\delta_n^{(\alpha,\beta)}, \tag{7.10.9}$$

where

$$G_n = \int_{-1}^{1}(1-x)^\alpha(1+x)^\beta P_n^{(\alpha,\beta)}(x)\, g(x)\, dx. \tag{7.10.10}$$

This is because the Jacobi polynomials possess the orthogonality property

$$\int_{-1}^{1}(1-x)^\alpha(1+x)^\beta P_m^{(\alpha,\beta)}(x)P_n^{(\alpha,\beta)}(x)\,dx = 0, \quad m\neq n,$$

$$= \delta_n^{(\alpha,\beta)}, \quad m=n, \tag{7.10.11}$$

where

$$\delta_n^{(\alpha,\beta)} = \frac{2^{\alpha+\beta+1}\Gamma(n+\alpha+1)\Gamma(n+\beta+1)}{n!(\alpha+\beta+2n+1)\Gamma(n+\alpha+\beta+1)}. \tag{7.10.12}$$

Hence, the solution of the problem may be expressed as

$$u(x,t) = \sum_{n=0}^{\infty} G_n(\delta_n^{(\alpha,\beta)})^{-1}\exp[\frac{-nt}{q}(n+\alpha+\beta+1)]P_n^{(\alpha,\beta)}(x). \qquad (7.10.13)$$

By way of illustration, suppose that $g(x) = 1+x$ and $\nu = 1$.Further, we take $\alpha = 1/2, \beta = -1/2$. Substituting these values into (7.10.10) and using the result

$$\int_{-1}^{1}(1-x)^{\alpha}(1+x)^{\beta}x^k\ P_n^{(\alpha,\beta)}(x) = 0 \text{ for } 0 \leq k \leq n,$$

$$= \mu_n^{\alpha,\beta} \text{ for } k = n, \qquad (7.10.14)$$

where $$\mu_n^{\alpha,\beta} = \frac{2^{1+\alpha+\beta+n}\Gamma(1+\alpha+n)\Gamma(1+\beta+n)}{\Gamma(2+\alpha+\beta+2n)}, \qquad (7.10.15)$$

we get the result

$$G_n = \pi/2 \quad \text{for } n = 0$$

$$= \pi/4 \quad \text{for } n = 1$$

$$= 0 \quad \text{for } n \geq 2. \qquad (7.10.16)$$

Now, $P_0^{(\alpha,\beta)}(x) = 1$ and $P_1^{(\alpha,\beta)}(x) = [(\alpha+\beta+2)x+(\alpha-\beta)]/2$, so that (7.10.13) gives

$$u(x,t) = 1/2 + (2x+1)/2\ e^{-2t} \qquad (7.10.17)$$

7.11 AN APPLICATION IN THE STUDY OF NON-LINEAR OSCILLATIONS

In the investigation of certain types of non-linear oscillations, a linear approximation which gives the greatest accuracy, in some sense, is made to the non-linear terms of the governing differential equation. This approximation is made over some interval [-A,A], say, of the dependent variable. This process, exemplified here by considering the free oscillations of the simple pendulum, enables us to compare the solution of the approximate differential equation with the exact solution of the original non-linear differential equation.

The equation of motion may be written

$$\frac{d^2\theta}{dt^2} + \omega_0^2\sin\theta = 0, \qquad (7.11.1)$$

so that one obtains the approximate solution

$$\theta^* = A\sin(\omega^* t + \phi), \qquad (7.11.2)$$

where ω^*, the approximate frequency, is to be determined. A quite general approach to this problem may be made by carrying out a linearisation of the term $\sin\theta$ using ultraspherical polynomials, given by

$$C_n^{(\alpha)}(x) = \frac{(2\alpha,n)}{n!}{}_2F_1(-n,n+2\alpha;\alpha+\frac{1}{2};\frac{1-x}{2}),\alpha > -\frac{1}{2}. \qquad (7.11.3)$$

These polynomials are orthogonal over the interval [-1,1] with respect to the weight function $(1-x^2)^{\alpha-1/2}$, and they may also be obtained from the Rodrigues formula

$$C_n^{(\alpha)}(x) = A_n^{(\alpha)}(1-x^2)^{-\alpha+1/2}(\frac{d}{dx})^n(1-x^2)^{n+\alpha-1/2}, \qquad (7.11.4)$$

where $A_n^{(\alpha)}$ is a normalisation factor given by

$$A_n^{(\alpha)} = \frac{(-1)^n\Gamma(\alpha+1/2)\Gamma(n+2\alpha)}{2^n\ n!\Gamma(2\alpha)\Gamma(n+\alpha+1/2)}. \qquad (7.11.5)$$

Then, $C_1^{(\alpha)}(x) = 2\alpha x$ and $C_n^{(0)}(x) = 0$ for $n \geq 1$. However, in the case of the Tchebycheff polynomials of the first kind from (7.11.4) with $\alpha = 0$, the normalising factor must be re-defined as

$$A_n^{(0)} = \frac{(-1)^n\ 2^n\ n!}{(2n)!}. \qquad (7.11.6)$$

This is consistent with

$$C_n^{(0)}(x) = T_n(x) = \cos(n\cos^{-1}x), \qquad (7.11.7)$$

where T_n is the Tchebycheff polynomial of the first kind. Other sets of ultraspherical (Gegenbauer) polynomials are the Legendre polynomials, for which $\alpha = 1/2$, and the Tchebycheff polynomials of the second, for which $\alpha = 1$. The Taylor series expansion of an analytic function about the origin corresponds to its expansion in ultraspherical polynomials for which $\alpha \to \infty$.

The ultraspherical polynomials on the interval [-A,A] are defined as the sets of polynomials orthogonal on this interval with respect to the weight function $(1-x^2/A^2)^{\alpha-1/2}$, where $\alpha > -1/2$. The normalisation is chosen to give rise to the polynomials $C_n^{(\alpha)}(x/A)$. Approximating $\sin\theta$ on the interval [-A,A] with the ultraspherical polynomial linearly, one obtains

$$\sin^*\theta = a_1^{(\alpha)}C_1^{(\alpha)}(\theta/A), \qquad (7.11.8)$$

where

$$a_1^{(\alpha)} = \int_{-A}^{A}(1-\theta^2/A^2)^{\alpha-1/2}\ C_1^{(\alpha)}(\theta/A)\ \sin\theta\ d\theta\ \times \qquad (7.11.9)$$

(continued)

(continued)

$$\times \left[\int_{-A}^{A} [C_1^{(\alpha)}(\theta/A)]^2 \, (1-\theta^2/A^2)^{\alpha-1/2} \, d\theta \right]^{-1}, \qquad (7.11.9)$$

which, after some reduction, becomes

$$a_1^{(\alpha)} = \Gamma(\alpha+2) J_{\alpha+1}(A)/[\alpha(A/2)^{\alpha}] \,, \qquad (7.11.10)$$

so that the required linear approximation of $\sin\theta$ becomes

$$\sin^*\theta = \Gamma(\alpha+2) J_{\alpha+1}(A)\theta/(A/2)^{\alpha+1} = \Lambda_{\alpha+1}(A)\theta, \qquad (7.11.11)$$

where

$$\Lambda_{\alpha}(A) = \Gamma(\alpha+1) J_{\alpha}(A)/(A/2)^{\alpha} \,. \qquad (7.11.12)$$

It was found by comparison with the exact solution of (7.11.1) in terms of elliptical integrals, that a good approximation to the solution for A in the closed interval $[0,\pi]$ is obtained when $\alpha = -0.21$. This method gives an amplitude-dependent solution to the problem under consideration which accords very well with reality. For more details see Denman and Howard (1964). An obvious extension is to a study on the above lines of, for example, the equation of the simple pendulum when a forcing function is present, and where an exact analytical solution is not known.

7.12 THE LAPLACE TRANSFORM OF THE PRODUCT OF SEVERAL MOLECULAR INTEGRALS

A molecular integral of frequent occurrence in quantum chemistry is the function $\beta_n(z)$ given by

$$\beta_n(z) = \int_{-1}^{1} t^n e^{-zt} \, dt, \quad n=0,1,2,.. \,. \qquad (7.12.1)$$

See Abramowitz and Stegun (1965) page 228.
It is of interest to develop the Laplace transform of the product of several such functions, that is

$$I^{\mu,p}_{n_1,n_2,..,n_r}(a_1,..,a_r) = \int_0^{\infty} e^{-ps} s^{\mu-1} \beta_{n_1}(a_1 s)..\beta_{n_r}(a_r s) \, ds, \qquad (7.12.2)$$

where p and μ are both positive. By an elementary change of the variable of integration of the right-hand member of (7.12.1), it may be shown that (Abramowitz and Stegun [1965] page 230)

$$\beta_n(z) = z^{-n-1}[\Gamma(n+1,-z) - \Gamma(n+1,z)]. \qquad (7.12.3)$$

Now,

$$\Gamma(a,x) = \Gamma(a) - \gamma(a,x) = \Gamma(a) - x^a/a \; {}_1F_1(a;a+1;-x), \qquad (7.12.4)$$

so that $\beta_n(z) = (n+1)^{-1}[{}_1F_1(n+1;n+2;-z)-{}_1F_1(n+1;n+2;z)]$.

$$\text{Hence,}\quad \prod_{j=1}^{r}(n_j+1)\, I^{\mu,p}_{n_1,n_2,..,n_r}(a_1,..,a_r) \tag{7.12.5}$$

$$= \int_0^\infty e^{-ps}\, s^{\mu-1} \prod_{j=1}^{r}[{}_1F_1(n_j+1;n_j+2;-a_js)-{}_1F_1(n_j+1;n_j+2;a_js)]ds \tag{7.12.6}$$

and this integral may be evaluated by using the formula

$$\int_0^\infty e^{-s}\, s^{a-1}\, {}_1F_1(b_1;c_1;x_1s)..{}_1F_1(b_r;c_r;x_rs)ds$$

$$= \Gamma(a)F_A^{(r)}(a,b_1,..,b_r;c_1,..,c_r;x_1,..,x_r). \tag{7.12.7}$$

We thus have, finally

$$p^\mu[\Gamma(\mu)]^{-1}\prod_{j=1}^{r}(n_j+1)\, I^{\mu,p}_{n_1,n_2,..,n_r}(a_1,..,a_r)$$

$$= \sum_{h_1,..,h_r=-1,\neq 0}^{1} (-1)^K\, F_A^{(r)}(\mu,n_1+1,..,n_r+1;n_1+2,..,n_r+2;\frac{h_1a_1}{p},..,\frac{h_ra_r}{p}), \tag{7.12.8}$$

where $K = \sum_{j=1}^{r} h_j+r$.

This is the result sought.

7.13 A PROBLEM IN COMMUNICATIONS ENGINEERING

The problem of the input to a limiter consisting of a desired sinusoidal signal, Gaussian noise and several interfering sinusoids leads to integrals of the general type

$$\int_0^\infty e^{-\sigma^2x^2/2}\, x^{k-1}\, J_{\nu_1}(a_1x)..J_{\nu_r}(a_rx)\, dx. \tag{7.13.1}$$

This integral can be written

$$[2^{\nu_1}\Gamma(\nu_1+1)..2^{\nu_r}\Gamma(\nu_r+1)]^{-1}\int_0^\infty x^{k+\nu_1+.+\nu_r-1}\, e^{-\sigma^2x^2/2}$$

$$\times\ {}_0F_1(-;\nu_1+1;-a_1^2x^2/4)..{}_0F_1(-;\nu_r+1;-a_r^2x^2/4)\ dx, \tag{7.13.2}$$

and if $z = \sigma^2x^2/2$, it becomes $\dfrac{2^{(k-\nu_1-.-\nu_r-2)/2}\,\sigma^{k+\nu_1+.+\nu_r}}{\Gamma(\nu_1+1)...\Gamma(\nu_r+1)}$

$$\times \int_0^\infty z^{(k+\nu_1+.+\nu_r-2)/2}e^{-z}{}_0F_1(-;\nu_1+1;-\sigma^2a_1^2z/2)..{}_0F_1(-;\nu_r+1;-\sigma^2a_r^2z/2)$$

$$\times\ dz. \tag{7.13.3}$$

Term-by term integration of the integrand yields the result

$$2^{(k-\nu_1-.-\nu_r-2)/2}\ \sigma^{k+\nu_1+.+\nu_r}\ \Gamma[(k+\nu_1+.+\nu_r)/2][\Gamma(\nu_1+1)..\Gamma(\nu_r+1)]^{-1}$$

$$\times\ \Psi_2^{(r)}(\frac{k+\nu_1+.+\nu_r}{2};\nu_1+1,.,\nu_r+1;-\sigma^2a_1^2/2,.,-\sigma^2a_r^2/2), \qquad (7.13.4)$$

provided that $k+\nu_1+.+\nu_r > 0$. A special case of frequent occurrence is

$$Q_n(a) = \int_0^\infty e^{-\sigma^2x^2/2}\ J_0(s_1x)..J_0(s_nx)J_1(ax)x^{-1}\ dx, \qquad (7.13.5)$$

which may readily be evaluated by means of (7.13.1) and (7.13.4) in the form

$$\frac{\sigma\sqrt{\pi}}{16}\ \Psi_2^{(n+1)}(1/2;1,.,1,2;-\sigma^2s_1^2/2,.,-\sigma^2s_n^2/2,-\sigma^2a^2/2). \qquad (7.13.6)$$

This confluent Lauricella function may be represented as a series which converges for all finite values of its variables, and provided that σ is not too large, it can be computed directly. See Mathai and Saxena (1973) page 255.

7.14 AN APPLICATION OF REPEATED INTEGRALS OF BESSEL FUNCTIONS

We consider a semi-infinite ladder network of impedences with mid-series termination, which is initially passive and to which a voltage $V(t)$ is applied at $t=0$. Let z' and z respectively be the generalised impedences of the series and shunt elements, let I_0 be the input current, I_n the current in the nth series element z', and V_n the voltage drop across the nth shunt element z, then, if $L(f)$ is the Laplace transform of the function f,

$$L(I_n) = L(V)[zz'(1+\frac{z'}{4z})]^{-1/2}\ [(1+\frac{z'}{4z})^{1/2}-(\frac{z'}{4z})^{1/2}]^{2n} \qquad (7.14.1)$$

and $$L(V_n) = L(V)(1+\frac{z'}{4z})^{-1/2}\ [(1+\frac{z'}{4z})^{1/2}-(\frac{z'}{4z})^{1/2}]^{2n-1}. \qquad (7.14.2)$$

We proceed to determine I_n and V_n for the high-pass filter with series elements of capacitance C and shunt elements of inductance L. Here

$$z' = 1/C_s; \quad z = L_s; \quad z'/(4z) = \omega^2/s^2, \qquad (7.14.3)$$

where $\omega^2 = 1/(4LC)$. For constant applied voltage, $V=E, L(V)=E/s$ and (7.14.1) and (7.14.2) become

$$L(I_n) = E\sqrt{(C/L)}[\sqrt{(s^2+\omega^2)} - \omega]^{2n}\ s^{-2n}\ (s^2+\omega^2)^{-1/2} \qquad (7.14.4)$$

and $L(V_n) = E[\surd(s^2+\omega^2)-\omega]^{2n-1}\ s^{1-2n}\ (s^2+\omega^2)^{-1/2}$. (7.14.5)

For convenience, we now define the 'high-pass filter function' $\Phi_n(t)$ by the relation

$$L[\Phi_n(t)] = [\surd(s^2+1)-1]^n\ s^{-n}\ (s^2+1)^{-1/2}, \quad (7.14.6)$$

where n is a positive integer. If n is even, equal to 2r, say, then we may expand the numerator of the right-hand member of (7.14.6) in the form

$$[\surd(s^2+1)-1]^{2r} = \sum_{m=0}^{r} \frac{r(r+m-1)!2^{2m}\ s^{2r-2m}}{(r-m)!\ (2m)!}$$

$$- \surd(s^2+1) \sum_{m=1}^{r} \frac{(r+m-1)!2^{2m-1}\ s^{2r-2m}}{(r-m)!\ (2m-1)!}. \quad (7.14.7)$$

Now, observing that $L(t^n/n!) = s^{-n-1}$, $n=0,1,2,..$, (7.14.8)

$$L[J_n(t)] = [\surd(s^2+1)-s]^n\ (s^2+1)^{-1/2}, \ n > -1 \quad (7.14.9)$$

and $L[Ji_{n,r}(t)] = [\surd(s^2+1)-s]^n\ s^{-r}\ (s^2+1)^{-1/2}$, $n > -1$, (7.14.10)

where $Ji_{n,r}(t)$ is the rth repeated integral of the Bessel function of order n, $J_n(t)$ given by

$$Ji_{n,r}(t) = \int_o^t dt..(r)..\int_o^t J_n(t)\ dt. \quad (7.14.11)$$

From (7.14.6), (7.14.7), (7.14.8) and (7.14.10), it is evident that

$$\Phi_{2r}(t) = \sum_{m=0}^{r} \frac{r(r+m-1)!2^{2m}}{(r-m)!(2m)!} Ji_{o,2m}(t) - \sum_{m=1}^{r} \frac{(r+m-1)!2^{2m-1}\ t^{2m-1}}{(r-m)![(2m-1)!]^2}. \quad (7.14.12)$$

Similar results, also involving repeated integrals of Bessel functions, arise in the solution of other related problems. See Jaeger (1946).

7.15 AN APPLICATION IN COMBINATORIAL ANALYSIS

Suppose that we have n events $A_1,..,A_n$ and let $P(A_{j1},..,A_{jk})$ be the probability that the events $A_{j1},..,A_{jk}$ occur jointly. The probability P_0 that none of $A_1,..,A_n$ occurs is given by the probabilistic version of the principle of inclusion and exclusion due to Poincaré:

$$P_0 = 1 - \sum_j P(A_j) + \sum_{j>k} P(A_j A_k) - \ldots \qquad (7.15.1)$$

When these events are symmetric, $P(A_{j1},.,A_{jk})$ depends only upon k; say it is ϕ_k. It may be shown that P_r, the probability that exactly r events occur, has integral representation

$$P_r(n) = \frac{(-1)^r}{r!} \int_0^\infty t^r [d^r/dt^r\, f(t)]\, d\alpha(t), \qquad (7.15.2)$$

where f(t) is the associated fundamental polynomial, and the weight function $\alpha(t)$ is of bounded variation on $(0,\infty)$.

In connection with the general derangement problem, where we have $k_1+.+k_n$ cards of which k_r are marked 'r', r=1,2,..,n, the cards marked 'r', or r-cards, are not to appear in any of P_r specified places with no places forbidden simultaneously to both r-cards and cards not so marked. Kaplansky proved that $\phi_k = k!$ [see Askey and Ismail (1976)], and that the number of these arrangements, say $C[{k_1,.,k_n \atop p_1,.,p_n}]$, is given by the integral

$$\int_0^\infty H(k_1,p_1,x)H(k_2,p_2,x)..H(k_n,p_n,x)e^{-x}\, dx, \qquad (7.15.3)$$

where
$$H(k,p,x) = x^k/k!\ {}_2F_0(-k,-p;-;-1/x). \qquad (7.15.4)$$

If $p \leqq k$, it follows that

$$H(k,p,x) = x^{k-p}(-1)^p \frac{p!}{k!} L_p^{k-p}(x) \qquad (7.15.5)$$

and if $p \geqq k$,

$$H(k,p,x) = (-1)^k L_k^{p-k}(x). \qquad (7.15.6)$$

So that, if $p_1 \leqq k_1,..,p_n \leqq k_n$,

$$C[{k_1,.,k_n \atop p_1,.,p_n}] = (-1)^{p_1+.+p_n} \int_0^\infty e^{-x}\, x^{k_1-p_1+.+k_n-p_n}$$
$$\times\ L_{p_1}^{k_1-p_1}(x)..L_{p_n}^{k_n-p_n}(x)\, dx. \qquad (7.15.7)$$

Recall that the generalised Laguerre polynomial may be expressed in hypergeometric form by means of the relation

$$L_n^\alpha(x) = \frac{(\alpha+1,n)}{n!}\ {}_1F_1(-n;\alpha+1;x), \qquad (7.15.8)$$

whence the right-hand member of (7.15.8) may be sritten

$$\frac{(-1)^{p_1+.+p_n}\,(k_1-p_1+1,p_1)..(k_n-p_n+1,p_n)}{p_1!\ldots p_n!}$$

$$\times \int_0^\infty e^{-x}\, x^{k_1-p_1+.+k_n-p_n}\; {}_1F_1\left({-p_1 \atop k_1-p_1+1};x\right)..{}_1F_1\left({-p_n \atop k_n-p_n+1};x\right)\,dx. \tag{7.15.9}$$

Hence, by the application of (5.2.5.44), we have

$$C\left[{k_1,.,k_n \atop p_1,.,p_n}\right] = (-1)^{p_1+.+p_n}\,(k_1-p_1+1,p_1)..(k_n-p_n+1,p_n)$$

$$\times\, (k_1-p_1+.+k_n-p_n)!\; F_A^{(n)}(k_1-p_1+.+k_n-p_n;-p_1,.,-p_n;$$

$$k_1-p_1+1,.,k_n-p_n+1;1,..,1), \tag{7.15.10}$$

which is a terminating form of the Lauricella function of the first kind. Integrals involving the product of several Laguerre polynomials occur in a number of other problems in combinatorial analysis. See Askey and Ismail (1976).

Part Two: Tables and Computer Programs

A
Tables of Hypergeometric Integrals

Introduction. In the following tables of hypergeometric integrals, it is inevitable that many cases of potential interest must be omitted on account of the restrictions imposed by lack of space. It had originally been hoped to include integrals involving those special functions which may be expressed in hypergeometric form. This aspect of the work has been postponed until a later date for the same reason.

The integrals are listed in each of the following sections in order of complexity of the integrand: the product of several hypergeometric functions each of one variable is treated as a multiple hypergeometric function for the purposes of classification. Unless otherwise indicated, it will be assumed that all quantities are real in what follows.

A.1 INTEGRALS OF EULER TYPE

A.1.1 Euler Integrals involving Single Hypergeometric Functions

	$f(u)$	$\int_0^1 f(u)\,du$
A.1.1.1	$u^{a-1}(1-u)^{b-1}$ $\times {}_0F_1(-;b;ux)$	$\frac{\Gamma(a)\Gamma(b)}{\Gamma(a+b)}\ {}_0F_1(-;a+b;x)$ $a,b > 0$
A.1.1.2	$u^{a-1}(1-u)^{b-1}$ $\times {}_0F_1(-;c;ux)$	$\frac{\Gamma(a)\Gamma(b)}{\Gamma(a+b)}\ {}_1F_2(a;a+b,c;x)$ $a,b > 0$
A.1.1.3	$u^{a-1}\ {}_0F_1(-;\frac{a}{2};u^2x)$	$\frac{1}{a}\ {}_0F_1(-;1+\frac{a}{2};x)$ $a > 0$
A.1.1.4	$u^{a-1}(1-u)^{b-1}$ $\times {}_0F_1(-;\frac{a}{2};u^2x)$	$\frac{\Gamma(a)\Gamma(b)}{\Gamma(a+b)}\ {}_1F_2\left[\begin{matrix}\frac{a+1}{2} & ; \\ \frac{a+b}{2},\ \frac{a+b+1}{2} & ;\end{matrix}\ x\right]$ $a,b > 0$

	$f(u)$	$\int_0^1 f(u)\,du$
A.1.1.5	$u^{a-1}(1-u)^{b-1}$ $\times {}_0F_1(-;c;u^2x)$	$\frac{\Gamma(a)\Gamma(b)}{\Gamma(a+b)}\ {}_2F_3\left[\begin{matrix}\frac{a}{2}, \frac{a+1}{2} & ; \\ \frac{a+b}{2}, \frac{a+b+1}{2}, c; & \end{matrix} x\right]$ $a,b > 0$
A.1.1.6	$u^{a-1}(1-u)^{a-1}$ $\times {}_0F_1(-;a;u[1-u]x)$	$\frac{\Gamma(a)\Gamma(a)}{\Gamma(2a+1)}\ {}_0F_1(-;a+\frac{1}{2};\frac{x}{4})$ $a > 0$
A.1.1.7	$u^{a-1}(1-u)^{b-1}$ $\times {}_0F_1(-;b;u[1-u]x)$	$\frac{\Gamma(a)\Gamma(b)}{\Gamma(a+b)}\ {}_1F_2\left[\begin{matrix} a & ; \\ \frac{a+b}{2}, \frac{a+b+1}{2} & ; \end{matrix} \frac{x}{4}\right]$ $a,b > 0$
A.1.1.8	$u^{a-1}(1-u)^{b-1}$ $\times {}_0F_1(-;c;u[1-u]x)$	$\frac{\Gamma(a)\Gamma(b)}{\Gamma(a+b)}\ {}_2F_3\left[\begin{matrix} a, b & ; \\ \frac{a+b}{2}, \frac{a+b+1}{2}, c & ; \end{matrix} \frac{x}{4}\right]$ $a,b > 0$
A.1.1.9	$u^{a-1}(1-u)^{b-1}$ $\times {}_1F_1(a+b;a;ux)$	$\frac{\Gamma(a)\Gamma(b)}{\Gamma(a+b)}\ e^x \quad a,b > 0$
A.1.1.10	$u^{a-1}(1-u)^{b-1}$ $\times {}_1F_1(c;a;ux)$	$\frac{\Gamma(a)\Gamma(b)}{\Gamma(a+b)}\ {}_1F_1(c;a+b;x) \quad a,b > 0$
A.1.1.11	$u^{a-1}(1-u)^{b-1}$ $\times {}_1F_1(c;d;ux)$	$\frac{\Gamma(a)\Gamma(b)}{\Gamma(a+b)}\ {}_2F_2\left[\begin{matrix} a, c; \\ a+b, d; \end{matrix} x\right] \quad a,b > 0$
A.1.1.12	u^{a-1} $\times {}_1F_1(\frac{a+1}{2};1+\frac{a}{2};u^2x)$	$e^x/a \quad a > 0$
A.1.1.13	$u^{a-1}(1-u)^{b-1}$ $\times {}_1F_1(\frac{a}{2};\frac{a+b+1}{2};u^2x)$	$\frac{\Gamma(a)\Gamma(b)}{\Gamma(a+b)}\ {}_1F_1(\frac{a+1}{2};\frac{a+b}{2};x) \quad a,b > 0$
A.1.1.14	$u^{a-1}(1-u)^{b-1}$ $\times {}_1F_1(c;\frac{a}{2};u^2x)$	$\frac{\Gamma(a)\Gamma(b)}{\Gamma(a+b)}\ {}_2F_2\left[\begin{matrix} \frac{a+1}{2}, c & ; \\ \frac{a+b}{2}, \frac{a+b+1}{2} & ; \end{matrix} x\right]$ $a,b > 0$

	$f(u)$	$\int_0^1 f(u)\,du$
A.1.1.15	$u^{a-1}(1-u)^{b-1}$ $\times {}_1F_1(c;d;u^2x)$	$\frac{\Gamma(a)\Gamma(b)}{\Gamma(a+b)}\,{}_3F_3\left[\begin{matrix}\frac{a}{2},\frac{a+1}{2}, c ; \\ \frac{a+b}{2},\frac{a+b+1}{2},d;\end{matrix} x\right]$ $a,b > 0$
A.1.1.16	$u^{a}(1-u)^{a-1}$ $\times {}_1F_1(a+\frac{1}{2};a;u[1-u]x)$	$\frac{\Gamma(a+1)\Gamma(a)}{\Gamma(2a+1)}\,e^{x/4}$ $a > 0$
A.1.1.17	$u^{a}(1-u)^{a-1}$ $\times {}_1F_1(b;a;u[1-u]x)$	$\frac{\Gamma(a+1)\Gamma(a)}{\Gamma(2a+1)}\,{}_1F_1(b;a+\frac{1}{2};\frac{x}{4})$ $a > 0$
A.1.1.18	$u^{a-1}(1-u)^{b-1}$ $\times {}_1F_1(c;d;u[1-u]x)$	$\frac{\Gamma(a)\Gamma(b)}{\Gamma(a+b)}\,{}_3F_3\left[\begin{matrix}c, a, b ; \\ d,\frac{a+b}{2},\frac{a+b+1}{2} ;\end{matrix}\frac{x}{4}\right]$ $a,b > 0$
A.1.1.19	$u^{a-1}(1-u)^{b-1}$ $\times {}_1F_1(c;b;u[1-u]x)$	$\frac{\Gamma(a)\Gamma(b)}{\Gamma(a+b)}\,{}_2F_2\left[\begin{matrix}c, a ; \\ \frac{a+b}{2},\frac{a+b+1}{2};\end{matrix}\frac{x}{4}\right]$ $a,b > 0$
A.1.1.20	$u^{a-1}(1-u)^{a-1}$ $\times {}_2F_1(c;d;\frac{c+d+1}{2};u)$	$\pi\frac{2^{1-2a}\,\Gamma(a)\Gamma(\frac{1+c+d}{2})\Gamma(\frac{1-c-d}{2}+a)}{\Gamma(\frac{1+c}{2})\Gamma(\frac{1+d}{2})\Gamma(\frac{1-c}{2}+a)\Gamma(\frac{1-d}{2}+a)}$ $a > 0,\quad 2a-c-d > -1$
A.1.1.21	$u^{c-1}(1-u)^{a-c}$ $\times {}_2F_1(a,a;c;\frac{u}{2})$	$\frac{\Gamma(c)\Gamma(a-c+1)}{\Gamma(a+1)}\,2^{a-1}a$ $\times[\psi(\frac{1+a}{2})-\psi(\frac{a}{2})]$ $a \neq -1,-2,..$ $c > 0 \quad a-c > -1$
A.1.1.22	$u^{a-1}(1-u)^{a-c}$ $\times {}_2F_1(b,1-b;c;u)$	$\frac{2^{1-2a}\,\Gamma(a)\Gamma(a-c+1)\Gamma(c)\,\pi}{\Gamma(\frac{b+c}{2})\Gamma(\frac{b-c+1}{2}+a)\Gamma(\frac{1-b+c}{2})}$ $\times 1/\Gamma(1+a-\frac{b+c}{2})$ $a > 0, a-c > -1$

	$f(u)$	$\int_0^1 f(u)\, du$
A.1.1.23	$(1-u)\,{}_2F_1(a,b;c;u)$ $a \neq 1,2;\ b \neq 1,2;$ $c-a-b > -2$	$\frac{(c-2,2)\Gamma(c-2)\Gamma(c-a-b+2)}{(a-2,2)(b-2,2)\Gamma(c-a)\Gamma(c-b)}$ $-\frac{(c-2,2)}{(a-2,2)(b-2,2)} - \frac{(c-1)}{(a-1)(b-1)}$
A.1.1.24	${}_2F_1(a,b;c;u)$ $a,b \neq 1,\ c-a-b > -1$	$\frac{(c-1)\Gamma(c-1)\Gamma(c-a-b+1)}{(a-1)(b-1)\Gamma(c-a)\Gamma(c-b)}$ $-\frac{(c-1)}{(a-1)(b-1)}$
A.1.1.25	$u^{a+b}\,{}_2F_1\left({-N,N+2b;\atop 2b+1\ ;}u\right)$ $a+b > -1$	$\frac{2b(a+b+1)N!(b-a+1,N)}{(b-a-1)(2b,N)(a+b+2,N)}$
A.1.1.26	$(1-u)^{h/2-2}$ $\times{}_2F_1\left({1-k,h/2+1;\atop h+k\ ;}-u\right)$	$h(h/2-1)(h+k)$ $h,h+2k > 2$
A.1.1.27	$u^{a-1}(1-u)^{b-1}$ $\times{}_2F_1\left({1+a/2,-N;\atop a/2\ ;}u\right)$	$\frac{\Gamma(a)\Gamma(b)(-b,N)(b-1,N)}{\Gamma(a+b+N)(1-b,N)}$ $a,b > 0,\ N=0,1,2,..$
A.1.1.28	$u^{a-1}(1-u)^{b-1}$ $\times{}_2F_1\left({-N,c\ ;\atop 1+c-b-N;}u\right)$	$\frac{\Gamma(b+N)\Gamma(a)\Gamma(a+b-c+N)\Gamma(b-c)}{\Gamma(a+b+N)\Gamma(a-b-c)\Gamma(b-c+N)}$ $a,b > 0,\quad N = 0,1,2,..$
A.1.1.29	$u^{c-1}(1-u)^{a/2+b/2-c}$ $\times{}_2F_1(a,b;c;u/2)$ $c > 0,\ \frac{a+b}{2}-c > -1$ $\frac{a+b}{2}+1 \neq 0,-1,-2,..$	$\frac{\Gamma(c)\Gamma(\frac{a+b}{2}-c+1)2\sqrt{\pi}}{(a-b)}$ $\times[\{\Gamma(\frac{a}{2})\Gamma(\frac{1+b}{2})\}^{-1}$ $-\{\Gamma(\frac{1+a}{2})\Gamma(\frac{b}{2})\}^{-1}]$
A.1.1.30	$u^{c-1}(1-u)^{a-c}$ $\times{}_2F_1(a,1;c;-u)$	$\frac{\Gamma(c)a(a-c)}{2}\{\psi(\frac{1+a}{2})-\psi(\frac{a}{2})\}$ $c > 0,\ a-c > -1$
A.1.1.31	$u^{c-1}(1-u)^{a-b-c}$ $\times{}_2F_1(a,b;c;-u)$	$\frac{\Gamma(c)\Gamma(1+a-b-c)}{\Gamma(1+a)\Gamma(1+a/2-b)}$ $c > 0,\ a-b-c > 1,\ b > 1$

	$f(u)$	$\int_0^1 f(u)\,du$
A.1.1.32	$u^{c-1}(1-u)^{a-b-c+1}$ $\times {}_2F_1(a,b;c;-u)$ $c>0,\ a-b-c>-1,$ $b<0$ or $a(b)=0,-1,-2,..$	$\frac{\Gamma(c)\Gamma(a-b-c+2)\sqrt{\pi}}{2^a(b-1)}$ $\times\{[\Gamma(\frac{a}{2})\Gamma(\frac{3+a}{2}-b)]^{-1}$ $-[\Gamma(\frac{1+a}{2})\Gamma(1+\frac{a}{2}-b)]^{-1}\}$
A.1.1.33	$u^{c-1}(1-u)^{d-1}$ $\times {}_2F_1(a,b;c;u)$	$\frac{\Gamma(c)\Gamma(d)\Gamma(c+d-a-b)}{\Gamma(c+d-a)\Gamma(c+d-b)}$ $c+d-a-b>0,\ c,d>0$
A.1.1.34	$(1-u)^{a/2+b/2-1/2-c}$ $\times u^{c-1}\,{}_2F_1\left(\begin{matrix}a,b;\frac{u}{2}\\ c\ ;\end{matrix}\right)$	$\frac{\Gamma(c)\Gamma(\frac{1+a+b}{2}-c)\Gamma(\frac{1}{2})}{\Gamma(\frac{1+a}{2})\Gamma(\frac{1+b}{2})}$ $c>0$ $a+b-2c>1$
A.1.1.35	$u^{c-1}(1-u)^{b-1}$ $\times {}_2F_1(a,1-a;c;\frac{u}{2})$	$\frac{\Gamma(c)\Gamma(b)\Gamma(\frac{c+b}{2})\Gamma(\frac{1+c+b}{2})}{\Gamma(c+b)\Gamma(\frac{c+b+a}{2})\Gamma(\frac{1+c+b-a}{2})}$ $b,c>0$
A.1.1.36	$u^{a-1}(1-u)^{2+a-2N}$ $\times {}_2F_1\left(\begin{matrix}3+2a-2N,-N;\\ 1/2\ \ ;\end{matrix}u^2\right)$ $a>0,3+a>2N$	$\frac{\Gamma(a)\Gamma(3+a-2N)}{\Gamma(3+2a-2N)}$ $\times\frac{(-1-\frac{a}{2},N)(\frac{a+1}{2},N)(-\frac{a+1}{2},N)}{(\frac{1}{2},N)(-\frac{a+2}{4},N)(-a-1,N)}$
A.1.1.37	$u^{a-1}(1-u)^{2b-2a-1}$ $\times {}_2F_1\left(\begin{matrix}b-a/2,b\ ;\\ 1+b-a/2;\end{matrix}u^2\right)$ $a>0,2b-2a>1$ $b-2a>-1$	$\frac{\Gamma(a)\Gamma(2b-2a)\Gamma(1+\frac{b}{2})\Gamma(1+b-\frac{a}{2})}{\Gamma(2b-a)\Gamma(1+b)\Gamma(1+\frac{b-a}{2})\Gamma(\frac{1}{2}+b-a)}$ $\times\frac{\Gamma(\frac{1}{2}+b-\frac{a}{2})\Gamma(\frac{1+b}{2}-a)}{\Gamma(\frac{1+b-a}{2})}$
A.1.1.38	$u^{a-1}(1-u)^{b-1}$ $\times {}_2F_1\left[\begin{matrix}\frac{a+b}{2},-N\ \ ;\\ \frac{a-b}{2}+1-N;\end{matrix}u^2\right]$ $a,b>0$	$\frac{\Gamma(a)\Gamma(b)(\frac{b+1}{2},N)(\frac{b}{2}+1)}{\Gamma(a+b)\ (\frac{a+b+1}{2},N)(\frac{b-a}{2},N)}$

	$f(u)$	$\int_0^1 f(u)\,du$
A.1.1.39	$u^{a-1}(1-u)^{b-1}$ $\times {}_2F_1\left[\begin{matrix}\frac{a+b}{2}, \frac{a+b+1}{2}; u^2\\ c\quad ;\end{matrix}\right]$	$\frac{\Gamma(a)\Gamma(b)\Gamma(c)\Gamma(c-a-b-1/2)}{\Gamma(a+b)\Gamma(c-\frac{a+b}{2})\Gamma(c-\frac{a+b+1}{2})}$ $a > 0, b > 0,\ c-a-b > -1/2$
A.1.1.40	$u^{c-1}(1-u)^{b-c-1}$ $\times {}_2F_1(a,b;c;ux)$	$\frac{\Gamma(c)\Gamma(b-c)}{\Gamma(b)}\ (1-x)^{-a}\quad \|x\| < 1,$ $c,\ b-c > 0$
A.1.1.41	$u^{e-1}(1-u)^{d-1}$ $\times {}_2F_1(a,b;c;ux)$	$\frac{\Gamma(e)\Gamma(d)}{\Gamma(e+d)}\ {}_3F_2\left(\begin{matrix}a,b\ e;\\ c,d+e;\end{matrix}x\right)$ $d,e > 0\quad \|x\| < 1$
A.1.1.42	$u^{c-1}(1-u)^{2a-c}$ $\times {}_2F_1\left(\begin{matrix}a,1/2+a;\\ c\quad ;\end{matrix}ux\right)$	$\frac{\Gamma(c)\Gamma(1+2a-c)}{\Gamma(1+2a)}\ 2^{2a}[1+\sqrt{(1-x)}]^{-2a}$ $c, 2a-c > 0\quad \|x\| < 1$
A.1.1.43	$u^{c-1}(1-u)^{2a-c-1}$ $\times {}_2F_1\left(\begin{matrix}a,1/2+a;\\ c\quad ;\end{matrix}ux\right)$	$\frac{\Gamma(c)\Gamma(2a-c)}{\Gamma(2a)(1-x)^{1/2}}\left[\frac{1+\sqrt{(1-x)}}{2}\right]^{1-2a}$ $c > 0,\ 2a-c > 0\quad \|x\| < 1$
A.1.1.44	$u^{c-1}(1-u)^{d-1}$ $\times {}_2F_1(a,b;c;ux)$	$\frac{\Gamma(c)\Gamma(d)}{\Gamma(c+d)}\ {}_2F_1(a,b;c+d;x)$ $c,d > 0\quad \|x\| < 1$
A.1.1.45	$u^{c-1}(1-u)^{-c-1/2}$ $\times {}_2F_1(a,1-a;c;-ux^2)$ $0 < c < 1/2,\ \|x\|<1$	$\frac{\Gamma(c)\Gamma(\frac{1}{2}-c)}{2\sqrt{\pi}\sqrt{(1+x^2)}}\{[\sqrt{(1+x^2)}+x]^{2a-1}$ $+[\sqrt{(1+x^2)}-x]^{2a-1}\}$
A.1.1.46	$u^{c-1}(1-u)^{-1/2-c}$ $\times {}_2F_1(a,-a;c;-ux^2)$ $0 < c < 1/2, \|x\| < 1$	$\frac{\Gamma(c)\Gamma(\frac{1}{2}-c)}{2\sqrt{\pi}}\ \{[\sqrt{(1+x^2)}+x]^{2a}$ $+[\sqrt{(1+x^2)}-x]^{2a}\}$
A.1.1.47	$u^{c-1}(1-u)^{1/2-c}$ $\times {}_2F_1\left(\begin{matrix}a,a+1/2;\\ c\quad ;\end{matrix}ux^2\right)$	$\frac{\Gamma(c)\Gamma(3/2-c)}{\sqrt{\pi}x(1-2a)}\{(1+x)^{1-2a}-(1-x)^{1-2a}\}$ $0 < c < 3/2,\ \|x\| < 1$
A.1.1.48	$(1-u)^{f-2}$ $\times {}_2F_1\left(\begin{matrix}-N,f+1;\\ 1-z\quad ;\end{matrix}\frac{z+f}{n+f}u\right)$	$\frac{(f-1)z(n+f)}{f(z-n)}\qquad f > 1$

	$f(u)$	$\int_0^1 f(u)\,du$
A.1.1.49	$u^{c-1}(1-u)^{-c-1/2}$ $\times {}_2F_1\left(\begin{matrix}a, a+1/2;\\ c\quad;\end{matrix} ux^2\right)$	$\frac{\Gamma(c)\Gamma(1/2-c)}{2\sqrt{\pi}}[(1+x)^{-2a}+(1-x)^{-2a}]$ $0 < c < 1/2,\quad \lvert x\rvert < 1$
A.1.1.50	$u^{d-1}(1-u)^{e-1}$ $\times {}_2F_1(a,b;c;xu^2)$	$\frac{\Gamma(d)\Gamma(e)}{\Gamma(d+e)}\,{}_4F_3\left[\begin{matrix}a, b, \frac{d}{2}, \frac{d+e}{2}\quad;\\ c, \frac{d+e}{2}, \frac{d+e+1}{2};\end{matrix} x\right]$ $d, e > 0\quad \lvert x\rvert < 1$
A.1.1.51	$u^{d-1}(1-u)^{e-1}$ $\times {}_2F_1(b, \frac{d+e}{2}; c; xu^2)$	$\frac{\Gamma(d)\Gamma(e)}{\Gamma(d+e)}\,{}_3F_2\left[\begin{matrix}b, \frac{d}{2}, \frac{d+1}{2};\\ c, \frac{d+e+1}{2};\end{matrix} x\right]$ $d, e > 0\ \lvert x\rvert < 1$
A.1.1.52	$u^{a-1}(1-u)^{b-1}$ $\times {}_2F_1\left[\begin{matrix}\frac{a+b}{2}, \frac{a+b+1}{2};\\ c\quad;\end{matrix} xu^2\right]$	$\frac{\Gamma(a)\Gamma(b)}{\Gamma(a+b)}\,{}_2F_1\left(\begin{matrix}a/2, a/2+1/2;\\ c\quad;\end{matrix} x\right)$ $a, b > 0\quad \lvert x\rvert < 1$
A.1.1.53	$u^{a-1}(1-u)^{b-1}$ $\times {}_2F_1\left[\begin{matrix}\frac{a+b}{2}, \frac{a+b+1}{2};\\ \frac{a+1}{2}\quad;\end{matrix} xu^2\right]$	$\frac{\Gamma(a)\Gamma(b)}{\Gamma(a+b)}(1-x)^{-a/2}$ $a, b > 0\quad \lvert x\rvert < 1$
A.1.1.54	$u^{a-1}(1-u)^{b-1}$ $\times {}_2F_1\left[\begin{matrix}\frac{a+b}{2}, \frac{a+b+1}{2};\\ 1/2\quad;\end{matrix} x^2u^2\right]$	$\frac{\Gamma(a)\Gamma(b)}{2\Gamma(a+b)}[(1+x)^{-a}+(1-x)^{-a}]$ $a, b > 0\quad \lvert x\rvert < 1$
A.1.1.55	$u^{a-1}(1-u)^{b-1}$ $\times {}_2F_1\left[\begin{matrix}\frac{a+b}{2}, \frac{a+b+1}{2};\\ 3/2\quad;\end{matrix} x^2u^2\right]$	$\frac{\Gamma(a)\Gamma(b)}{\Gamma(a+b)\ 2x(1-a)}$ $\times[(1+x)^{1-a}-(1-x)^{1-a}]$ $a, b > 0\quad \lvert x\rvert < 1$
A.1.1.56	$u^{-1/2}(1-u)^{b-1}$ $\times {}_2F_1\left[\begin{matrix}\frac{1+2b}{4}, \frac{3+2b}{4};\\ 1/2\quad;\end{matrix} -x^2u^2\right]$ $b > 0\quad \lvert x\rvert < 1$	$\frac{\Gamma(b)\sqrt{\pi}}{2\Gamma(1/2+b)(1+x^2)}$ $\times\{[\sqrt{(1+x^2)}+x]^{-1/2}$ $+[\sqrt{(1+x^2)}-x]^{-1/2}\}$

	$f(u)$	$\int_0^1 f(u)\,du$
A.1.1.57	$u^{a-1}(1-u)^{b-1}$ $\times {}_C F_D((c);(d);xu^m)$ $a,b>0 \quad \lvert x\rvert<1$	$\frac{\Gamma(a)\Gamma(b)}{\Gamma(a+b)}$ $\times {}_{C+m}F_{D+m}\left[\begin{matrix}(c),\frac{a}{m},..,\frac{a+m-1}{m}; \\ (d),\frac{a+b}{m},..,\frac{a+b+m-1}{m};\end{matrix}\; x\right]$
A.1.1.58	$u^{1/2-a}(1-u)^{a-1/2}$ $\times {}_2F_1\left[\begin{matrix}a,\ 1\ ; \\ 3/2-a;\end{matrix}\; 4u(1-u)x^2\right]$	$\frac{\Gamma(\frac{3}{2}-a)\Gamma(a+\frac{1}{2})}{2x(1-2a)}$ $\times\ [(1+x)^{1-2a}-(1-x)^{1-2a}]$ $-1/2<a<3/2 \quad \lvert x\rvert>1$
A.1.1.59	$u^{-1/2+3a}(1-u)^{a-1/2}$ $\times {}_2F_1\left[\begin{matrix}a,2a+\frac{1}{2}; \\ 3a+\frac{1}{2}\ ;\end{matrix}\; 4u(1-u)x\right]$	$\frac{\Gamma(\frac{1}{2}+3a)\Gamma(\frac{1}{2}+a)}{\Gamma(1+4a)}2^{2a}[1+\sqrt{(1-x)}]^{-2a}$ $a>1/2 \quad \lvert x\rvert<1$
A.1.1.60	$u^{d-1}(1-u)^{e-1}$ $\times {}_2F_1(a,b;c;u[1-u]x)$ $d,e>0 \quad \lvert x\rvert<4$	$\frac{\Gamma(d)\Gamma(e)}{\Gamma(d+e)}$ $\times {}_4F_3\left[\begin{matrix}a\ ,b\ ,d\ ,e\ ; \\ c,\frac{d+e}{2},\frac{d+e+1}{2};\end{matrix}\; x/4\right]$
A.1.1.61	$u^{d-1}(1-u)^{e-1}$ $\times {}_0F_3(-;a,b,c;u[1-u]x)$ $d,e>0$	$\frac{\Gamma(d)\Gamma(e)}{\Gamma(d+e)}$ $\times {}_2F_5\left[\begin{matrix}d,\ e\ ; \\ \frac{d+e}{2},\frac{d+e+1}{2},a,b,c;\end{matrix}\; x/4\right]$
A.1.1.62	$u^{d-1}(1-u)^{e-1}$ $\times {}_2F_1(a,b;d;u[1-u]x)$ $d,e>0 \quad \lvert x\rvert<4$	$\frac{\Gamma(d)\Gamma(e)}{\Gamma(d+e)}{}_3F_2\left[\begin{matrix}a\ ,\ b\ ,\ e; \\ \frac{d+e}{2},\frac{d+e+1}{2};\end{matrix}\; \frac{x}{4}\right]$
A.1.1.63	$u^{d-1}(1-u)^{e-1}$ $\times {}_2F_1\left[\begin{matrix}a,\frac{d+e}{2}; \\ d\ ;\end{matrix}\; u(1-u)x\right]$	$\frac{\Gamma(d)\Gamma(e)}{\Gamma(d+e)}{}_2F_1\left[\begin{matrix}a,e\ ; \\ \frac{d+e+1}{2};\end{matrix}\; \frac{x}{4}\right]$ $d,e>0 \quad \lvert x\rvert<4$

	$f(u)$	$\int_0^1 f(u)\,du$
A.1.1.64	$u^{a-1}(1-u)^{b-1} \times {}_2F_1\left[\begin{matrix}\frac{a+b+1}{2}, \frac{a+b}{2}; \\ a \quad ; \\ u(1-u)x\end{matrix}\right]$	$\frac{\Gamma(a)\Gamma(b)}{\Gamma(a+b)}(1-x/4)^{-b}$ $a,b > 0 \quad \|x\| < 4$
A.1.1.65	$u^{a-1}(1-u)^{b-1} \times {}_2F_1\left[\begin{matrix}N+a-\frac{1}{2}, -N; \\ a \quad ; \\ 4u(1-u)\end{matrix}\right]$	$\frac{\Gamma(a)\Gamma(b)\,(\frac{a-b}{2},N)\,(\frac{a-b+1}{2},N)}{\Gamma(a+b)\,(\frac{a+b}{2},N)\,(\frac{a+b+1}{2},N)}$ $a,b > 0$
A.1.1.66	$u^{a-1}(1-u)^{b-1} \times {}_2F_1\left[\begin{matrix}\frac{a+3b-3}{2}, \frac{a+b}{2}; \\ a \quad ; \\ 4u(1-u)\end{matrix}\right]$	$\frac{\Gamma(a)\Gamma(b)}{\Gamma(a+b)}2^{(3-3b-a)/2}$ $\times \{[\Gamma(\frac{3b+a-3}{4})\Gamma(\frac{a-b+3}{4})]^{-1}$ $-[\Gamma(\frac{a+3b-1}{4})\Gamma(\frac{a-b+1}{4})]^{-1}\}$ $a,b > 0$
A.1.1.67	$u^{a-1}(1-u)^{b-1} \times {}_2F_1\left[\begin{matrix}\frac{a+3b-1}{2}, \frac{a+b}{2}; \\ a \quad ; \\ 4u(1-u)\end{matrix}\right]$	$\frac{\Gamma(a)\Gamma(b)\Gamma(\frac{a+b+1}{2})\sqrt{\pi}\;2^{(1-a-3b)/2}}{\Gamma(a+b)\Gamma(\frac{3+a-b}{4})\Gamma(\frac{a+3b+1}{4})}$ $a > 0,\ b > 1$
A.1.1.68	$u^{a-1}(1-u)^{b-1} \times {}_2F_1(c,a+b;a;\ 4u[1-u])$	$\frac{\Gamma(a)\Gamma(b)\Gamma(\frac{a+b+1}{2})\Gamma(\frac{a-c+1}{2})}{\Gamma(a+b)\Gamma(\frac{a+b+1}{2}-c)\Gamma(\frac{a-b+1}{2})}$ $a,b > 0 \quad a-b-2c > -1$
A.1.1.69	$u^{3a-5/2}(1-u)^{a-1/2} \times {}_2F_1\left[\begin{matrix}a, 2a-1/2; \\ 3a-3/2 \ ; \\ 4u(1-u)x\end{matrix}\right]$	$\frac{\Gamma(3a-3/2)\Gamma(a+1/2)\,2^{2a-1}}{\Gamma(4a-1)\sqrt{(1-x)}}$ $\times\ [1+\sqrt{(1-x)}]^{1-2a}$ $a > 1/2 \quad \|x\| < 1$

	$f(u)$	$\int_0^1 f(u)\,du$
A.1.1.70	$u^{a-1}(1-u)^{b-1}$ $\times {}_2F_1(b,[a+3b-1]/2;a;$ $4u[1-u])$ $a,b > 0 \quad a-5b > -1$	$\frac{\Gamma(a)\Gamma(b)\Gamma(\frac{2+a+3b}{4})\Gamma(\frac{a+b}{2})\Gamma(\frac{a+b+1}{2})}{\Gamma(a+b)\Gamma(\frac{1+a+3b}{2})\Gamma(\frac{1+a-b}{4})\Gamma(\frac{3+a-b}{4})}$ $\times\frac{\Gamma(\frac{1+a-5b}{4})}{\Gamma(\frac{a-b}{2})}$
A.1.1.71	$u^{b-1/2}(1-u)^{b-1}$ $\times {}_2F_1(a,1-a;1/2+b;$ $4u[1-u])$	$\frac{\pi 2^{1-2b}\Gamma(b)\Gamma(b+1/4)\Gamma(b+3/4)}{\Gamma(\frac{1}{2}+2b)\Gamma(\frac{a+b}{2}+\frac{3}{8})\Gamma(\frac{a+b}{2}+\frac{1}{8})}$ $\times 1/\Gamma(b/2-a/2+5/8) \qquad b > 0$
A.1.1.72	$u^{c-1}(1-u)^{1/2-c}$ $\times {}_2F_1(a,1-a;c;$ $u\sin^2 x)$	$\frac{\Gamma(c)\Gamma(3/2-c)\sin[(2a-1)x]}{2\sqrt{\pi}(2a-1)\ \sin x}$ $0 < c < 3/2$
A.1.1.73	$u^{c-1}(1-u)^{1/2-c}$ $\times {}_2F_1(a,2-a;c;$ $u\sin^2 x)$	$\frac{\Gamma(c)\Gamma(3/2-c)\sin[(2a-2)x]}{2\sqrt{\pi}(a-1)\ \sin(2x)}$ $0 < c < 3/2$
A.1.1.74	$u^{c-1}(1-u)^{-1/2-c}$ $\times {}_2F_1(a,-a;c;$ $u\sin^2 x)$	$\frac{\Gamma(c)\Gamma(1/2-c)\ \cos(2ax)}{\sqrt{\pi}}$ $0 < c < 1/2$
A.1.1.75	$u^{c-1}(1-u)^{-1/2-c}$ $\times {}_2F_1(a,1-a;1/2;$ $u\sin^2 x)$	$\frac{\Gamma(c)\Gamma(1/2-c)\ \cos[(2a-1)x]}{\sqrt{\pi}\ \cos x}$ $0 < c < 1/2$
A.1.1.76	$u^{c-1}(1-u)^{-1/2-c}$ $\times {}_2F_1(a,1/2+a;c;$ $-u\tan^2 x)$	$\frac{\Gamma(c)\Gamma(1/2-c)}{\sqrt{\pi}}(\cos x)^{2a}\cos(2ax)$ $0 < c < 1/2 \quad \lvert\tan x\rvert < 1$
A.1.1.77	$u^{a}(1-u)^{-a}$ $\times {}_2F_1(a,1;1+a;$ $4u[1-u]\sin^2 x)$	$\frac{\Gamma(1+a)\Gamma(1-a)\sin[(2a-1)x]}{(2a-1)\ \sin x}$ $-1 < a < 1$

	$f(u)$	$\int_0^1 f(u)\,du$
A.1.1.78	$u^{1/2}(1-u)^{b-1} \times {}_2F_1\left[\begin{matrix}\frac{3+2b}{4},\frac{5+2b}{4}; \\ 3/2 \quad ;\end{matrix} u^2\sin^2 x\right]$	$-\frac{2\Gamma(3/2)\Gamma(b)\sin(x/2)}{\Gamma(3/2+b)\sin(2x)}$ $b > 0$
A.1.1.79	$u^{-1/2}(1-u)^{b-1} \times {}_2F_1\left[\begin{matrix}\frac{1+2b}{4},\frac{3+2b}{4}; \\ 3/2 \quad ;\end{matrix} u^2\sin^2 x\right]$	$\frac{2\sqrt{\pi}\Gamma(b)\sin(x/2)}{\Gamma(1/2+b)\sin(2x)}$ $b > 0$
A.1.1.80	$u^{a-1}(1-u)^{b-1} \times {}_1F_2(a+b;c,a;ux)$	$\frac{\Gamma(a)\Gamma(b)}{\Gamma(a+b)}\ {}_0F_1(-;c;x)$ $a,b > 0$
A.1.1.81	$u^{c-1}(1-u)^{d-1} \times {}_1F_2(a;c,b;ux)$	$\frac{\Gamma(c)\Gamma(d)}{\Gamma(c+d)}\ {}_1F_2(a;c+d,b;x)$ $c,d > 0$
A.1.1.82	$u^{d-1}(1-u)^{e-1} \times {}_1F_2(a;b,c;ux)$	$\frac{\Gamma(d)\Gamma(e)}{\Gamma(d+e)}\ {}_2F_3\left(\begin{matrix}d,a \quad ; \\ d+e,b,c;\end{matrix} x\right)$ $d,e > 0$
A.1.1.83	$u^{b}(1-u)^{b-1} \times {}_1F_2(a;b+1,b;u[1-u]x)$	$\frac{\Gamma(b+1)\Gamma(b)}{\Gamma(2b+1)}\ {}_0F_1(-;b+1/2;x/4)$ $b > 0$
A.1.1.84	$u^{b-1}(1-u)^{c-1} \times {}_1F_2(a;b,c;u[1-u]x)$	$\frac{\Gamma(b)\Gamma(c)}{\Gamma(b+c)}\ {}_1F_2(a;\frac{b+c}{2},\frac{b+c+1}{2};x/4)$ $b,c > 0$

	$f(u)$	$\int_0^1 f(u)\, du$
A.1.1.85	$u^{c-1}(1-u)^{d-1}$ $\times {}_1F_2(a;b,d;u[1-u]x)$	$\frac{\Gamma(c)\Gamma(d)}{\Gamma(c+d)}\, {}_2F_3\left[\begin{matrix} c, & a & ; \\ \frac{c+d}{2}, & \frac{c+d+1}{2}, b: \end{matrix} x/4\right]$ $c,d > 0$
A.1.1.86	$u^{d-1}(1-u)^{e-1}$ $\times {}_1F_2(a;b,c;u[1-u]x)$	$\frac{\Gamma(d)\Gamma(e)}{\Gamma(d+e)}\, {}_3F_4\left[\begin{matrix} d, & e, & a & ; \\ \frac{d+e}{2}, & \frac{d+e+1}{2}, & b,c; \end{matrix} \frac{x}{4}\right]$ $d,e > 0$
A.1.1.87	$u^{a-1}(1-u)^{b-1}$ $\times {}_1F_2(\frac{a+b}{2};\frac{a}{2},\frac{a+1}{2};u^2x)$	$\frac{\Gamma(a)\Gamma(b)}{\Gamma(a+b)}\, {}_0F_1(-;[a+b+1]/2;x)$ $a,b > 0$
A.1.1.88	$u^{a-1}(1-u)^{b-1}$ $\times {}_1F_2(\frac{a+b}{2};\frac{a}{2},c;u^2x)$	$\frac{\Gamma(a)\Gamma(b)}{\Gamma(a+b)}\, {}_1F_2(\frac{a+1}{2};\frac{a+b+1}{2},c;x)$ $a,b > 0$
A.1.1.89	$u^{d-1}(1-u)^{e-1}$ $\times {}_1F_2(\frac{d+e}{2};b,c;u^2x)$	$\frac{\Gamma(d)\Gamma(e)}{\Gamma(d+e)}\, {}_2F_3\left[\begin{matrix} \frac{d}{2}, & \frac{d+1}{2} & ; \\ \frac{d+e+1}{2}, & b,c; & \end{matrix} x\right]$ $d,e > 0$
A.1.1.90	$u^{d-1}(1-u)^{e-1}$ $\times {}_1F_2(a;b,c;u^2x)$	$\frac{\Gamma(d)\Gamma(e)}{\Gamma(d+e)}\, {}_3F_4\left[\begin{matrix} \frac{d}{2}, & \frac{d+1}{2}, & a & ; \\ \frac{d+e}{2}, & \frac{d+e+1}{2}, & b,c; & \end{matrix} x\right]$ $d,e > 0$
A.1.1.91	$u^{d-1}(1-u)^{c-d-1}$ $\times {}_3F_2\left(\begin{matrix} a,b,c & ; \\ a-b+2,d; & \end{matrix} -u\right)$ $d,c-d > 0;\ b < 3$ unless $a(b)=0,-1,-2,..$	$\frac{\Gamma(d)\Gamma(c-d)\Gamma(a-b+2)\sqrt{\pi}}{\Gamma(c)(b-1)2^a}$ $\times \{[\Gamma(\frac{a}{2})\Gamma(\frac{3+a-2b}{2})]^{-1}$ $- [\Gamma(\frac{1+a}{2})\Gamma(\frac{2+a-2b}{2})]^{-1}\}$
A.1.1.92	$u^{d-1}(1-u)^{c-d-1}$ $\times {}_3F_2\left(\begin{matrix} a, & b, & c & ; \\ [a+b+2]/2, & d; & \end{matrix} \frac{u}{2}\right)$ $d,c-d > 0$	$\frac{2\sqrt{\pi}\Gamma(d)\Gamma(c-d)\Gamma(\frac{a+b+2}{2})}{\Gamma(c)\Gamma(a-b)}$ $\times\{[\Gamma(a/2)\Gamma(a/2+1/2)]^{-1}$ $-[\Gamma(b/2)\Gamma(a/2+1/2)]^{-1}\}$

	$f(u)$	$\int_0^1 f(u)\,du$
A.1.1.93	$u^{e-1}(1-u)^{c-e-1} \times {}_3F_2\left({a,b,c; \atop d,e\ ;}u\right)$	$\frac{\Gamma(e)\Gamma(c-e)\Gamma(d)\Gamma(d-a-b)}{\Gamma(c)\Gamma(d-a)\Gamma(d-b)}$ $e, c-e > 0;\ d-a-b > 0$ unless $a(b) = 0,-1,-2,..$
A.1.1.94	$u^{d-1}(1-u)^{a+b-c-d-N} \times {}_3F_2\left({a,b,-N; \atop c,d\ ;}u\right)$	$\frac{\Gamma(d)\Gamma(a+b-c-d-N+1)(c-a,N)(c-b,N)}{\Gamma(a+b-c-N+1)(c,N)(c-a-b,N)}$ $d > 0 \quad a+b-c-d-N > -1$
A.1.1.95	$u^{b-1}(1-u)^{a-2b} \times {}_3F_2\left[{a, 1+\frac{a}{2}, -N; \atop \frac{a}{2}, 2+2b-N;}u\right]$	$\frac{\Gamma(b)\Gamma(1+a-2b)(a-2b-1,N)(-b-1,N)}{\Gamma(1+a-b+N)(a/2-1/2-b,N)}$ $\times \frac{(a/2+1/2-b,N)}{(-2b-1,N)} \quad b > 0, a-2b > -1$
A.1.1.96	$u^{e-1}(1-u)^{2c-d-e} \times {}_3F_2\left({a,1-a,c; \atop d,e\ ;}u\right)$	$\frac{2^{1-2c}\pi\Gamma(e)\Gamma(d)\Gamma(1+2c-d-e)}{\Gamma(\frac{a}{2}+\frac{1}{2}+c-\frac{d}{2})\Gamma(\frac{a+d}{2})\Gamma(d-c-\frac{1}{2})}$ $\times\ 1/\Gamma(b/2+d/2) \quad e > 0;$ $2c-d-e > -1$ and $c > 0$ unless $c=0,-1,-2,...$
A.1.1.97	$u^{d-1}(1-u)^{c-d-1} \times {}_3F_2\left({a,b,c \atop [a+b+1]/2,d;}\frac{u}{2}\right)$	$\frac{\Gamma(d)\Gamma(c-d)\Gamma(\frac{1+a+b}{2})\sqrt{\pi}}{\Gamma(c)\Gamma(\frac{1+a}{2})\Gamma(\frac{1+b}{2})} \quad d, c-d > 0$
A.1.1.98	$u^{c-1}(1-u)^{a-2c} \times {}_3F_2\left({a,1+a/2,b; \atop a/2,1+a+b;}u\right)$	$\frac{\Gamma(c)\Gamma(1+a-2c)\Gamma(1/2+a/2)\Gamma(1+a-b)}{\Gamma(1+a)\Gamma(1/2+a/2-b)\Gamma(1/2+a/2-c)}$ $\times\frac{\Gamma(1/2+a/2-b-c)}{\Gamma(1+a-b-c)} \quad c, a-2c+1 > 0$ $a-2b-2c > -1$ unless $b=0,-1,..$
A.1.1.99	$u^{b-1}(1-u)^{a-2b} \times {}_3F_2\left({a,1+a/2,-N; \atop a/2,1+2b-N;}u\right)$	$\frac{\Gamma(b)\Gamma(1+a-2b)(a-2b,N)(-b,N)}{\Gamma(1+a-b+N)\ (-2b,N)}$ $b, a-2b+1 > 0$
A.1.1.100	$u^{a/2-1/2}(1-u)^{a/2-1/2} \times {}_3F_2\left({a/2,b+N,-N\ ; \atop b/2,b/2+1/2;}u\right)$	$\frac{\Gamma(\frac{1+a}{2})(b-a,N)}{\Gamma(1+\frac{a}{2})(b,N)2^a\sqrt{\pi}} \quad a > -1$

	$f(u)$	$\int_0^1 f(u)\, du$
A.1.1.101	$u^{c-1}(1-u)^{b-c-1} \times {}_3F_2\left({a,1+a/2,-N; \atop a/2,\ c\ ;}u\right)$	$\dfrac{\Gamma(c)\Gamma(b-c)\,(a-b,N)\,(b-a-1,N)}{\Gamma(b+N)\quad(1+a-b,N)}$ $c,b-c > 0$
A.1.1.102	$u^{c-1}(1-u)^{2b-c-N} \times {}_3F_2\left({a,b,-N\ ; \atop 1+a-b,c;}u\right)$	$\dfrac{\Gamma(c)\Gamma(1+2b-c-N)\,(a-2b,N)\,(-b,N)}{\Gamma(1+2b-N)\,(1+a-b,N)\,(-2b,N)}$ $\times\ \dfrac{(1+a/2-b,N)}{(a/2-b,N)}\quad c,2b-c-N+1 > 0$
A.1.1.103	$u^{d-1}(1-u)^{2c-d-1} \times {}_3F_2\left({a,\ \ b,\ \ \ c\ \ ; \atop 1/2+a/2+b/2,d;}u\right)$	$\dfrac{\Gamma(d)\Gamma(2c-d)\Gamma(c+\frac{1}{2})\,\Gamma(\frac{1}{2}+\frac{a}{2}+\frac{b}{2})\sqrt{\pi}}{\Gamma(2c)\Gamma(\frac{1}{2}+\frac{a}{2})\Gamma(\frac{1+b}{2})\Gamma(\frac{1-a}{2}+c)}$ $\times\dfrac{\Gamma(1/2-a/2-b/2+c)}{\Gamma(1/2-b/2+c)},\ d,2c-d > 0$ and $a+b+2c < 1/2$ unless $b=0,-1,-2,..$
A.1.1.104	$u^{c-1}(1-u)^{a-2c} \times {}_3F_2\left({a,1+a/2,b; \atop a/2,1+a-b;}-u\right)$	$\dfrac{\Gamma(c)\Gamma(1+a-2c)\Gamma(1+a-b)}{\Gamma(1+a)\Gamma(1+a-b-c)}$ $c,a-2c+1 > 0$, and $a-2b-2c > -2$ unless $b=0,-1,-2,...$
A.1.1.105	$u^{d-1}(1-u)^{b-d-1} \times {}_3F_2\left({a,1-a,b; \atop c,\ d\ \ ;}u/2\right)$	$\dfrac{\Gamma(d)\Gamma(b-d)\Gamma(c)\Gamma(1/2)\,2^{1-c}}{\Gamma(b)\Gamma(c/2+a/2)\Gamma(c/2-a/2+1/2)}$ $d,b-d > 0$
A.1.1.106	$u^{d-1}(1-u)^{c-d-1} \times {}_3F_2\left({a,b,c\ \ ; \atop 1+a-b,d;}-u\right)$	$\dfrac{\Gamma(d)\Gamma(c-d)\Gamma(1+a-b)\Gamma(1+a/2)}{\Gamma(c)\Gamma(1+a/2-b)\Gamma(1+a)}$ $a > 0$ unless $a(b)=0,-1,-2,..$ and $d,c-d > 0$
A.1.1.107	$u^{d-1}(1-u)^{c-d-1} \times {}_3F_2\left({a,a+1/2,c; \atop 1+2a,\ d\ \ ;}ux\right)$	$\dfrac{\Gamma(d)\Gamma(c-d)\,2^{2a}}{\Gamma(c)\,[1+\sqrt{(1-x)}]^{2a}}$ $d,c-d > 0\quad \lvert x\rvert < 1$
A.1.1.108	$u^{c-1}(1-u)^{b-c-1} \times {}_3F_2\left({a,1/2+a,b; \atop 2a,\ \ c\ \ ;}ux\right)$	$\dfrac{\Gamma(c)\Gamma(b-c)\,2^{2a-1}\,[1+\sqrt{(1-x)}]^{1-2a}}{\Gamma(b)\ \sqrt{(1-x)}}$ $c,b-c > 0\quad \lvert x\rvert < 1$

	$f(u)$	$\int_0^1 f(u)\,du$
A.1.1.109	$u^{c-1}(1-u)^{b-c-1}$ $\times {}_3F_2\left({a,1/2+a,b; \atop 3/2, \quad c \;;} ux^2\right)$	$\frac{\Gamma(c)\Gamma(b-c)}{\Gamma(b)2x(1-2a)}[(1+x)^{1-2a}-(1-x)^{1-2a}]$ $c, b-c > 0 \qquad \lvert x\rvert < 1$
A.1.1.110	$u^{c-1}(1-u)^{b-c-1}$ $\times {}_3F_2\left({-a,a,b; \atop 1/2,d \;;} -ux^2\right)$	$\frac{\Gamma(c)\Gamma(b-c)}{2\;\Gamma(b)}$ $\times\{[\sqrt{(1+x^2)}+x]^{2a}+[\sqrt{(1+x^2)}-x]^{2a}\}$ $c, b-c > 0 \qquad \lvert x\rvert < 1$
A.1.1.111	$u^{c-1}(1-u)^{b-c-1}$ $\times {}_3F_2\left({a,1-a,b; \atop 3/2, \;c \;;} u\sin^2 x\right)$	$\frac{\Gamma(c)\Gamma(b-c)}{\Gamma(b)} \frac{\sin[(2a-1)x]}{(2a-1)\sin x}$ $c, b-c > 0$
A.1.1.112	$u^{e-1}(1-u)^{c-e-1}$ $\times {}_3F_2\left({a,b,c; \atop d,e \;;} ux\right)$	$\frac{\Gamma(e)\Gamma(c-e)}{\Gamma(c)}\;{}_2F_1(a,b;d;x)$ $e, c-e > 0 \qquad \lvert x\rvert < 1$
A.1.1.113	$u^{e-1}(1-u)^{f-1}$ $\times {}_3F_2\left({a,b,c; \atop d,e \;;} ux\right)$	$\frac{\Gamma(e)\Gamma(f)}{\Gamma(e+f)}\;{}_3F_2\left({a,b,c; \atop d,e+f;} x\right)$ $e, f > 0 \quad \lvert x\rvert < 1$
A.1.1.114	$u^{f-1}(1-u)^{g-1}$ $\times {}_3F_2\left({a,b,c; \atop d,\; e;} ux\right)$	$\frac{\Gamma(f)\Gamma(g)}{\Gamma(f+g)}\;{}_4F_3\left({a,b,c,f; \atop d,e,f+g;} x\right)$ $f, g > 0 \qquad \lvert x\rvert < 1$
A.1.1.115	$u^{c-1}(1-u)^{d-1}$ $\times {}_0F_3(-;a,b,c;ux)$	$\frac{\Gamma(c)\Gamma(d)}{\Gamma(c+d)}\;{}_0F_3(-;a,b,c+d;x)$ $c, d > 0$
A.1.1.116	$u^{d-1}(1-u)^{e-1}$ $\times {}_0F_3(-;a,b,c;ux)$	$\frac{\Gamma(d)\Gamma(e)}{\Gamma(d+e)}\;{}_1F_4(d;d+e,a,b,c;x)$ $d, e > 0$
A.1.1.117	$u^{b-1}(1-u)^{c-1}$ $\times {}_0F_3(-;a,b,c;u[1-u]x)$	$\frac{\Gamma(b)\Gamma(c)}{\Gamma(b+c)}\;{}_0F_3(-;a,\frac{b+c}{2},\frac{b+c+1}{2};\frac{x}{4})$ $b, c > 0$
A.1.1.118	$u^{c-1}(1-u)^{d-1}$ $\times {}_0F_3(-;a,b,c;u[1-u]x)$	$\frac{\Gamma(c)\Gamma(d)}{\Gamma(c+d)}\;{}_1F_4(d;\frac{c+d}{2},\frac{c+d+1}{2},a,b;\frac{x}{4})$ $c, d > 0$

	$f(u)$	$\int_0^1 f(u)\,du$
A.1.1.119	$u^{e-1}(1-u)^{f-1}$ $\times {}_2F_3\left({a,e+f; \atop c,d,e;}ux\right)$	$\frac{\Gamma(e)\Gamma(f)}{\Gamma(e+f)}\,{}_1F_2(a;c,d;x)$ $e,f>0$
A.1.1.120	$u^{e-1}(1-u)^{f-1}$ $\times {}_2F_3(a,b;c,d,e;ux)$	$\frac{\Gamma(e)\Gamma(f)}{\Gamma(e+f)}\,{}_2F_3(a,b;c,d,e+f;x)$ $e,f>0$
A.1.1.121	$u^{f-1}(1-u)^{g-1}$ $\times {}_2F_3(a,b;c,d,e;ux)$	$\frac{\Gamma(f)\Gamma(g)}{\Gamma(f+g)}\,{}_3F_4\left({a,b,\ f\ ; \atop c,d,e,f+g;}x\right)$ $f,g>0$
A.1.1.122	$u^{a-1}(1-u)^{b-1}$ $\times {}_2F_3\left[{\frac{a+b}{2},\frac{a+b+1}{2}; \atop c,a,b\ ;}u[1-u]x\right]$	$\frac{\Gamma(a)\Gamma(b)}{\Gamma(a+b)}\,{}_0F_1(-;c;x/4)$ $a,b>0$
A.1.1.123	$u^{d-1}(1-u)^{e-1}$ $\times {}_2F_3(\frac{d+e}{2},b;c,d,e;$ $u[1-u]x)$	$\frac{\Gamma(d)\Gamma(e)}{\Gamma(d+e)}\,{}_1F_2(b;\frac{d+e+1}{2},c;x/4)$ $d,e>0$
A.1.1.124	$u^{d-1}(1-u)^{e-1}$ $\times {}_2F_3(a,b;c,d,e;$ $u[1-u]x)$	$\frac{\Gamma(d)\Gamma(e)}{\Gamma(d+e)}\,{}_2F_3(a,b;\frac{d+e}{2},\frac{d+e+1}{2},c;\frac{x}{4})$ $d,e>0$
A.1.1.125	$u^{e-1}(1-u)^{f-1}$ $\times {}_2F_3(a,b;c,d,f;$ $u[1-u]x)$	$\frac{\Gamma(e)\Gamma(f)}{\Gamma(e+f)}\,{}_3F_4\left({e,\ a,\ b\ ; \atop \frac{e+f}{2},\frac{e+f+1}{2},c,d;}\frac{x}{4}\right)$ $e,f>0$
A.1.1.126	$u^{f-1}(1-u)^{g-1}$ $\times {}_2F_3(a,b;c,d,e;$ $u[1-u]x)$	$\frac{\Gamma(f)\Gamma(g)}{\Gamma(f+g)}\,{}_4F_5\left({f,\ g,\ a,\ b\ ; \atop \frac{f+g}{2},\frac{f+g+1}{2},c,d,e;}\frac{x}{4}\right)$ $f,g>0$
A.1.1.127	$u^{e-1}(1-u)^{c-e-1}$ $\times {}_4F_3\left({a,b,-N,c\ ; \atop d,1+a+b-d-N,e;}u\right)$	$\frac{\Gamma(e)\Gamma(c-e)\,(d-a,N)\,(d-b,N)}{\Gamma(c)\,(d,N)\,(d-a-b,N)}$ $e,c-e>0$

	$f(u)$	$\int_0^1 f(u)\,du$
A.1.1.128	$u^{d-1}(1-u)^{c-d-1} \times {}_4F_3\left(\begin{matrix} a, 1+\frac{a}{2}, -N, c; \\ \frac{a}{2}, b, c; \end{matrix} u\right)$	$\frac{\Gamma(d)\Gamma(c-d)(a-b,N)(b-a-1,N)}{\Gamma(c)\ (b,N)(1+a-b,N)}$ $d, c-d > 0$
A.1.1.129	$u^{d-1}(1-u)^{a-c-d} \times {}_4F_3\left(\begin{matrix} a, 1+\frac{a}{2}, b, c \\ \frac{a}{2}, 1+a-b, d; \end{matrix} -u\right)$	$\frac{\Gamma(d)\Gamma(1+a-c-d)\Gamma(1+a-b)}{\Gamma(1+a)\Gamma(1+a-b-c)}$ $d, 1+a-c-d > 0;\ a-2b-2c > -2$ unless $a(b,c)=0,-1,-2,\ldots$
A.1.1.130	$u^{c-1}(1-u)^{1+2b-c-N} \times {}_4F_3\left(\begin{matrix} a, 1+\frac{a}{2}, b, -N; \\ \frac{a}{2}, 1+a-b, c; \end{matrix} u\right)$ $c, 2b-c-N+2 > 0$	$\frac{\Gamma(c)\Gamma(2+2b-c-N)(a-2b-1,N)}{\Gamma(2+2b-N)(1+a-b,N)(-2b-1,N)} \times \frac{(-b-1,N)(\frac{a+1}{2}-b,N)}{(\frac{a-1}{2}-b,N)}$
A.1.1.131	$u^{e-1}(1-u)^{d-e-1} \times {}_4F_3\left(\begin{matrix} a, b, c, d; \\ \frac{1+a+b}{2}, 2c, e; \end{matrix} u\right)$ $e, d-e > 0$ and $c-a/2-b/2 > -\frac{1}{2}$ unless $a(b,c)=0,-1,-2,\ldots$	$\frac{\Gamma(e)\Gamma(d-e)\Gamma(\frac{1}{2})\Gamma(c+\frac{1}{2})\Gamma(\frac{1+a+b}{2})}{\Gamma(d)\Gamma(\frac{1+a}{2})\Gamma(\frac{1+b}{2})\Gamma(\frac{1-a}{2}+c)} \times \frac{\Gamma(\frac{1-a-b}{2}+c)}{\Gamma(\frac{1-b}{2}+c)}$
A.1.1.132	$u^{c-1}(1-u)^{a-c} \times {}_4F_3\left(\begin{matrix} a/2, 1/2+a/2, b+N, -N; \\ b/2, b/2+1/2, b; \end{matrix} u\right)$	$\frac{\Gamma(c)\Gamma(1+a-c)(b-a,N)}{\Gamma(1+a)(b,N)}$ $c, a-c > 0$
A.1.1.133	$u^{c-1}(1-u)^{b-c-1} \times {}_4F_3\left(\begin{matrix} a+e, 1+\frac{a+e}{2}, \frac{e}{2}, b; \\ 1+a+\frac{e}{2}, 1/2+e/2, d; \end{matrix} ux\right)$	$\frac{\Gamma(c)\Gamma(b-c)}{\Gamma(b)\ (1-x)^e}\ {}_2F_1\left(\begin{matrix} a, -e/2; \\ 1+a+e/2; \end{matrix} x\right)$ $c, b-c > 0 \quad \lvert x\rvert < 1$

	$f(u)$	$\int_0^1 f(u)\,du$
A.1.1.134	$u^{g-1}(1-u)^{d-g-1}$ $\times {}_4F_3\left({a,b,c,d; \atop e,f,g\ ;}ux\right)$	$\frac{\Gamma(g)\Gamma(d-g)}{\Gamma(d)}\ {}_3F_2\left({a,b,c; \atop e,f\ ;}x\right)$ $g,d-g > 0 \quad \lvert x\rvert < 1$
A.1.1.135	$u^{g-1}(1-u)^{h-1}$ $\times {}_4F_3\left({a,b,c,d; \atop e,f,g\ ;}ux\right)$	$\frac{\Gamma(g)\Gamma(h)}{\Gamma(g+h)}\ {}_4F_3\left({a,b,c,d; \atop e,f,g+h;}x\right)$ $g,h > 0 \quad \lvert x\rvert < 1$
A.1.1.136	$u^{h-1}(1-u)^{j-i}$ $\times {}_4F_3\left({a,b,c,d; \atop e,f,g\ ;}ux\right)$	$\frac{\Gamma(h)\Gamma(j)}{\Gamma(h+j)}\ {}_5F_4\left({a,b,c,d,h; \atop e,f,g,h+j;}x\right)$ $h,j > 0 \quad \lvert x\rvert < 1$
A.1.1.137	$u^{d-1}(1-u)^{c-d-1}$ $\times {}_5F_4\left({a/2,a/2+1/2,\ b+N,-N,c; \atop b/2,b/2+1/2\ \ 1+a,\ \ d\ ;}u\right)$	$\frac{\Gamma(d)\Gamma(c-d)(b-a,N)}{\Gamma(c)\ (b,N)}$ $d,c-d > 0$
A.1.1.138	$u^{d-1}(1-u)^{c-d-1}$ $\times {}_5F_4\left({a,1+a/2,b,\ -N,c\ ; \atop a/2,1+a-b,\ 2+2b-N,d;}u\right)$	$\frac{\Gamma(d)\Gamma(c-d)(a-2b-1,N)(-b-1,N)}{\Gamma(c)(1+a-b,N)(-2b-1,N)}$ $\times\frac{(a/2+1/2-b,N)}{(a/2-1/2-b,N)}$ $d,c-d > 0$
A.1.1.139	$u^{d-1}(1-u)^{c-d-1}$ $\times {}_5F_4\left({a,1+a/2,b,\ -N,c\ \ ; \atop a/2,1+a-b,\ 1+2b-N,d;}u\right)$	$\frac{\Gamma(d)\Gamma(c-d)(a-2b,N)(-b,N)}{\Gamma(c)(1+a-b,N)(-2b,N)}$ $d,c-d > 0$
A.1.1.140	$u^{j-1}(1-u)^{e-j-1}$ $\times {}_5F_4\left({a,b,c,d,e; \atop f,g,h,j\ ;}ux\right)$	$\frac{\Gamma(j)\Gamma(e-j)}{\Gamma(e)}\ {}_4F_3\left({a,b,c,d; \atop f,g,h\ ;}x\right)$ $j,e-j > 0 \quad \lvert x\rvert < 1$
A.1.1.141	$u^{j-1}(1-u)^{k-1}$ $\times {}_5F_4\left({a,b,c,d,e; \atop f,g,h,j\ ;}ux\right)$	$\frac{\Gamma(j)\Gamma(k)}{\Gamma(j+k)}\ {}_5F_4\left({a,b,c,d,e; \atop f,g,h,j+k;}x\right)$ $j,k > 0 \quad \lvert x\rvert < 1$
A.1.1.142	$u^{k-1}(1-u)^{p-1}$ $\times {}_5F_4\left({a,b,c,d,e; \atop f,g,h,j\ ;}ux\right)$	$\frac{\Gamma(k)\Gamma(p)}{\Gamma(k+p)}\ {}_6F_5\left({a,b,c,d,e,k; \atop f,g,h,j,k+p;}x\right)$ $k,p > 0 \quad \lvert x\rvert < 1$

	$f(u)$	$\int_0^1 f(u)\,du$
A.1.1.143	$u^{e-1}(1-u)^{a-2e} \times {}_6F_5\left(\begin{matrix} a,1+a/2,b,\ e\ ,\ 1+2a-b-c-e+N,-N; \\ a/2,1+a-b,1+a-c,\ b+c+e-a-N,1+a+N; \end{matrix} u\right)$	$\dfrac{\Gamma(e)\Gamma(1+a-2e)(1+a,N)(1+a-b-c,N)}{\Gamma(1+a-e)(1+a-b,N)(1+a-c,N)} \times \dfrac{(1+a-b-e,N)(1+a-c-e,N)}{(1+a-e,N)(1+a-b-c-e,N)}$ $e,a-2e+1 > 0$
A.1.1.144	$u^{f-1}(1-u)^{e-f-1} \times {}_6F_5\left(\begin{matrix} a,1+a/2,b,\ c\ ,\ d,\ e\ ; \\ a/2,1+a-b,1+a-c,\ 1+a-d,f; \end{matrix} u\right)$	$\dfrac{\Gamma(f)\Gamma(e-f)\Gamma(1+a-b)\Gamma(1+a-c)}{\Gamma(e)\Gamma(1+a)\Gamma(1+a-b-c)\Gamma(1+a-b-d)} \times \dfrac{\Gamma(1+a-d)\Gamma(1+a-b-c-d)}{\Gamma(1+a-c-d)}$ $f,e-f > 0;\ \ a-b-c-d+1 > 0$ unless $a(b,c,d)=0,-1,-2,..$
A.1.1.145	$u^{a-d}(1-u)^{a+d-e-1} \times {}_6F_5\left(\begin{matrix} a,1+a/2,b,1-b,\ d,\ 1-d\ ; \\ a/2,1+a-b,a+b,\ 1+a-d,e; \end{matrix} u\right)$	$\dfrac{\Gamma(e)\Gamma(a+b-e)\Gamma(1+a-b)\Gamma(a+b)}{\Gamma(a)\Gamma(1+a)\Gamma(a+b-d)\Gamma(2+a-b-d)} \times \dfrac{\Gamma(1+a-d)\Gamma(a/2)\Gamma(a/2)}{\Gamma(1+a/2-b/2-d/2)\Gamma(a/2+b/2+d/2)}$ $a-d+1,a+d-e > 0;\ a > 0$ unless $a(b,d)=0,-1,-2,..$
A.1.1.146	$u^{2a-d+N}(1-u)^{d-a-N-1} \times {}_6F_5\left(\begin{matrix} a,1+a/2,d/2,a-d,\ 1/2+d/2,-N\ ; \\ a/2,1+a-d/2,1+d,\ a+1/2-d/2,1+a+N; \end{matrix} u\right)$	$\dfrac{\Gamma(1+2a-d)\Gamma(d-a-N)\Gamma(1+a+N)}{(1+a-d,N)} \times (1+2a-2d,N)$ $2a-d+N-1,\ d-a-N > 0$
A.1.1.147	$u^{c-1}(1-u)^{d-1} \times {}_AF_B(\ (a);(b);ux)$	$\dfrac{\Gamma(c)\Gamma(d)}{\Gamma(c+d)}\ {}_{A+1}F_{B+1}\left(\begin{matrix} c,(a)\ ; \\ c+d,(b); \end{matrix} x\right)$ $c,d > 0$
A.1.1.148	$u^{c-1}(1-u)^{d-1} \times {}_AF_B\left(\begin{matrix}(a); \\ (b);\end{matrix} u^m[1-u]^n x\right)$ $c,d > 0$. If $p=q+1$, $\lvert x\rvert < 1$. No restriction on x if $p \leq q$. No restriction on x,p,q if an 'a' parameter is a non-positive integer.	$\dfrac{\Gamma(c)\Gamma(d)}{\Gamma(c+d)} \times {}_{A+m+n}F_{B+m+n}\left[\begin{matrix} (a),\frac{c}{m},..,\frac{c+m-1}{m},\ \frac{d}{n},..,\frac{d+n-1}{n}; \\ (b),\frac{c+d}{m+n},.....,\ \frac{c+d+m+n-1}{m+n}; \end{matrix} \frac{m^m n^n\, x}{(m+n)^{m+n}}\right]$

	$f(u)$	$\int_0^1 f(u)\,du$
A.1.1.149	$u^{c-1}(1-u)^{d-1}$ $\times G^{m,n}_{p,q}\left(ux\,\middle\vert\,\begin{matrix}(a)\\(b)\end{matrix}\right)$ $c,d > 0$	$\Gamma(d)$ $\times G^{m,n+1}_{p+1,q+1}\left[x\,\middle\vert\,\begin{matrix}a_1,..,a_n,1-c,a_{n+1},..,a_p\\b_1,..,b_m,1-c-d,b_{m+1},..,b_q\end{matrix}\right]$
A.1.1.150	$u^{c-1}(1-u)^{d-1}$ $\times G^{m,n}_{p,q}\left(u[1-u]x\,\middle\vert\,\begin{matrix}(a)\\(b)\end{matrix}\right)$ $c,d > 0$	$\sqrt{\pi}\,2^{1-c-d}$ $\times G^{m,n+2}_{p+2,q+2}\left[\frac{x}{4}\,\middle\vert\,\begin{matrix}a_1,..,a_n,1-c,1-d,a_{n+1},....,a_p\\b_1,..,b_m,1-c/2-d/2,1/2-c/2-d/2,b_{m+1},..,b_q\end{matrix}\right]$

A.1.2 Euler Integrals involving Double Hypergeometric Functions

	$f(u)$	$\int_0^1 f(u)\,du$
A.1.2.1	$u^{a-1}(1-u)^{b-1}$ $\times {}_2F_1\left(\begin{matrix}c,-N\,;\\c-b-N+1;\end{matrix}u\right)$ $\times {}_DF_E\left(\begin{matrix}(d);\\(e);\end{matrix}ux\right)$	$\frac{\Gamma(b+N)\Gamma(a)\ (a+b-c,N)}{\Gamma(a+b+N)\ (b-c,N)}$ $\times {}_DF_E\left(\begin{matrix}(d),a,a+b-c+N\,;\\(e),a+b-c,a+b+N;\end{matrix}x\right)$ $a,b > 0$, ${}_DF_E$ convergent over the range of integration
A.1.2.2	$u^{f-1}(1-u)^{g-1}$ $\times {}_2F_1\left(\begin{matrix}a,-N\,;\\a-g-N+1;\end{matrix}u\right)$ $\times G^{m,n}_{p,q}\left(ux\,\middle\vert\,\begin{matrix}(d)\\(e)\end{matrix}\right)$ $f,g > 0$	$\frac{\Gamma(f)\Gamma(g+N)\Gamma(g-a)}{\Gamma(f+g)\Gamma(g-a+N)}$ $\times G^{m,n+2}_{p+2,q+2}\left[x\,\middle\vert\,\begin{matrix}d_1,..,d_n,1-f-g,1+a-f-g-N,d_{n+1},..,d_p\\e_1,..,e_m,1-f-g-N,1+a-f-g,e_{m+1},....,e_q\end{matrix}\right]$

	$f(u)$	$\int_0^1 f(u)\,du$
A.1.2.3	$u^{a-1}(1-u)^{b-1} \times {}_2F_1\left({c,-N \atop c-b-N+1};u\right) \times F^{F:G;G'}_{H:K;K'}\left[{(f):(g);(g'); \atop (h):(k);(k');}ux,y\right]$	$\dfrac{\Gamma(b+N)\Gamma(a)(a+b-c,N)}{\Gamma(a+b+N)(b-c,N)} \times F^{F:G+2;G'}_{H:K+2;K'}\left[{(f):(g),a+b-c+n,a\ ;(g'); \atop (h):(k),a+b-c,a+b+n;(k');}x,y\right]$ a,b > 0; the series of the integrand must converge over the range of integration
A.1.2.4	$u^{a-1}(1-u)^{b-1} \times {}_2F_1\left({c,-N \atop c-b-N+1};u\right) \times F^{F:G;G'}_{H:K;K'}\left[{(f):(g);(g'); \atop (h):(k);(k');}ux,uy\right]$	$\dfrac{\Gamma(b+N)\Gamma(a)(a+b-c,N)}{\Gamma(a+b+N)(b-c,N)} \times F^{F+2:G;G'}_{H+2:K;K'}\left[{(f),a,a+b-c+N:(g);(g'); \atop (h),a+b-c,a+b+N:(k);(k');}x,y\right]$ a,b > 0; series cgt. over the range of integration
A.1.2.5	$u^{a-1}(1-u)^{b-1} \times {}_CF_D(\ (c);(d);ux) \times {}_EF_F(\ (e);(f);uy)$	$\dfrac{\Gamma(a)\Gamma(b)}{\Gamma(a+b)} F^{1:C;E}_{1;D;F}\left[{a\ :(c);(e); \atop a+b:(d);(f);}x,y\right]$ a,b > 0;series cgt. over the range of integration.
A.1.2.6	$u^{a-1}(1-u)^{b-1} \times {}_CF_D(\ (c);(d);ux) \times {}_EF_F((e);(f);[1-u]y)$ a,b > 0; series	$\dfrac{\Gamma(a)\Gamma(b)}{\Gamma(a+b)} F^{0:C+1;E+1}_{1:\ D\ ;\ F}\left[{-\ :(c),a;(e),b; \atop a+b:\ (d)\ ;(f)\ ;}x,y\right]$ cgt. over range of integration
A.1.2.7	$u^{a}(1-u)^{a-1} \times {}_CF_D((c);(d);xu[1-u]) \times {}_EF_F((e);(f);yu[1-u])$	$\dfrac{\Gamma(a)\Gamma(a+1)}{\Gamma(2a+1)} F^{1:C;E}_{1:D;F}\left[{a\ ;(c);(e); \atop a+1/2;(d);(f);}x/4,y/4\right]$ a > 0; series cgt. over the range of integration

	$f(u)$	$\int_0^1 f(u)\,du$
A.1.2.8	$u^{a-1}(1-u)^{b-1}$ $\times {}_CF_D((c);(d);xu[1-u])$ $\times {}_EF_F((e);(f);yu[1-u])$	$\frac{\Gamma(a)\Gamma(b)}{\Gamma(a+b)} F^{2:C;E}_{2:D;F}\left[\begin{matrix} a\ ,\ b & :(c); & (e); \\ \frac{a+b}{2},\frac{a+b+1}{2} & :(d); & (f); \end{matrix} x/4,y/4\right]$ a,b > 0; series cgt. over the range of integration
A.1.2.9	$u^{a-1}(1-u)^{b-1}$ $\times {}_CF_D(\ (c);(d);xu^m)$ $\times {}_EF_F(\ (e);(f);yu^m)$	$\frac{\Gamma(a)\Gamma(b)}{\Gamma(a+b)} F^{m:C;E}_{m:D;F}\left[\begin{matrix} \frac{a}{m},..,\frac{a+m-1}{m} & : & (c);(e); \\ \frac{a+b}{m},..,\frac{a+b+m-1}{m} & : & (d);(f); \end{matrix} x,y\right]$ a,b > 0; series cgt. over the range of integration
A.1.2.10	$u^{a-1}(1-u)^{b-1}$ $\times {}_CF_D((c);(d);xu^m[1-u]^n)$ $\times {}_EF_G((e);(f);yu^m[1-u]^n)$ a,b > 0;series cgt over the range of integration	$\frac{\Gamma(a)\Gamma(b)}{\Gamma(a+b)} F^{m+n:C;E}_{m+n:D;F}$ $\left[\begin{matrix} \frac{a}{m},..,\frac{a+m-1}{m},\frac{b}{n},..,\frac{b+n-1}{n} & ;(c); & (e); \\ \frac{a+b}{m+n},...,\frac{a+b+m+n-1}{m+n} & ;(d); & (f); \end{matrix} \frac{m^m x}{(m+n)^{m+n}},\frac{n^n y}{(m+n)^{m+n}}\right]$
A.1.2.11	$u^{a-1}(1-u)^{c-1}$ $\times\Phi_2(b,b';a;ux,uy)$	$\frac{\Gamma(a)\Gamma(c)}{\Gamma(a+c)}\ \Phi_2(b,b';a+c;x,y)$ a,c > 0
A.1.2.12	$u^{a-1}(1-u)^{d-1}$ $\times\Phi_2(b,b';c;ux,uy)$	$\frac{\Gamma(a)\Gamma(d)}{\Gamma(a+d)} F^{1:1;1}_{2:0;0}\left(\begin{matrix} a & :b;b'; \\ c,a+d: & -\ ; \end{matrix} x,y\right)$ a,d > 0
A.1.2.13	$u^{a}(1-u)^{a-1}$ $\times\Phi_2(a+\frac{1}{2},b;c;xu[1-u],y)$	$\frac{\Gamma(a+1)\Gamma(a)}{\Gamma(2a+1)}\ \Phi_2(a,b;c;x/4,y)$ a > 0
A.1.2.14	$u^{a-1}(1-u)^{d-1}$ $\times\Phi_2(\frac{a+d}{2},b;c;$ $xu[1-u],y)$ a,d > 0	$\frac{\Gamma(a)\Gamma(d)}{\Gamma(a+d)}$ $\times F^{0:2;1}_{1:1;0}\left[\begin{matrix} -:a,d & ;b; \\ c:\frac{a+d+1}{2} & ;-; \end{matrix} x/4,y\right]$

	$f(u)$	$\int_0^1 f(u)\,du$
A.1.2.15	$u^{a-1}(1-u)^{d-1}$ $\times\Phi_2(b,b';c;xu[1-u],y)$ $a,d>0$	$\frac{\Gamma(a)\Gamma(d)}{\Gamma(a+d)}F^{0:3;1}_{1:2;0}\left[\begin{matrix}-: a,\ b,\ d\ ; & b'; \\ c: \frac{a+d}{2}, \frac{a+d+1}{2}; & -\ ;\end{matrix}\ \frac{x}{4}, y\right]$
A.1.2.16	$u^{a-1}(1-u)^{a-1}$ $\times\Phi_1(a+1/2,b;a;$ $xu[1-u],yu[1-u])$	$\frac{\Gamma(a+1)\Gamma(a)}{\Gamma(2a+1)}e^{x/4}(1-y/4)^{-b}$ $a>0;\ \lvert y\rvert<4$ unless $b=0,-1,-2,..$
A.1.2.17	$u^{f-1}(1-u)^{g-1}$ $\times\Phi_1(a,b;c;$ $xu[1-u],yu[1-u])$ $f,g>0;\ \lvert y\rvert<4$ unless $a(b)=0,-1,..$	$\frac{\Gamma(f)\Gamma(g)}{\Gamma(f+g)}F^{3:0;1}_{3:0;0}\left[\begin{matrix}a,\ f,\ g\ :-; & b; \\ c, \frac{f+g}{2}, \frac{f+g+1}{2}:-; & -;\end{matrix}\ \frac{x}{4}, \frac{y}{4}\right]$
A.1.2.18	$u^{c}(1-u)^{c-1}$ $\times\Phi_2(b,b';c;xu[1-u],$ $yu[1-u])$	$\frac{\Gamma(c+1)\Gamma(c)}{\Gamma(2c+1)}$ $\times\Phi_2(b,b';c+\frac{1}{2};\frac{x}{4},\frac{y}{4})$ $c>0$
A.1.2.19	$u^{c-1}(1-u)^{d-1}$ $\times\Phi_2(b,b';d;xu[1-u],$ $yu[1-u])$ $c,d>0$	$\frac{\Gamma(c)\Gamma(d)}{\Gamma(c+d)}F^{1:1;1}_{2:0;0}\left[\begin{matrix}c\ : & b;b'; \\ \frac{c+d}{2}, \frac{c+d+1}{2}: & -;-\ ;\end{matrix}\ x/4, y/4\right]$
A.1.2.20	$u^{f-1}(1-u)^{g-1}$ $\times\Phi_2(b,b';c;xu[1-u],$ $yu[1-u])$ $f,g>0$	$\frac{\Gamma(f)\Gamma(g)}{\Gamma(f+g)}F^{2:1;1}_{3:0;0}\left[\begin{matrix}f,\ g\ : & b;b'; \\ c, \frac{f+g}{2}, \frac{f+g+1}{2}: & -;-\ ;\end{matrix}\ x/4, y/4\right]$
A.1.2.21	$u^{f-1}(1-u)^{g-1}$ $\times\Psi_2(a;f,c;ux,y)$	$\frac{\Gamma(f)\Gamma(g)}{\Gamma(f+g)}\Psi_2(a;f+g,c;x,y)$ $f,g>0$
A.1.2.22	$u^{f-1}(1-u)^{g-1}$ $\times\Psi_2(a;c,c';ux,y)$	$\frac{\Gamma(f)\Gamma(g)}{\Gamma(f+g)}F^{1:1;0}_{0:2;1}\left[\begin{matrix}a: & f\ ;-; \\ -: & f+g,c;c';\end{matrix}\ x, y\right]$ $f,g>0$
A.1.2.23	$u^{f-1}(1-u)^{g-1}$ $\times\Psi_2(f+g;c,c';ux,uy)$	$\frac{\Gamma(f)\Gamma(g)}{\Gamma(f+g)}\Psi_2(f;c,c';x,y)$ $f,g>0$

	$f(u)$	$\int_0^1 f(u)$
A.1.2.24	$u^{f-1}(1-u)^{g-1}$ $\times\Psi_2(a;c,c';ux,uy)$	$\frac{\Gamma(f)\Gamma(g)}{\Gamma(f+g)} F^{2:0;0}_{1:1;1}\left[\begin{matrix} a,f:-; & -; \\ f+g:c;c'; \end{matrix} x,y)\right]$ $f,g > 0$
A.1.2.25	$u^{f}(1-u)^{f-1}$ $\times\Psi_2(a;f,c;x[1-u]u,y)$	$\frac{\Gamma(f+1)\Gamma(f)}{\Gamma(2f+1)}\ \Psi_2(a;f+\frac{1}{2},c;\frac{x}{4},y)$ $f > 0$
A.1.2.26	$u^{f-1}(1-u)^{g-1}$ $\times\Psi_2(a;c,c';xu[1-u],y)$ $f,g > 0$	$\frac{\Gamma(f)\Gamma(g)}{\Gamma(f+g)} F^{1:2;0}_{0:3;1}\left[\begin{matrix} a: & f,\ g & ; & - ; \\ -:c, & \frac{f+g}{2}, & \frac{f+g+1}{2}; & c'; \end{matrix} x/4,y\right]$
A.1.2.27	$u^{f-1}(1-u)^{g-1}$ $\times\Psi_2(f+g;f,g;ux,$ $y[1-u])$	$\frac{\Gamma(f)\Gamma(g)}{\Gamma(f+g)}\ e^{x+y}$ $f,g > 0$
A.1.2.28	$u^{g}(1-u)^{g-1}$ $\times\Psi_2(g+1/2;c,c';$ $xu[1-u],yu[1-u])$	$\frac{\Gamma(g+1)\Gamma(g)}{\Gamma(2g+1)}\ \Psi_2(g;c,c';x/4,y/4)$ $g > 0$
A.1.2.29	$u^{f-1}(1-u)^{g-1}$ $\times\Psi_2(f/2+g/2;c,c';$ $xu[1-u],yu[1-u])$	$\frac{\Gamma(f)\Gamma(g)}{\Gamma(f+g)} F^{2:0;0}_{1:1;1}\left[\begin{matrix} f,g & :-;-; \\ \frac{f+g+1}{2} & :c;c'; \end{matrix} \frac{x}{4},\frac{y}{4}\right]$ $f,g > 0$
A.1.2.30	$u^{f-1}(1-u)^{g-1}$ $\times\Psi_2(a;c,c';xu[1-u],$ $yu[1-u])$	$\frac{\Gamma(f)\Gamma(g)}{\Gamma(f+g)} F^{3:0;0}_{2:1;1}\left[\begin{matrix} a,f,g & :-;- ; \\ \frac{f+g}{2},\frac{f+g+1}{2} & : c;c'; \end{matrix} x/4.y/4\right]$ $f,g > 0$
A.1.2.31	$u^{c-1}(1-u)^{1+a-b-b'-c}$ $\times F_1(a,b,b';c;u,-u)$ $0 < c < 2+a-b-b'$	$\frac{\Gamma(c)\Gamma(2+a-b-b'-c)\Gamma(2-b')\Gamma(1/2)}{\Gamma(2+b-b')2^{a}(b'-1)}$ $\times\{[\Gamma(a/2)\Gamma(3/2+a/2-b')]^{-1}$ $-[\Gamma(a/2+1/2)\Gamma(1+a/2-b')]^{-1}\}$
A.1.2.32	$u^{f-1}(1-u)^{g-1}$ $\times F_1(a,b,1-a;f;u,u/2)$ $f,g > 0$	$\frac{\Gamma(f)\Gamma(g)\Gamma(f+g-a-b)\sqrt{\pi}2^{1-f-g-b}}{\Gamma(f+g-a)\Gamma(\frac{a-b+f+g}{2})\Gamma(\frac{1-b+f+g-a}{2})}$

	$f(u)$	$\int_0^1 f(u)\, du$
A.1.2.33	$u^{c-1}(1-u)^{\frac{a+b'+1}{2}+b-c}$ $\times F_1(a,b,b';c;u,u/2)$	$\dfrac{\Gamma(c)\Gamma(\frac{a+b'+1}{2}+b-c)\Gamma(\frac{1-a+b'}{2})\sqrt{\pi}}{\Gamma(\frac{1-a+b'}{2}+b)\Gamma(\frac{1+a}{2})\Gamma(\frac{1+b'}{2})}$ $0 < c < 1/2+b'/2+a/2+b$
A.1.2.34	$u^{f-1}(1-u)^{g-1}$ $\times F_1(a,b,b';c;xu,y)$	$\frac{\Gamma(f)\Gamma(g)}{\Gamma(f+g)} F^{1:2;1}_{1:1;0}\left(\begin{matrix}a:b,f;b';\\c:f+g;-\ ;\end{matrix}x,y\right)$ $f,g > 0;\ \|x\|,\|y\| < 1$
A.1.2.35	$u^{b-1}(1-u)^{c-1}$ $\times F_2(a,b+c,b';b,a;$ $xu,y)$	$\frac{\Gamma(c)\Gamma(b)}{\Gamma(c+b)}(1-x)^{b'-a}(1-x-y)^{-b'}$ $b,c > 0;\ \|x\|+\|y\| < 1$
A.1.2.36	$u^{f-1}(1-u)^{g-1}$ $\times F_2(a,f+g,b;f,c;xu,y)$	$\frac{\Gamma(f)\Gamma(g)}{\Gamma(f+g)}(1-x)^{-a}\,{}_2F_1\left(\begin{matrix}a,b;\\c\ ;\end{matrix}\frac{y}{1-x}\right)$ $f,g > 0;\ \|x\| + \|y\| < 1$
A.1.2.37	$u^{f-1}(1-u)^{g-1}$ $\times F_2(a,b,b';f,c;ux,y)$	$\frac{\Gamma(f)\Gamma(g)}{\Gamma(f+g)} F_2(a,b,b';f+g,c;x,y)$ $f,g > 0;\ \|x\| + \|y\| < 1$
A.1.2.38	$u^{f-1}(1-u)^{g-1}$ $\times F_2(a,b,b';c,c';ux,y)$	$\frac{\Gamma(f)\Gamma(g)}{\Gamma(f+g)} F^{1:2;1}_{0:2;1}\left(\begin{matrix}a:b,f\ \ ;b';\\-:c,f+g;c';\end{matrix}x,y\right)$ $f,g > 0;\ \|x\| + \|y\| < 1$
A.1.2.39	$u^{f-1}(1-u)^{g-1}$ $\times F_3(a,a',f+g,b;c;$ $ux,y)$	$\frac{\Gamma(f)\Gamma(g)}{\Gamma(f+g)} F_3(a,a',f,b;c;x,y)$ $f,g > 0;\ \|x\|,\|y\| < 1$
A.1.2.40	$u^{f-1}(1-u)^{g-1}$ $\times F_3(a,a',b,b';c;$ $ux,y)$	$\frac{\Gamma(f)\Gamma(g)}{\Gamma(f+g)} F^{0:3;2}_{1:1;0}\left(\begin{matrix}-:a,b,f;a',b';\\c:\ f+g\ ;\ \ -\ \ ;\end{matrix}x,y\right)$ $f,g > 0; \|x\|,\|y\| < 1$
A.1.2.41	$u^{f-1}(1-u)^{g-1}$ $\times F_4(a,b;c,c';ux,y)$	$\frac{\Gamma(f)\Gamma(g)}{\Gamma(f+g)} F^{2:2;1}_{0:1;0}\left(\begin{matrix}a,b:\ \ \ f\ ;\ -\ ;\\-\ :f+g,c;c';\end{matrix}x,y\right)$ $f,g > 0; \|\sqrt{x}\|+\|\sqrt{y}\| < 1$
A.1.2.42	$u^{f-1}(1-u)^{g-1}$ $\times F_1(a,\frac{f+g}{2},b;c;$ $xu[1-u],y)$	$\frac{\Gamma(f)\Gamma(g)}{\Gamma(f+g)} F^{1:2;1}_{1:1;0}\left(\begin{matrix}a:f,g\ ;b\ ;\\c:\frac{f+g+1}{2};-;\end{matrix}\frac{x}{4},y\right)$ $f,g > 0; \|x\| < 4, \|y\| < 1$

	$f(u)$	$\int_0^1 f(u)\,du$
A.1.2.43	$u^{f-1}(1-u)^{g-1}$ $\times F_1(a,b,b';c;xu[1-u],y)$ $f,g>0$	$\frac{\Gamma(f)\Gamma(g)}{\Gamma(f+g)} F^{1:3;1}_{1:2;0}\left(\begin{matrix} a:f,g,b & ;b'; \\ c:\frac{f+g}{2},\frac{f+g+1}{2} & ;-; \end{matrix} x/4,y\right)$ $\lvert x\rvert<4,\ \lvert y\rvert<1$
A.1.2.44	$u^{f}(1-u)^{f-1}$ $\times F_2(a,f+\frac{1}{2},b;f,c;xu[1-u],y)$	$\frac{\Gamma(f)\Gamma(f+1)}{\Gamma(2f+1)}\ {}_2F_1(a,b;c;\frac{4y}{1-x})$ $f>0;\ \lvert x/4\rvert+\lvert y\rvert<1$
A.1.2.45	$u^{f-1}(1-u)^{g-1}$ $\times F_2(a,\frac{f+g}{2},b;g,c;ux[1-u],y)$	$\frac{\Gamma(f)\Gamma(g)}{\Gamma(f+g)} F_2(a,f,b;\frac{f+g+1}{2},c;\frac{x}{4},y)$ $f,g>0;\ \lvert x/4\rvert+\lvert y\rvert<1$
A.1.2.46	$u^{f-1}(1-u)^{g-1}$ $\times F_2(a,\frac{f+g}{2},b;c,c';xu[1-u],y)$ $f,g>0;$	$\frac{\Gamma(f)\Gamma(g)}{\Gamma(f+g)} F^{1:2;1}_{0:2;1}\left(\begin{matrix} a:\ f,\ g & ;b\ ; \\ -:c,\frac{f+g+1}{2} & ;c'; \end{matrix} x/4,y\right)$ $\lvert x/4\rvert+\lvert y\rvert<1$
A.1.2.47	$u^{f-1}(1-u)^{g-1}$ $\times F_3(a,a',b,b';c;xu[1-u],y)$ $f,g>0$	$\frac{\Gamma(f)\Gamma(g)}{\Gamma(f+g)} F^{0:4;2}_{1:2;0}\left(\begin{matrix} -:f,g,a,b\ ; & a',b'; \\ c:\frac{f+g}{2},\frac{f+g+1}{2}; & -\ ; \end{matrix} \frac{x}{4},y\right)$ $\lvert x\rvert<4,\lvert y\rvert<1$
A.1.2.48	$u^{f}(1-u)^{f-1}$ $\times F_4(a,b;f,c;xu[1-u],y)$	$\frac{\Gamma(f+1)\Gamma(f)}{\Gamma(2f+1)} F_4(a,b;f+\frac{1}{2},c;\frac{x}{4},y)$ $f>0;\ \lvert\sqrt{x}/2\rvert+\lvert y\rvert<1$
A.1.2.49	$u^{f-1}(1-u)^{g-1}$ $\times F_4(a,b;g,c;xu[1-u],y)$ $f,g>0$	$\frac{\Gamma(f)\Gamma(g)}{\Gamma(f+g)} F^{2:1;0}_{0:2;1}\left[\begin{matrix} a,b:\ f\ ; & -; \\ -\ :\frac{f+g}{2},\frac{f+g+1}{2}; & c; \end{matrix} x/4,y\right]$ $\left[\lvert\sqrt{x}/2\rvert+\lvert y\rvert<1\right]$
A.1.2.50	$u^{f-1}(1-u)^{g-1}$ $\times F_4(a,b;c,c';xu[1-u],y)$ $f,g>0$	$\frac{\Gamma(f)\Gamma(g)}{\Gamma(f+g)} F^{2:2;0}_{0:3;1}\left[\begin{matrix} a,b:\ f,\ g\ ; & -; \\ -\ :c,\frac{f+g}{2},\frac{f+g+1}{2}; & c'; \end{matrix} x/4,y\right]$ $\left[\lvert\sqrt{x}/2\rvert+\lvert y\rvert<1\right]$

	$f(u)$	$\int_0^1 f(u)\,du$
A.1.2.51	$u^{f-1}(1-u)^{g-1}$ $\times F_3(\frac{f+g}{2},a,\frac{f+g+1}{2},b;$ $c;ux[1-u],y)$	$\frac{\Gamma(f)\Gamma(g)}{\Gamma(f+g)}F_3(f,a,g,b;c;x/4,y)$ $f,g>0;\ \lvert x\rvert<4,\ \lvert y\rvert<1$
A.1.2.52	$u^{f-1}(1-u)^{g-1}$ $\times F_3(\frac{f+g}{2},a,b,b';c;$ $xu[1-u],y)$ $f,g>0;\ \lvert x\rvert<4$	$\frac{\Gamma(f)\Gamma(g)}{\Gamma(f+g)}F^{0:3;2}_{1:1;0}\left(\begin{matrix}-:f,g,b; & a,b'; \\ c:\frac{f+g+1}{2}; & -; \end{matrix}x/4,y\right)$ $\lvert y\rvert<1$
A.1.2.53	$u^{f-1}(1-u)^{\frac{a}{2}+\frac{b}{2}-\frac{1}{2}-f}$ $\times F_1(a,b,\frac{a-b+1}{2};f;$ $ux,u[2x-1])$	$\frac{\Gamma(f)\Gamma(\frac{1+a+b}{2}-f)\Gamma(1/2)}{\Gamma(\frac{1+a}{2})\Gamma(\frac{1+b}{2})(-2x)^a}$ $0<f<(a+b+1)/2$
A.1.2.54	$u^{f-1}(1-u)^{b-a-f}$ $\times F_1(a,1-a,b;f;$ $ux,u[2x-1])$	$\frac{\Gamma(f)\Gamma(1+b-a-f)\Gamma(1/2)2^{a-b}}{\Gamma(\frac{1+b}{2})\Gamma(1-a-\frac{b}{2})(-2x)^a}$ $f,b-a-f+1>0$
A.1.2.55	$u^{f-1}(1-u)^{1/2-f}$ $\times F_1(a,1/2+a,1-a;f;$ $ux,uy)$ $0<f<3/2$	$\frac{\Gamma(f)\Gamma(3/2-f)(1-y)^{1-a}}{2\Gamma(3/2)(x-y)(1-2a)}$ $\times\{[1+\surd(\frac{x-y}{1-y})]^{1-2a}$ $+[1-\surd(\frac{x-y}{1-y})]^{1-2a}\}$
A.1.2.56	$u^{f-1}(1-u)^{2a-f}$ $\times F_1(a,\frac{1}{2}+a,\frac{1}{2}+a;f;$ $ux,uy)$	$\frac{\Gamma(f)\Gamma(1+2a-f)2^{2a}}{\Gamma(1+2a)(1-y)^a}[1+\surd(1-\frac{x-y}{1-y})]^{-2a}$ $0<f<1+2a$
A.1.2.57	$u^{f-1}(1-u)^{-f-1/2}$ $\times F_1(a,\frac{1}{2}+a,-a;f;$ $ux,uy)$	$\frac{\Gamma(f)\Gamma(1/2-f)}{2\Gamma(1/2)(1-y)^a}\{[1+\surd(\frac{x-y}{1-y})]^{-2a}$ $+[1-\surd(\frac{x-y}{1-y})]^{-2a}\}$ $0<f<1/2$
A.1.2.58	$u^{f-1}(1-u)^{g-1}$ $\times F_1(f+g,b,b';f;$ $ux,uy)$	$\frac{\Gamma(f)\Gamma(g)}{\Gamma(f+g)}(1-x)^{-b}(1-y)^{-b'}$ $f,g>0;\ \lvert x\rvert,\lvert y\rvert<1$

	$f(u)$	$\int_0^1 f(u)\,du$
A.1.2.59	$u^{f-1}(1-u)^{g-1}$ $\times F_1(a,f,g;f;ux,uy)$	$\frac{\Gamma(f)\Gamma(g)}{\Gamma(f+g)}(1-y)^{-a}\,{}_2F_1\left(\begin{matrix}a,f;\\f+g;\end{matrix}\frac{x-y}{1-y}\right)$ $f,g>0;\ \|x-y\|<\|1-y\|$
A.1.2.60	$u^{f-1}(1-u)^{g-1}$ $\times F_1(a,b,b';f;ux,uy)$	$\frac{\Gamma(f)\Gamma(g)}{\Gamma(f+g)}F_1(a,b,b';f+g;x,y)$ $f,g>0;\ \|x\|,\|y\|<1$
A.1.2.61	$u^{f-1}(1-u)^{g-1}$ $\times F_1(a,b,b';c;ux,uy)$	$\frac{\Gamma(f)\Gamma(g)}{\Gamma(f+g)}F^{2:1;1}_{2:0;0}\left(\begin{matrix}f,\ c:b;b';\\f+g,c:-;\ -\ ;\end{matrix}x,y\right)$ $f,g>0;\ \|x\|,\|y\|<1$
A.1.2.62	$u^{f-1}(1-u)^{g-1}$ $\times F_2(f+g,b,b';c,c';ux,uy)$	$\frac{\Gamma(f)\Gamma(g)}{\Gamma(f+g)}F_2(f,b,b';c,c';x,y)$ $f,g>0;\ \|x\|+\|y\|<1$
A.1.2.63	$u^{f-1}(1-u)^{g-1}$ $\times F_2(a,b,b';c,c';ux,uy)$	$\frac{\Gamma(f)\Gamma(g)}{\Gamma(f+g)}F^{2:1;1}_{1:1;1}\left(\begin{matrix}f,g:b;b';\\f+g;c;c';\end{matrix}x,y\right)$ $f,g>0;\ \|x\|+\|y\|<1$
A.1.2.64	$u^{f-1}(1-u)^{g-1}$ $\times F_3(a,a',b,b';c;ux,uy)$	$\frac{\Gamma(f)\Gamma(g)}{\Gamma(f+g)}F_3(a,a',b,b';f+g;x,y)$ $f,g>0;\ \|x\|,\|y\|<1$
A.1.2.65	$u^{f-1}(1-u)^{g-1}$ $\times F_3(a,a',b,b';c;ux,uy)$	$\frac{\Gamma(f)\Gamma(g)}{\Gamma(f+g)}F^{1:2;2}_{2:0;0}\left(\begin{matrix}f\ :a,b;a',b';\\f+g,c:\ -\ ;\ -\ ;\end{matrix}x,y\right)$ $f,g>0;\|x\|,\|y\|<1$
A.1.2.66	$u^{f-1}(1-u)^{g-1}$ $\times F_4(f+g,b;c,c';ux,uy)$	$\frac{\Gamma(f)\Gamma(g)}{\Gamma(f+g)}F_4(f,b;c,c';x,y)$ $f,g>0;\ \|\sqrt{x}\|+\|\sqrt{y}\|<1$
A.1.2.67	$u^{f-1}(1-u)^{g-1}$ $\times F_4(a,b;c,c';ux,uy)$	$\frac{\Gamma(f)\Gamma(g)}{\Gamma(f+g)}F^{3:0;0}_{1:1;1}\left(\begin{matrix}f,a,b:-;-;\\f+g\ \ :c;c';\end{matrix}x,y\right)$ $f,g>0;\ \|\sqrt{x}\|+\|\sqrt{y}\|<1$
A.1.2.68	$u^{f-1}(1-u)^{g-1}$ $\times F_1(f+g,b,b';c;ux,[1-u]y)$	$\frac{\Gamma(f)\Gamma(g)}{\Gamma(f+g)}F_3(b,b',f,g;c;x,y)$ $f,g>0;\ \|x\|,\|y\|<1$

	$f(u)$	$\int_0^1 f(u)\,du$
A.1.2.69	$u^{f-1}(1-u)^{g-1}$ $\times F_1(a,b,b';c;$ $ux,[1-u]y)$	$\frac{\Gamma(f)\Gamma(g)}{\Gamma(f+g)} F^{1:2;2}_{2:0;0}\left(\begin{matrix} a & :b,f;b',g; \\ c,f+g: & - \;;\; - \;; \end{matrix} x,y\right)$ $f,g > 0;\ \lvert x\rvert,\lvert y\rvert < 1$
A.1.2.70	$u^{f-1}(1-u)^{g-1}$ $\times F_2(f+g,b,b';c,c';$ $ux,[1-u]y)$	$\frac{\Gamma(f)\Gamma(g)}{\Gamma(f+g)}\,{}_2F_1(b,f;c;x)$ $\times{}_2F_1(b',g;c';y)$ $f,g > 0;\ \lvert x\rvert,\lvert y\rvert < 1$
A.1.2.71	$u^{f-1}(1-u)^{g-1}$ $\times F_2(a,b,b';c,c';$ $ux,[1-u]y)$	$\frac{\Gamma(f)\Gamma(g)}{\Gamma(f+g)} F^{1:2;2}_{1:1;1}\left(\begin{matrix} a & :b,f;b',g; \\ f+g; & c \;;\; c' \;; \end{matrix} x,y\right)$ $f,g > 0;\ \lvert x\rvert,\lvert y\rvert < 1$
A.1.2.72	$u^{f-1}(1-u)^{g-1}$ $\times F_2(f+g,b,b';f,g;$ $ux,[1-u]y)$	$\frac{\Gamma(f)\Gamma(g)}{\Gamma(f+g)}(1-x)^{-b}(1-y)^{-b'}$ $f,g > 0;\ \lvert x\rvert,\lvert y\rvert < 1$
A.1.2.73	$u^{f-1}(1-u)^{g-1}$ $\times F_2(a,b,b';f,g;$ $ux,[1-u]y)$	$\frac{\Gamma(f)\Gamma(g)}{\Gamma(f+g)} F_1(a,b,b';f+g;x,y)$ $f,g > 0;\ \lvert x\rvert,\lvert y\rvert < 1$
A.1.2.74	$u^{f-1}(1-u)^{g-1}$ $\times F_4(f+g,b;f,g;$ $ux,[1-u]y)$	$\frac{\Gamma(f)\Gamma(g)}{\Gamma(f+g)}(1-x-y)^{-b}$ $f,g > 0;\ \lvert x+y\rvert < 1$
A.1.2.75	$u^{f-1}(1-u)^{g-1}$ $\times F_4(a,b;f,g;$ $ux,[1-u]y)$	$\frac{\Gamma(f)\Gamma(g)}{\Gamma(f+g)}\,{}_2F_1(a,b;f+g;x+y)$ $f,g > 0;\ \lvert x+y\rvert < 1$
A.1.2.76	$u^{f-1}(1-u)^{g-1}$ $\times F_4(f+g,b;c,c';$ $ux,[1-u]y)$	$\frac{\Gamma(f)\Gamma(g)}{\Gamma(f+g)} F_2(b,f,g;c,c';x,y)$ $f,g > 0;\ \lvert x\rvert+\lvert y\rvert < 1$
A.1.2.77	$u^{f-1}(1-u)^{g-1}$ $\times F_4(a,b;c,c';$ $ux,[1-u]y)$	$\frac{\Gamma(f)\Gamma(g)}{\Gamma(f+g)} F^{2:1;1}_{1:1;1}\left(\begin{matrix} a,b;f;g\;; \\ f+g:c;c'; \end{matrix} x,y\right)$ $f,g > 0;\ \lvert\sqrt{x}\rvert+\lvert\sqrt{y}\rvert < 1$

	$f(u)$	$\int_0^1 f(u)\ du$
A.1.2.78	$u^{f-1}(1-u)^{g-1}$ $\times F_1(\frac{f+g}{2},g,\frac{f-g+1}{2};g;$ $4x[1-u],4[2x-1]$ $\times u[1-u])$	$\frac{\Gamma(f)\Gamma(g)\Gamma(\frac{1+f+g}{2})\Gamma(\frac{1}{2})(-2x)^{-f}}{\Gamma(f+g)\Gamma(\frac{1+f}{2})\Gamma(\frac{1+g}{2})}$ $f,g > 0$
A.1.2.79	$u^{f-1}(1-u)^{g-1}$ $\times F_1(\frac{f+g}{2},b,1-f;g;$ $4u[1-u],2u[1-u])$ $f,g > 0$	$\frac{\Gamma(f)\Gamma(g)\Gamma(\frac{f+g+1}{2})\Gamma(\frac{g-f+1}{2}-b)\Gamma(\frac{1}{2})}{\Gamma(\frac{g+1}{2})\Gamma(f+\frac{g-b+1}{2})\Gamma(1+\frac{g-b}{2})}$ $\times 2^{1/2-f/2-g/2+b}$
A.1.2.80	$u^{f-1}(1-u)^{g-1}$ $\times F_1(\frac{f+g}{2},1-f,f+\frac{g}{2}-\frac{1}{2};g;$ $4xu[1-u],$ $4[2x-1]u[1-u])$	$\frac{\Gamma(f)\Gamma(g)\Gamma(\frac{1}{2})\Gamma(\frac{3-f-g}{2})(-x)^{-f}}{\Gamma(f+g)\Gamma(\frac{1+2f+g}{4})\Gamma(\frac{5-6f-g}{4})}$ $\times 2^{1/2-3f/2-g/2}$ $f,g > 0$
A.1.2.81	$u^{f-1}(1-u)^{g-1}$ $\times F_1(\frac{f+g}{2},\frac{g-3f-1}{2},\frac{1}{2}+f;$ $g;4u[1-u],$ $4yu[1-u])$	$\frac{\Gamma(f)\Gamma(g)\Gamma(\frac{1+f+g}{2})\Gamma(1+f)}{\Gamma(f+g)\Gamma(1+2f)\Gamma(\frac{1-f+g}{2})}2^{2f}$ $\times[1+\sqrt{(1-y)}]^{-2f}$ $f,g > 0;\ \lvert y\rvert < 1$
A.1.2.82	$u^{f-1}(1-u)^{g-1}$ $\times F_1(\frac{f+g}{2},\frac{1}{2}+f,\frac{1}{2}+f;$ $1+2f;4xu[1-u],$ $4yu[1-u])$	$\frac{\Gamma(f)\Gamma(g)}{\Gamma(f+g)}(1-y)^{-2f}2^{2f}$ $\times[1+\sqrt{(1-\frac{x-y}{1-y})}]^{-2f}$ $f,g > 0;\ \lvert x-y\rvert < \lvert 1-y\rvert$
A.1.2.83	$u^{f-1}(1-u)^{g-1}$ $\times F_1(\frac{f+g}{2},b,b';g;$ $4u[1-u],yu[1-u])$ $f,g > 0;\ \lvert y\rvert < 4$	$\frac{\Gamma(f)\Gamma(g)\Gamma(\frac{f+g+1}{2})\Gamma(\frac{g-f+1}{2}-b)}{\Gamma(f+g)\Gamma(\frac{g-f+1}{2})\Gamma(\frac{f+g+1}{2}-b)}$ $\times {}_2F_1([f+g]/2,b;[f+g+1]/2;y/4)$

	$f(u)$	$\int_0^1 f(u)\,du$
A.1.2.84	$u^{f-1}(1-u)^{g-1}$ $\times F_1(\frac{f+g}{2},\frac{f}{2},\frac{g+1}{2};g;$ $4x[1-u],4y[1-u])$	$\frac{\Gamma(f)\Gamma(g)}{\Gamma(f+g)}(1-y)^{-f/2-g/2}$ $\times {}_2F_1(\frac{f+g}{2},\frac{f}{2};\frac{f+g+1}{2};\frac{x-y}{1-y})$ $f,g > 0;\ \|x-y\| < \|1-y\|$
A.1.2.85	$u^{f-1}(1-u)^{g-1}$ $\times F_1(\frac{f+g}{2},b,b';g;$ $xu[1-u],yu[1-u])$	$\frac{\Gamma(f)\Gamma(g)}{\Gamma(f+g)}F_1(f,b,b';\frac{f+g+1}{2};\frac{x}{4},\frac{y}{4})$ $f,g > 0;\ \|x\|,\|y\| < 4$
A.1.2.86	$u^{f-1}(1-u)^{g-1}$ $\times F_1(\frac{f+g}{2},b,b';c;$ $xu[1-u],yu[1-u])$	$\frac{\Gamma(f)\Gamma(g)}{\Gamma(f+g)}F^{2:1;1}_{2:0;0}\left(\begin{matrix}f,\ g & :b;b'; \\ c,\frac{f+g+1}{2} & ;-;-;\end{matrix}\frac{x}{4},\frac{y}{4}\right)$ $f,g > 0;\ \|x\|,\|y\| < 4$
A.1.2.87	$u^{f-1}(1-u)^{g-1}$ $\times F_1(a,b,b';c;$ $xu[1-u],yu[1-u])$ $f,g > 0;$	$\frac{\Gamma(f)\Gamma(g)}{\Gamma(f+g)}F^{3:1;1}_{3:0;0}\left(\begin{matrix}a,\ f,\ g & : b;b'; \\ c,\frac{f+g}{2},\frac{f+g+1}{2} & : -;- ;\end{matrix}\frac{x}{4},\frac{y}{4}\right)$ $\|x\|,\|y\| < 4$
A.1.2.88	$u^{f}(1-u)^{f-1}$ $\times F_2(f+\frac{1}{2},b,b';c,c';$ $xu[1-u],yu[1-u])$	$\frac{\Gamma(f)\Gamma(g)}{\Gamma(f+g)}F_2(f,b,b';c,c';\frac{x}{4},\frac{y}{4})$ $f > 0;\ \|x\|+\|y\| < 4$
A.1.2.89	$u^{f-1}(1-u)^{g-1}$ $\times F_2(\frac{f+g}{2},b,b';c,c';$ $xu[1-u],yu[1-u])$	$\frac{\Gamma(f)\Gamma(g)}{\Gamma(f+g)}F^{2:1;1}_{1:1;1}\left(\begin{matrix}f,\ g & :b;b'; \\ \frac{f+g+1}{2} & :c;c';\end{matrix}\frac{x}{4},\frac{y}{4}\right)$ $f,g > 0;\ \|x\|+\|y\| < 4$
A.1.2.90	$u^{f-1}(1-u)^{g-1}$ $\times F_2(a,b,b';c,c';$ $xu[1-u],yu[1-u])$ $f,g > 0;$	$\frac{\Gamma(f)\Gamma(g)}{\Gamma(f+g)}F^{3:1;1}_{2:1;1}\left(\begin{matrix}a,\ f,\ g & :b;b'; \\ \frac{f+g}{2},\frac{f+g+1}{2} & :c;c';\end{matrix}x/4,y/4\right)$ $\|x\|+\|y\| < 4$
A.1.2.91	$u^{f}(1-u)^{f-1}$ $\times F_3(a,a',b,b';f;$ $ux[1-u],uy[1-u])$	$\frac{\Gamma(f+1)\Gamma(f)}{\Gamma(2f+1)}F_3(a,a',b,b';f+\frac{1}{2};\frac{x}{4},\frac{y}{4})$ $f > 0;\ \|x\|,\|y\| < 4$

	$f(u)$	$\int_0^1 f(u)\,du$
A.1.2.92	$u^{f-1}(1-u)^{g-1}$ $\times F_3(a,a',b,b';g;$ $xu[1-u],yu[1-u])$ $f,g > 0$	$\frac{\Gamma(f)\Gamma(g)}{\Gamma(f+g)} F^{1:2;2}_{2:0;0}\left(\begin{matrix} f & :a,b; & a',b'; \\ \frac{f+g}{2},\frac{f+g+1}{2} & : - & - ; \end{matrix}\ \frac{x}{4},\frac{y}{4}\right)$ $\lvert x\rvert,\lvert y\rvert < 4$
A.1.2.93	$u^{f-1}(1-u)^{g-1}$ $\times F_3(a,a',b,b';c;$ $xu[1-u],yu[1-u])$ $f,g > 0;\ \lvert x\rvert,\lvert y\rvert < 4$	$\frac{\Gamma(f)\Gamma(g)}{\Gamma(f+g)} F^{2:2;2}_{3:0;0}\left(\begin{matrix} f,\ g & : a,b; & a',b'; \\ c,\frac{f+g}{2},\frac{f+g+1}{2} & : - ; & - ; \end{matrix}\ \frac{x}{4},\frac{y}{4}\right)$
A.1.2.94	$u^{f-1}(1-u)^{g-1}$ $\times F_4(\frac{f+g}{2},\frac{f+g+1}{2};c,c';$ $xu[1-u],yu[1-u])$	$\frac{\Gamma(f)\Gamma(g)}{\Gamma(f+g)} F_4(f,g;c,c';\frac{x}{4},\frac{y}{4})$ $f,g > 0;\ \lvert\sqrt{x}\rvert+\lvert\sqrt{y}\rvert < 2$
A.1.2.95	$u^{f-1}(1-u)^{g-1}$ $\times F_4(a,b;c,c';$ $xu[1-u],yu[1-u])$ $f,g > 0;$	$\frac{\Gamma(f)\Gamma(g)}{\Gamma(f+g)} F^{4:0;0}_{2:1;1}\left(\begin{matrix} f,g,a,b & :-;-; \\ \frac{f+g}{2},\frac{f+g+1}{2} & :c;c'; \end{matrix}\ x/4,y/4\right)$ $\lvert\sqrt{x}\rvert+\lvert\sqrt{y}\rvert < 2$
A.1.2.96	$u^{g-1}(1-u)^{g-1}$ $\times H_3(a,b;g;$ $xu[1-u],yu[1-u])$	$\frac{\Gamma(f)\Gamma(g)}{\Gamma(f+g)} H_3(a,b;g+\frac{1}{2};\frac{x}{4},\frac{y}{4})$ $f,g > 0$; series must be cgt.
A.1.2.97	$u^{f-1}(1-u)^{g-1}$ $\times F^{A:B;B'}_{C:D;D'}\left(\begin{matrix} (a):(b); & (b'); \\ (c):(d); & (d'); \end{matrix}\ xu^m,y\right)$ $f,g > 0;$	$\frac{\Gamma(f)\Gamma(g)}{\Gamma(f+g)} F^{A:B+m;B'}_{C:D+m;D'}\left(\begin{matrix} (a):(b), \frac{f}{m},..,\frac{f+m-1}{m}; (b') ; \\ (c):(d), \frac{f+g}{m},..,\frac{f+g+m-1}{m}; (d') ; \end{matrix}\ x,y\right)$ series must be cgt.
A.1.2.98	$u^{f-1}(1-u)^{g-1}$ $\times F^{A:B;B'}_{C:D;D'}\left(\begin{matrix} (a):(b); & (b'); \\ (c):(d); & (d'); \end{matrix}\ xu^m(1-u)^n,y\right)$ $f,g > 0$; series cgt.	$\frac{\Gamma(f)\Gamma(g)}{\Gamma(f+g)} F^{A:B+m+n;B'}_{C:D+m+n;D'}\left(\begin{matrix} (a):(b), \frac{f}{m},..,\frac{f+m-1}{m},\frac{g}{n},..,\frac{g+n-1}{n}; (b'); \\ (c):(d), \frac{f+g}{m+n},...,\frac{f+g+m+n-1}{m+n} ; (d'); \end{matrix}\ m^m n^n x/[m+n]^{m+n},y\right)$

	$f(u)$	$\int_0^1 f(u)\,du$
A.1.2.99	$u^{f-1}(1-u)^{g-1} \times F^{A:B;B'}_{C:D;D'}\left[\begin{matrix}(a):(b);(b');\\(c):(d);(d');\end{matrix}\, xu^m, y(1-u)^m\right]$ $f,g > 0$; series must be cgt.	$\frac{\Gamma(f)\Gamma(g)}{\Gamma(f+g)} F^{A\ :B+m;B'+m}_{C+m;\ D\ ;\ D'}\left[\begin{matrix}(a) :(b),\frac{f}{m},..,\frac{f+m-1}{m}; (b'),\frac{g}{m},..,\frac{g+m-1}{m};\\(c), \frac{f+g}{m+n},..,\frac{f+g+m+n-1}{m+n}:(d) ; (d) ;\end{matrix}\, x,y\right]$
A.1.2.100	$u^{f-1}(1-u)^{g-1} \times F^{A:B;B'}_{C:D;D'}\left[\begin{matrix}(a):(b);(b');\\(c):(d);(d');\end{matrix}\, xu^m, yu^m\right]$ $f,g > 0$;	$\frac{\Gamma(f)\Gamma(g)}{\Gamma(f+g)} F^{A+m:B;B'}_{C+m:D;D'}\left[\begin{matrix}(a), \frac{f}{m},...,\frac{f+m-1}{m}:(b);(b');\\(c), \frac{f+g}{m},..\frac{f+g+m-1}{m}:(d);(d');\end{matrix}\, x,y\right]$ series must be cgt.

A.1.3 Euler Integrals involving Multiple Hypergeometric Functions

	$f(u)$	$\int_0^1 f(u)\,du$
A.1.3.1	$u^{f-1}(1-u)^{g-1} \times {}_{A'}F_{B'}\left[\begin{matrix}(a');\\(b');\end{matrix}\, x_1 u\right] .. \times {}_{A^{(r)}}F_{B^{(r)}}\left[\begin{matrix}(a^{(r)});\\(b^{(r)});\end{matrix}\, x_r u\right]$	$\frac{\Gamma(f)\Gamma(g)}{\Gamma(f+g)} F^{1:A',..,A^{(r)}}_{1:B',..,B^{(r)}}\left[\begin{matrix}f\ :(a');..;(a^{(r)});\\f+g:(b');..;(b^{(r)});\end{matrix}\, x_1,..,x_r\right]$

$f,g > 0$; all series must be convergent over the range of integration.

	$f(u)$	$\int_0^1 f(u)\, du$
A.1.3.2	$u^{f-1}(1-u)^{g-1}$ $\times {}_{A'}F_{B'}\left[\begin{matrix}(a');\\(b');\end{matrix} x_1 u(1-u)\right]$ $\ldots \times {}_{A^{(r)}}F_{B^{(r)}}\left[\begin{matrix}(a^{(r)});\\(b^{(r)});\end{matrix} x_r u(1-u)\right]$	$\frac{\Gamma(f)\Gamma(g)}{\Gamma(f+g)} F^{2:A';..;A^{(r)}}_{2:B';..;B^{(r)}}\left[\begin{matrix} f,\ g: (a');..;(a^{(r)}); \\ \frac{f+g}{2},\frac{f+g+1}{2}: (b');..;(b^{(r)}); \end{matrix} \frac{x_1}{4},\ldots,\frac{x_r}{4}\right]$ $f,g > 0$; all series cgt.
A.1.3.3	$u^{f-1}(1-u)^{g-1}$ $\times {}_2F_1(a,-N;a-g-N+1;u)$ $\times F_1(f+g-a,b,b';f+g;ux,uy)$	$\frac{(g,N)(f+g-a,N)}{(g-a,N)(f+g,N)}$ $\times F_1(f+g-a+N,b,b';f+g+N;x,y)$ $f,g > 0;\ \|x\|,\|y\| < 1$
A.1.3.4	$u^{f-1}(1-u)^{g-1}$ $\times {}_2F_1(a,-N;a-g+N-1;u)$ $\times F_2(c,f+g-a,b;f+g\ ,d;ux,y)$	$\frac{(g,N)(f+g-a,N)}{(g-a,N)(f+g,N)} F_2(c,f+g-a+N,b;f+g+N,d;x,y)$ $f,g > 0;\ \|x\|+\|y\| < 1$
A.1.3.5	$u^{f-1}(1-u)^{g-1}$ $\times {}_2F_1(a,-N;a-g+N-1;u)$ $\times F_4(f+g-a,f+g+N;d,d';ux,uy)$	$\frac{(g,N)(f+g-a,N)}{(g-a,N)(f+g,N)}$ $\times F_4(f+g-a+N,f+g;d,d';x,y)$ $f,g > 0;\ \|\sqrt{x}\|+\|\sqrt{y}\| < 1$
A.1.3.6	$u^{f-1}(1-u)^{g-1}$ $\times {}_2F_1(a,-N;a-g-N+1;u)$ $\times F_A^{(r)}(b,f+g-a,c_2,..,c_r;f+g\ ,d_2,..,d_r;ux_1,x_2,..,x_r)$	$\frac{(g,N)(f+g-a,N)}{(g-a,N)(f+g,N)}$ $\times F_A^{(r)}(b,f+g-a+n,c_2,..,c_r;f+g+N,d_2,..,d_r;x_1,..,x_r)$ $f,g > 0;\ \|x_1\|+..+\|x_r\| < 1$

	$f(u)$	$\int_0^1 f(u)\,du$
A.1.3.7	$u^{f-1}(1-u)^{g-1}$ $\times {}_2F_1(a,-N;a-g-N+1;u)$ $\times F_B^{(r)}(f+g-a,b_2,..,b_r,$ $f+g+N,c_2,..,c_r;d;$ $ux_1,x_2,..,x_r)$	$\frac{(g,N)(f+g-a,N)}{(g-a,N)(f+g,N)}F_B^{(r)}(f+g-a+N,$ $b_2,..,b_r,f+g,c_2,..,c_r;d;x_1,..,x_r)$ $f,g>0;\ \lvert x_1\rvert,...,\lvert x_r\rvert<1$
A.1.3.8	$u^{f-1}(1-u)^{g-1}$ $\times {}_2F_1(a,-N;a-g-N+1;u)$ $\times F_D^{(r)}(f+g-a,c_1,..,c_r;$ $f+g;ux_1,..,ux_r)$	$\frac{(g,N)(f+g-a,N)}{(g-a,N)(f+g,N)}F_D^{(r)}(f+g-a+N,$ $c_1,..,c_r;f+g+N;x_1,..,x_r)$ $f,g>0;\ \lvert x_1\rvert,...,\lvert x_r\rvert<1$
A.1.3.9	$u^{f-1}(1-u)^{g-1}$ $\times {}_2F_1(a,-N;a-g-N+1;u)$ $\times F_C^{(r)}(f+g-a,f+g+N;$ $d_1,..,d_r;ux_1,..,ux_r)$	$\frac{(g,N)(f+g-a,N)}{(g-a,N)(f+g,N)}F_C^{(r)}(f+g-a+N,$ $f+g;d_1,..,d_r;x_1,..,x_r)$ $f,g>0;\lvert\sqrt{x_1}\rvert+..+\lvert\sqrt{x_r}\rvert<1$
A.1.3.10	$u^{a-1}(1-u)^{b-1}$ $\times {}_2F_1(c,-N;c-b-N+1;u)$ $\times F^{F:G';.;G^{(r)}}_{H:K';.;K^{(r)}}\left[\begin{matrix}(f): \\ (h):\end{matrix}\right.$ $(g');.;(g^{(r)});$ $(k');.;(k^{(r)});$ $\left.ux_1x_2,..,x_r\right]$	$\frac{\Gamma(b+N)\Gamma(a)(a+b-c,N)}{\Gamma(a+b+N)\ (b-c,N)}$ $\times F^{F:G'+2;G'',..,G^{(r)}}_{H:K'+2;K'',..,K^{(r)}}\left[\begin{matrix}(f):(g'), \\ (h):(k'),\end{matrix}\right.$ $a+b-c+N,a\ ;(g'');.;(g^{(r)});$ $a+b-c,a+b+N;(k'');.;(k^{(r)});$ $\left.x_1,..,x_r\right]$ $f,g>0$; series must be cgt.
A.1.3.11	$u^{f-1}(1-u)^{g-1}$ $\times\Phi_2^{(r)}(f+g,b_2,..,b_r;$ $c;ux_1,x_2,..,x_r)$	$\frac{\Gamma(f)\Gamma(g)}{\Gamma(f+g)}\Phi_2^{(r)}(f,b_2,..,b_r;c;$ $x_1,..,x_r)$ $f,g>0$

	$f(u)$	$\int_0^1 f(u)\,du$
A.1.3.12	$u^{a-1}(1-u)^{b-1}$ $\times {}_2F_1(c,-N;c-b-N+1;u)$ $\times F^{F:G';.;G^{(r)}}_{H:K';.;K^{(r)}}\left[\begin{matrix}(f): & (g');.;(g^{(r)}); \\ (h): & (k');.;(k^{(r)});\end{matrix} ux_1,..,ux_r\right]$	$\frac{\Gamma(b+N)\Gamma(a)}{\Gamma(a+b+N)}\frac{(a+b-c,N)}{(b-c,N)}$ $\times F^{F+2:G';.;G^{(r)}}_{H+2:K';.;K^{(r)}}\left[\begin{matrix}(f),a+b-c+N,a: & (g');.;(g^{(r)}); \\ (h),a+b-c,a+b+N: & (k');.;(k^{(r)});\end{matrix} x_1,..,x_r\right]$ $f,g > 0$; all series cgt.
A.1.3.13	$u^f(1-u)^{f-1}$ $\times\Phi_2^{(r)}(f+1/2,b_2,..,b_r;$ $c;\ x_1u[1-u],x_2,..,x_r)$	$\frac{\Gamma(f+1)\Gamma(f)}{\Gamma(2f+1)}$ $\times\Phi_2^{(r)}(f,b_2,..,b_r;c;\frac{x_1}{4},x_2,..,x_r)$ $f > 0$
A.1.3.14	$u^f(1-u)^{f-1}$ $\times\Phi_2^{(r)}(b_1,..,b_r;f;$ $x_1u[1-u],..,x_ru[1-u])$	$\frac{\Gamma(f+1)\Gamma(f)}{\Gamma(2f+1)}$ $\times\Phi_2^{(r)}(b_1,..,b_r;f+1/2;\frac{x_1}{4},..,\frac{x_r}{4})$ $f > 0$
A.1.3.15	$u^{f-1}(1-u)^{g-1}$ $\times\Phi_2^{(r)}(b_1,..,b_r;f;$ $ux_1,..,ux_r)$	$\frac{\Gamma(f)\Gamma(g)}{\Gamma(f+g)}\Phi_2^{(r)}(b_1,..,b_r;f+g;$ $x_1,..,x_r)$ $f,g > 0$
A.1.3.16	$u^{f-1}(1-u)^{g-1}$ $\times\Psi_2^{(r)}(a;f,d_2,..,d_r;$ $ux_1,x_2,..,x_r)$	$\frac{\Gamma(f)\Gamma(g)}{\Gamma(f+g)}\Psi_2^{(r)}(a;f+g,d_2,..,d_r;$ $x_1,..,x_2)$ $f,g > 0$
A.1.3.17	$u^f(1-u)^{f-1}$ $\times\Psi_2^{(r)}(a;f+1,d_2,..,d_r;$ $x_1u[1-u],x_2,..,x_r)$	$\frac{\Gamma(f+1)\Gamma(f)}{\Gamma(2f+1)}\Psi_2^{(r)}(a;f+\frac{1}{2},d_2,..,d_r;$ $x_1/4,x_2,..,x_r)$ $f > 0$
A.1.3.18	$u^{f-1}(1-u)^{g-1}$ $\times\Psi_2^{(r)}(f+g;c_1,..,c_r;$ $ux_1,..,ux_r)$	$\frac{\Gamma(f)\Gamma(g)}{\Gamma(f+g)}\Psi_2^{(r)}(f;c_1,..,c_r;x_1,..,x_r)$ $f,g > 0$

	$f(u)$	$\int_0^1 f(u)\,du$
A.1.3.19	$u^f(1-u)^{f-1}$ $\times\Psi_2^{(r)}(f+1/2;d_1,..,d_r;$ $x_1u[1-u],..,x_ru[1-u])$	$\frac{\Gamma(f+1)\Gamma(f)}{\Gamma(2f+1)}\Psi_2^{(r)}(f;d_1,..,d_r;$ $x_1/4,..,x_r/4)$ $f>0$
A.1.3.20	$u^{f-1}(1-u)^{g-1}$ $\times\Psi_2^{(r)}(f+g;d_1,..,d_r;$ $ux_1,...,ux_k,$ $[1-u]x_{k+1},..,[1-u]x_r)$	$\frac{\Gamma(f)\Gamma(g)}{\Gamma(f+g)}\Psi_2^{(r)}(f;d_1,..,d_k;x_1,..,x_k)$ $\times\Psi_2^{(r-k)}(g;d_{k+1},..,d_r;x_{k+1},..,x_r)$ $f,g>0$
A.1.3.21	$u^{f-1}(1-u)^{g-1}$ $\times F_A^{(r)}(a,f+g,b_2,..,b_r;$ $f,d_2,..,d_r;ux_1,x_2,..,x_r)$	$\frac{\Gamma(f)\Gamma(g)}{\Gamma(f+g)}(1-x_1)^{-a}F_A^{(r-1)}(a,b_2,..,b_r;$ $d_2,..,d_r;x_2/[1-x_1],..,x_r/[1-x_1])$ $f,g>0;\lvert x_1\rvert+.+\lvert x_r\rvert<1$
A.1.3.22	$u^{f-1}(1-u)^{g-1}$ $\times F_A^{(r)}(a,f+g,b_2,..,b_r;$ $d_1,..,d_r;ux_1,x_2,..,x_r)$	$\frac{\Gamma(f)\Gamma(g)}{\Gamma(f+g)}F_A^{(r)}(a,f,b_2,..,b_r;$ $d_1,..,d_r;x_1,..,x_r)$ $f,g>0;\lvert x_1\rvert+.+\lvert x_r\rvert<1$
A.1.3.23	$u^{f-1}(1-u)^{g-1}$ $\times F_B^{(r)}(f+g,a_2,..,a_r;$ $b_1,..,b_r;c;ux_1,x_2,..,x_r)$	$\frac{\Gamma(f)\Gamma(g)}{\Gamma(f+g)}F_B^{(r)}(f,a_2,..,a_r,$ $b_1,..,b_r;c;x_1,..,x_r)$ $f,g>0;\lvert x_1\rvert,..,\lvert x_r\rvert<1$
A.1.3.24	$u^{f-1}(1-u)^{g-1}$ $\times F_C^{(r)}(a,b;f,d_2,..,d_r;$ $ux_1,x_2,..,x_r)$	$\frac{\Gamma(f)\Gamma(g)}{\Gamma(f+g)}F_C^{(r)}(a,b;f+g,d_2,..,d_r;$ $x_1,..,x_r)$ $f,g>0;\lvert\sqrt{x_1}\rvert+.+\lvert\sqrt{x_r}\rvert<1$
A.1.3.25	$u^{f-1}(1-u)^{g-1}$ $\times F_D^{(r)}(a,f+g,b_2,..,b_r;$ $c;ux_1,x_2,..,x_r)$	$\frac{\Gamma(f)\Gamma(g)}{\Gamma(f+g)}F_D^{(r)}(a,f,b_2,..,b_r;c;$ $x_1,..,x_r)$ $f,g>0;\lvert x_1\rvert,..,\lvert x_r\rvert<1$
A.1.3.26	$u^f(1-u)^{f-1}$ $\times F_D^{(r)}(a,f+\frac{1}{2},b_2,..,b_r;$ $c;ux_1[1-u],x_2,..,x_r)$	$\frac{\Gamma(f)\Gamma(f+1)}{\Gamma(2f+1)}F_D^{(r)}(a,f,b_2,..,b_r;$ $c:\frac{x_1}{4},x_2,..,x_r)$ $f,g>0;\lvert x_1\rvert<4,\lvert x_2\rvert,..,\lvert x_r\rvert<1$

	$f(u)$	$\int_0^1 f(u)\,du$
A.1.3.27	$u^f(1-u)^{f-1}$ $\times F_A^{(r)}(a,f+\frac{1}{2},b_2,..,b_r;$ $f,d_2,..,d_r;x_1u[1-u],$ $x_2,..,x_r)$	$\frac{\Gamma(f+1)\Gamma(f)}{\Gamma(2f+1)}(1-x_1/4)^{-a}$ $\times F_A^{(r-1)}(a,b_2,..,b_r;d_2,..,d_r;$ $4x_2/[4-x_1],..,4x_r/[4-x_1])$ $f>0;\lvert x_1/4\rvert+\lvert x_2\rvert+.+\lvert x_r\rvert<1$
A.1.3.28	$u^{f-1}(1-u)^{g-1}$ $\times F_A^{(r)}(a,\frac{f+g}{2},b_2,..,b_r;$ $g,d_2,..,d_r;x_1u[1-u],$ $x_2,..,x_r)$	$\frac{\Gamma(f)\Gamma(g)}{\Gamma(f+g)}F_A^{(r)}(a,f,b_2,..,b_r;\frac{f+g+1}{2},$ $d_2,..,d_r;x_1/4,x_2,..,x_r)$ $f,g>0;\lvert x_1/4\rvert+\lvert x_2\rvert+.+\lvert x_r\rvert<1$
A.1.3.29	$u^{f-1}(1-u)^{g-1}$ $\times F_B^{(r)}(\frac{f+g}{2},a_2,..,a_r,$ $\frac{f+g+1}{2},b_2,..,b_r;c;$ $x_1u[1-u],x_2,..,x_r)$	$\frac{\Gamma(f)\Gamma(g)}{\Gamma(f+g)}F_B^{(r)}(f,a_2,..,a_r,$ $g,b_2,..,b_r;c;x_1/4,x_2,..,x_r)$ $f,g>0;\lvert x_1\rvert<4,\lvert x_1\rvert,..,\lvert x_r\rvert<1$
A.1.3.30	$u^f(1-u)^{f-1}$ $\times F_C^{(r)}(a,b;f+1,d_2,..,$ $d_r;x_1u[1-u],x_2,..,x_r)$	$\frac{\Gamma(f+1)\Gamma(f)}{\Gamma(2f+1)}F_C^{(r)}(a,b;f+\frac{1}{2},d_2,..,d_r;$ $x_1/4,x_2,..,x_r)$ $f>0;\lvert\sqrt{x_1}/2\rvert+\lvert\sqrt{x_2}\rvert+.+\lvert\sqrt{x_r}\rvert<1$
A.1.3.31	$u^{f-1}(1-u)^{g-1}$ $\times F_A^{(r)}(f+g,b_1,..,b_r;$ $d_1,..,d_r;ux_1,..,ux_r)$	$\frac{\Gamma(f)\Gamma(g)}{\Gamma(f+g)}F_A^{(r)}(f,b_1,..,b_r;d_1,..,d_r;$ $x_1,..,x_r)$ $f,g>0;\ \lvert x_1\rvert+.+\lvert x_r\rvert<1$
A.1.3.32	$u^{f-1}(1-u)^{g-1}$ $\times F_B^{(r)}(a_1,..,a_r,b_1,..,b_r;$ $f;ux_1,..,ux_r)$	$\frac{\Gamma(f)\Gamma(g)}{\Gamma(f+g)}F_B^{(r)}(a_1,..,a_r,b_1,..,b_r;$ $f+g;x_1,..,x_r)$ $f,g>0;\lvert x_1\rvert,..,\lvert x_r\rvert<1$
A.1.3.33	$u^{f-1}(1-u)^{g-1}$ $\times F_C^{(r)}(f+g,b;d_1,..,d_r;$ $ux_1,..,ux_r)$ $f,g>0$	$\frac{\Gamma(f)\Gamma(g)}{\Gamma(f+g)}F_C^{(r)}(f,b;d_1,..,d_r;$ $x_1,..,x_r)$ $\lvert\sqrt{x_1}\rvert+.+\lvert\sqrt{x_r}\rvert<1$

	$f(u)$	$\int_0^1 f(u)\,du$
A.1.3.34	$u^{f-1}(1-u)^{g-1}$ $\times F_D^{(r)}(f+g,b_1,..,b_r;f;$ $ux_1,..,ux_r)$	$\frac{\Gamma(f)\Gamma(g)}{\Gamma(f+g)}(1-x_1)^{-b_1}..(1-x_r)^{-b_r}$ $f,g > 0;\lvert x_1\rvert,..,\lvert x_r\rvert < 1$
A.1.3.35	$u^{f-1}(1-u)^{g-1}$ $\times F_D^{(r)}(f+g,b_1,..,b_r;c;$ $ux_1,..,ux_r)$	$\frac{\Gamma(f)\Gamma(g)}{\Gamma(f+g)}F_D^{(r)}(f,b_1,..,b_r;c;$ $x_1,..,x_r)$ $f,g > 0;\lvert x_1\rvert,...,\lvert x_r\rvert < 1$
A.1.3.36	$u^f(1-u)^{f-1}$ $\times F_A^{(r)}(f+\frac{1}{2},b_1,..,b_r;$ $d_1,..,d_r;x_1u[1-u],$ $..,x_ru[1-u])$	$\frac{\Gamma(f+1)\Gamma(f)}{\Gamma(2f+1)}F_A^{(r)}(f,b_1,..,b_r;d_1,..,d_r;$ $x_1/4,..,x_r/4)$ $f > 0;\lvert x_1\rvert+.+\lvert x_r\rvert < 4$
A.1.3.37	$u^f(1-u)^{f-1}$ $\times F_B^{(r)}(b_1,..,b_r,d_1,..,$ $d_r;x_1u[1-u],...,$ $x_ru[1-u])$	$\frac{\Gamma(f+1)\Gamma(f)}{\Gamma(2f+1)}F_B^{(r)}(b_1,..,b_r,d_1,..,d_r;$ $f;x_1/4,..,x_r/4)$ $f > 0;\lvert x_1\rvert,..,\lvert x_r\rvert < 4$
A.1.3.38	$u^f(1-u)^{f-1}$ $\times F_C^{(r)}(b,f;d_1,..,d_r;$ $x_1u[1-u],..,x_ru[1-u])$	$\frac{\Gamma(f+1)\Gamma(f)}{\Gamma(2f+1)}F_C^{(r)}(b,f;d_1,..,d_r;$ $x_1/4,..,x_r/4)$ $f > 0;\lvert\sqrt{x_1}\rvert+.+\lvert\sqrt{x_r}\rvert < 2$
A.1.3.39	$u^{f-1}(1-u)^{g-1}$ $\times F_C^{(r)}(\frac{f+g}{2},\frac{f+g+1}{2};$ $d_1,..,d_r;x_1u[1-u],$ $..,x_ru[1-u])$	$\frac{\Gamma(f)\Gamma(g)}{\Gamma(f+g)}F_C^{(r)}(f,g;d_1,..,d_r;$ $x_1/4,..,x_r/4)$ $f,g > 0;\lvert\sqrt{x_1}\rvert+.+\lvert\sqrt{x_r}\rvert < 2$
A.1.3.40	$u^f(1-u)^{f-1}$ $\times F_D^{(r)}(f+\frac{1}{2},b_1,..,b_r;f;$ $x_1u[1-u],..,x_ru[1-u])$	$\frac{\Gamma(f)\Gamma(f+1)}{\Gamma(2f+1)}(1-\frac{x_1}{4})^{-b_1}...$ $(1-\frac{x_r}{4})^{-b_r}$ $f > 0;\lvert x_1\rvert,..,\lvert x_r\rvert < 4$

	$f(u)$	$\int_0^1 f(u)\,du$
A.1.3.41	$u^{f-1}(1-u)^{g-1}$ $\times F_D^{(r)}(\frac{f+g}{2},b_1,..,b_r;$ $g;x_1u[1-u],..,$ $x_ru[1-u])$	$\frac{\Gamma(f)\Gamma(g)}{\Gamma(f+g)}F_D^{(r)}(f,b_1,..,b_r;\frac{f+g+1}{2};$ $x_1/4,..,x_r/4)$ $f,g > 0; \|x_1\|,..,\|x_r\| < 4$
A.1.3.42	$u^{f-1}(1-u)^{g-1}$ $\times F_A^{(r)}(f+g,b_1,..,b_r;$ $d_1,..,d_r;ux_1,..,ux_k,$ $[1-u]x_{k+1},..,[1-u]x_r)$ $f,g > 0;$	$\frac{\Gamma(f)\Gamma(g)}{\Gamma(f+g)}F_A^{(k)}(f,b_1,..,b_k;d_1,..,d_k;$ $x_1,..,x_k)$ $\times F_A^{(r-k)}(g,b_{k+1},..,b_r;d_{k+1},..,d_r;$ $x_{k+1},..,x_r)$ $\|x_1\|+.+\|x_k\| < 1;$ $\|x_{k+1}\|+.+\|x_r\| < 1$
A.1.3.43	$u^{f-1}(1-u)^{g-1}$ $\times F_C^{(r)}(f+g,b;d_1,..,d_r;$ $ux_1,..,ux_k,$ $[1-u]x_{k+1},..,[1-u]x_r)$	$\frac{\Gamma(f)\Gamma(g)}{\Gamma(f+g)}\,{}^{(k)}_{(1)}E_C^{(r)}(b,f,g;d_1,..,d_r;$ $x_1,..,x_r)$ $f,g > 0$; both series must be cgt. over the range of integration.
A.1.3.44	$u^{f-1}(1-u)^{g-1}$ $\times F_D^{(r)}(f+g,b_1,..,b_r;$ $c;ux_1,..,ux_k,$ $[1-u]x_{k+1},..,[1-u]x_r)$	$\frac{\Gamma(f)\Gamma(g)}{\Gamma(f+g)}\,{}^{(k)}_{(2)}E_D^{(r)}(f,g,b_1,..,b_r;$ $c;x_1,..,x_r)$ $f,g > 0$; both series must be cgt. over the range of integration.

A.2 DEFINITE INTEGRALS

A.2.1 Definite Integrals involving Single Hypergeometric Functions

	$f(u)$	$\int_0^z f(u)\,du$
A.2.1.1	$u^{f-1}\,{}_0F_1(-;f;ux)$	$z^f/f\ {}_0F_1(-;f+1;zx)$ $f > 0$
A.2.1.2	$u^{f-1}\,{}_0F_1(-;\frac{f}{2};xu^2)$	$z^f/f\ {}_0F_1(-;f/2+1;xz^2)$ $f > 0$
A.2.1.3	$u^{f-1}\,{}_1F_1(f+1;f;ux)$	$z^f/f\ e^{xz}\quad ;f > 0$
A.2.1.4	$u^{f-1}\,{}_1F_1(\frac{f}{2}+1;\frac{f}{2};xu^2)$	$z^f/f\ \exp(xz^2)$ $f > 0$
A.2.1.5	$u^{f-1}\,{}_1F_1(a;f;ux)$	$z^f/f\ {}_1F_1(a;f+1;xz)$ $f > 0$
A.2.1.6	$u^{f-1}\,{}_1F_1(f+1;b;ux)$	$z^f/f\ {}_1F_1(f;b;ux)$ $f > 0$
A.2.1.7	$u^{f-1}\,{}_1F_1(a;\frac{f}{2};xu^2)$	$z^f/f\ {}_1F_1(a;f/2+1;xz^2)$ $f > 0$
A.2.1.8	$u^{f-1}\,{}_1F_1(\frac{f}{2}+1;b;xu^2)$	$z^f/f\ {}_1F_1(f/2;b;xz^2)$ $f > 0$
A.2.1.9	$u^{f-1}\,{}_1F_1(\frac{f}{m}+1;\frac{f}{m};xu^m)$	$z^f/f\ \exp(xz^m)$ $f > 0$
A.2.1.10	$u^{f-1}\,{}_2F_1\left({a,f+1; \atop c\ ;}xu\right)$	$z^f/f\ {}_2F_1(a,f;b;xz)$ $f > 0\quad \lvert z\rvert \le 1;\ \lvert x\rvert < 1$
A.2.1.11	$u^{f-1}\,{}_2F_1\left({a,b; \atop f/2;}xu^2\right)$	$z^f/f\ {}_2F_1(a,b;\frac{f}{2}+1;xz^2)$ $f > 0;\ \lvert z\rvert \le 1;\ \lvert x\rvert < 1$
A.2.1.12	$u^{f-1}\,{}_2F_1\left({a,b; \atop f\ ;}ux\right)$	$z^f/f\ {}_2F_1(a,b;f+1;xz)$ $f > 0;\ \lvert z\rvert \le 1;\ \lvert x\rvert < 1$

	$f(u)$	$\int_0^z f(u)\,du$
A.2.1.13	$u^{f-1}\ {}_2F_1\left({a,b; \atop c\ ;}ux\right)$	$z^f/f\ {}_3F_2\left({a,b,f; \atop c,f+1;}xz\right)$ $f > 0;\ \lvert z\rvert \leq 1;\ \lvert x\rvert < 1$
A.2.1.14	$u^{f-1}\ {}_2F_1\left({a,b; \atop c;}xu^2\right)$	$z^f/f\ {}_3F_2\left({a,b,f/2; \atop c,1+f/2;}xz^2\right)$ $f > 0;\ \lvert z\rvert \leq 1;\ \lvert x\rvert < 1$
A.2.1.15	$u^{f-1}\ {}_AF_B\left({(a); \atop (b);}xu^m\right)$ $A \leq B+1$ unless ${}_AF_B$ terminates.	$z^f/f\ {}_{A+1}F_{B+1}\left({(a),f/m\ ; \atop (b),f/m+1;}xz^m\right)$ $f > 0;\ \lvert z\rvert \leq 1$ $\lvert x\rvert < 1$ if $A=B+1$. No restriction on x if $A \leq B$ or if ${}_AF_B$ terminates.
A.2.1.16	$u^{f-1}(1-u)^{g-1}$ $\times {}_AF_B\left({(a); \atop (b);}ux\right)$ $\lvert x\rvert < 1$ if $A=B+1$. No restriction on x if $A \leq B$ or if ${}_AF_B$ terminates.	$z^f/f\ F^{1:A;1}_{1:B;0}\left({f\ :(a);1-g; \atop f+1:(b);\ -\ ;}xz,z\right)$ $f,g > 0;\ \lvert z\rvert \leq 1;\ A \leq B+1$ unless ${}_AF_B$ terminates.
A.2.1.17	$u^{f-1}\ e^{-u}$ $\times {}_AF_B\left({(a); \atop (b);}ux\right)$	$z^f/f\ F^{1:A;0}_{1:B;0}\left({f\ :(a);-; \atop f+1:(b);-;}xz,-z\right)$ $f > 0;\ A \leq B$ unless ${}_AF_B$ terminates.
A.2.1.18	$u^{f-1}\ G^{m,n}_{p,q}\left(x\,\middle\vert\,{(a) \atop (b)}\right)$ $f > 0$	$z^f\ G^{m\ ,n+1}_{p+1,q+1}\left(zx\,\middle\vert\,{a_1,..,a_n,1-f,\ a_{n+1},..,a_p \atop b_1,..,b_m,\ -f,\ b_{m+1},..,b_q}\right)$

A.2.2 Definite Integrals involving Double Hypergeometric Functions

	$f(u)$	$\int_0^z f(u)\, du$
A.2.2.1	$u^{f-1}\Phi_2(\frac{f}{m}+1,b;c;xu^m,y)$	$z^f/f\ \Phi_2(\frac{f}{m},b;c;xz^m,y)$ $f>0$
A.2.2.2	$u^{f-1}\times\Phi_2(b,b';\frac{f}{m};xu^m,yu^m)$	$z^f/f\ \Phi_2(b,b';\frac{f}{m}+1;xz^m,yz^m)$ $f>0$
A.2.2.3	$u^{f-1}\Psi_2(\frac{f}{m}+1;c,c';xu^m,yu^m)$	$z^f/f\ \Psi_2(\frac{f}{m};c,c';xz^m,yz^m)$ $f>0$
A.2.2.4	$u^{f-1}\times\Psi_2(a;\frac{f}{m},c;xu^m,y)$	$z^f/f\ \Psi_2(a;\frac{f}{m}+1,c;xz^m,y)$ $f>0$
A.2.2.5	$u^{f-1}\ F_1(a,\frac{f}{m}+1,b;c;xu^m,y)$	$z^f/f\ F_1(a,\frac{f}{m},b;c;xz^m,y)$ $f>0$ $\quad \lvert z^m x\rvert,\lvert y\rvert<1$
A.2.2.6	$u^{f-1}\ F_1(\frac{f}{m}+1,b,b';\frac{f}{m};xu^m,yu^m)$	$z^f/f(1-z^m x)^{-b}(1-z^m y)^{-b'}$ $f>0$ $\quad \lvert xz^m\rvert,\lvert yz^m\rvert<1$
A.2.2.7	$u^{f-1}\ F_1(\frac{f}{m}+1,b,b';c;xu^m,yu^m)$	$z^f/f\ F_1(\frac{f}{m},b,b';c;xz^m,yz^m)$ $f>0$ $\quad \lvert xz^m\rvert,\lvert yz^m\rvert<1$
A.2.2.8	$u^{f-1}\ F_1(a,b,b';\frac{f}{m};xu^m,yu^m)$	$z^f/f\ F_1(a,b,b';\frac{f}{m}+1;xz^m,yz^m)$ $f>0;$ $\quad \lvert xz^m\rvert,\lvert yz^m\rvert<1$
A.2.2.9	$u^{f-1}\ F_2(a,\frac{f}{m}+1,b;\frac{f}{m},c;xu^m,y)$	$z^f/f\ (1-xz^m)^{-a}\ {}_2F_1\left({a,b;\atop c\ ;}\frac{y}{1-xz^m}\right)$ $f>0;$ $\quad \lvert y\rvert+\lvert xz^m\rvert<1$
A.2.2.10	$u^{f-1}\ F_2(a,\frac{f}{m}+1,b;c,c';xu^m,y)$	$z^f/f\ F_2(a,\frac{f}{m},b;c,c';xz^m,y)$ $f>0;$ $\quad \lvert y\rvert+\lvert xz^m\rvert<1$

	$f(u)$	$\int_0^z f(u)\,du$
A.2.2.11	$u^{f-1}\ F_2(a,b,b';\ \frac{f}{m},c;xu^m,y)$	$z^f/f\ F_2(a,b,b';1+\frac{f}{m},c;xz^m,y)$ $f > 0 \qquad \lvert y\rvert+\lvert xz^m\rvert < 1$
A.2.2.12	$u^{f-1}\ F_2(\frac{f}{m}+1,b,b';\ c,c';xu^m,yu^m)$	$z^f/f\ F_2(\frac{f}{m},b,b';c,c';xz^m,yz^m)$ $f > 0 \qquad \lvert xz^m\rvert+\lvert yz^m\rvert < 1$
A.2.2.13	$u^{f-1}\ F_3(\frac{f}{m}+1,a,b,b';\ c;xu^m,y)$	$z^f/f\ F_3(\frac{f}{m},a,b,b';c;xz^m,y)$ $f > 0 \qquad \lvert y\rvert,\lvert xz^m\rvert < 1$
A.2.2.14	$u^{f-1}\ F_3(a,a';b,b';\ \frac{f}{m};xu^m,yu^m)$	$z^f/f\ F_3(a,a',b,b';\frac{f}{m}+1;xz^m,yz^m)$ $f > 0 \qquad \lvert xz^m\rvert,\lvert yz^m\rvert < 1$
A.2.2.15	$u^{f-1}\ F_4(a,b;\frac{f}{m},c;\ xu^m,y)$	$z^f/f\ F_4(a,b;\frac{f}{m}+1,c;xz^m,y)$ $f > 0 \qquad \lvert\sqrt{(xz^m)}\rvert+\lvert\sqrt{y}\rvert < 1$
A.2.2.16	$u^{f-1}\ F_4(\frac{f}{m}+1,b;c,c';\ u^mx,u^my)$ $f > 0$	$z^f/f\ F_4(\frac{f}{m},b;c,c';xz^m,yz^m)$ $\lvert\sqrt{(xz^m)}\rvert+\lvert\sqrt{(yz^m)}\rvert < 1$
A.2.2.17	$u^{f-1}\ e^{-u}$ $\times F^{A:B;B'}_{C:D;D'}\left(\begin{matrix}(a):(b);(b');\\(c):(d);(d');\end{matrix}\ ux,y\right)$	$\frac{z^f}{f}F^{(3)}\left(\begin{matrix}-::(a):-:\ f\ ;(b);(b');-;\\-::(c):-:f+1;(d);(d');-;\end{matrix}\ xz,y,-z\right)$ $f > 0$; series must be cgt.
A.2.2.18	$u^{f-1}(1-u)^{g-1}$ $\times F^{A:B;B'}_{C:D;D'}\left(\begin{matrix}(a):(b);(b');\\(c):(d);(d');\end{matrix}\ ux,y\right)$	$\frac{z^f}{f}F^{(3)}\left(\begin{matrix}-::(a):-:\ f\ ;(b);(b');1-g;\\-::(c):-:f+1;(d);(d');-\ ;\end{matrix}\ xz,y,z\right)$ $f,g > 0; \lvert z\rvert \leq 1$; series must be cgt.
A.2.2.19	$u^{f-1}\ e^{-u}$ $\times F^{A:B;B'}_{C:D;D'}\left(\begin{matrix}(a):(b);(b');\\(c):(d);(d');\end{matrix}\ ux,uy\right)$ $f > 0$	$\frac{z^f}{f}F^{(3)}\left(\begin{matrix}f\ ::(a):-:-;(b);(b');-;\\f+1::(c):-:-;(d);(d');-;\end{matrix}\ xz,yz,-z\right)$ series must be cgt.

	f(u)	$\int_0^z f(u)\,du$
A.2.2.20	$u^{f-1}(1-u)^{g-1}$ $\times F^{A:B;B'}_{C:D;D'}\left(\begin{matrix}(a):(b);(b');\\(c):(d);(d');\end{matrix}\ ux,uy\right)$ f,g > 0	$\frac{z^f}{f}F^{(3)}\left(\begin{matrix}f::(a):-:-;(b);(b');\\f+1::(c):-:-;(d);(d');\end{matrix}\ \begin{matrix}1-g;\\-;\end{matrix}xz,yz,z\right)$ $\lvert z\rvert \leq 1$; series must be cgt.

A.2.3 Definite Integrals involving Multiple Hypergeometric Functions

	f(u)	$\int_0^z f(u)\,du$
A.2.3.1	u^{f-1} $\times\Phi_2^{(r)}(\frac{f}{m}+1,b_2,..,b_r;c;$ $u^m x_1,x_2,..,x_r)$	$\frac{z^f}{f}\Phi_2^{(r)}(\frac{f}{m},b_2,..,b_r;c;z^m x_1,$ $x_2,..,x_r)$ f > 0
A.2.3.2	u^{f-1} $\times\Phi_2^{(r)}(b_1,..,b_r;\frac{f}{m};$ $u^m x_1,..,u^m x_r)$	$\frac{z^f}{f}\Phi_2^{(r)}(b_1,..,b_r;\frac{f}{m}+1;$ $z^m x_1,..,z^m x_r)$ f > 0
A.2.3.3	u^{f-1} $\times\Psi_2^{(r)}(a;\frac{f}{m},c_2,..,c_r;$ $u^m x_1,x_2,..,x_r)$	$\frac{z^f}{f}\Psi_2^{(r)}(a;\frac{f}{m}+1,c_2,..,c_r;$ $z^m x_1,x_2,..,x_r)$ f > 0
A.2.3.4	u^{f-1} $\times\Psi_2^{(r)}(\frac{f}{m}+1;c_1,..,c_r;$ $u^m x_1,..,u^m x_r)$	$\frac{z^f}{f}\Psi_2^{(r)}(\frac{f}{m};c_1,..,c_r;$ $z^m x_1,..,z^m x_r)$ f > 0
A.2.3.5	u^{f-1} $\times F_A^{(r)}(a,b_1,..,b_r;\frac{f}{m},$ $d_2,..,d_r;u^m x_1,x_2,..,x_r)$	$\frac{z^f}{f}F_A^{(r)}(a,b_1,..,b_r;\frac{f}{m}+1,$ $d_2,..,d_r;z^m x_1,x_2,..,x_r)$ $f > 0;\lvert z^m x_1\rvert+\lvert x_2\rvert+.+\lvert x_r\rvert < 1$

	$f(u)$	$\int_0^z f(u)\,du$
A.2.3.6	$u^{f-1} \times F_A^{(r)}(a,\frac{f}{m}+1,b_2,..,b_r; d_1,..,d_r; u^m x_1,x_2,..,x_r)$	$\frac{z^f}{f} F_A^{(r)}(a,\frac{f}{m},b_2,..,b_r;d_1,..,d_r; z^m x_1,x_2,..,x_r)$ $f > 0; \|z^m x_1\|+\|x_2\|+.+\|x_r\| < 1$
A.2.3.7	$u^{f-1} \times F_A^{(r)}(a,\frac{f}{m}+1,b_2,..,b_r; \frac{f}{m},d_2,..,d_r;u^m x_1, x_2,..,x_r)$ $f > 0$	$\frac{z^f}{f}(1-z^m x_1)^{-a} \times F_A^{(r-1)}(a,b_2,..,b_r;d_2,..,d_r; x_2/[1-z^m x_1],.., x_r/[1-z^m x_1])$ $\|z^m x_1\|+\|x_2\|+.+\|x_r\| < 1$
A.2.3.8	$u^{f-1} \times F_A^{(r)}(\frac{f}{m}+1,b_1,..,b_r; c_1,..,c_r;u^m x_1,..,u^m x_r)$	$\frac{z^f}{f} F_A^{(r)}(\frac{f}{m},b_1,..,b_r;c_1,..,c_r; z^m x_1,...,z^m x_r)$ $f > 0; \|z^m x_1\|+.+\|z^m x_r\| < 1$
A.2.3.9	$u^{f-1} \times F_B^{(r)}(\frac{f}{m}+1,a_2,..,a_r, b_1,..,b_r;c;u^m x_1,x_2, ..,x_r)$	$\frac{z^f}{f} F_B^{(r)}(\frac{f}{m},a_2,..,a_r,b_1,..,b_r;c; z^m x_1,x_2,..,x_r)$ $f > 0; \|z^m x_1\|, \|x_2\|,.., \|x_r\|<1$
A.2.3.10	$u^{f-1} \times F_B^{(r)}(a_1,..,a_r,b_1,..,b_r; \frac{f}{m};u^m x_1,..,u^m x_r)$	$\frac{z^f}{f} F_B^{(r)}(a_1,..,a_r,b_1,..,b_r;\frac{f}{m}+1; z^m x_1,..,z^m x_r)$ $f > 0; \|z^m x_1\|,..,\|z^m x_r\| < 1$
A.2.3.11	$u^{f-1} \times F_D^{(r)}(\frac{f}{m}+1,b_1,..,b_r; c;u^m x_1,..,u^m x_r)$	$\frac{z^f}{f} F_D^{(r)}(\frac{f}{m},b_1,..,b_r;c; z^m x_1,..,z^m x_r)$ $f > 0; \|z^m x_1\|,..,\|z^m x_r\| < 1$

	$f(u)$	$\int_0^z f(u)\,du$
A.2.3.12	$u^{f-1} \times F_D^{(r)}(a,\frac{f}{m}+1,b_2,..,b_r; c;u^m x_1,x_2,..,x_r)$	$\frac{z^f}{f} F_D^{(r)}(a,\frac{f}{m},b_2,..,b_r;c; u^m x_1,x_2,..,x_r)$ $f > 0; \lvert z^m x_1 \rvert, \lvert x_2 \rvert,.., \lvert x_r \rvert < 1$
A.2.3.13	$u^{f-1} \times F_C^{(r)}(a,b;\frac{f}{m},c_2,..,c_r; u^m x_1,x_2,..,x_r)$ $f > 0$	$\frac{z^f}{f} F_C^{(r)}(a,b;\frac{f}{m}+1,c_2,..,c_r; z^m x_1,x_2,..,x_r)$ $\lvert \sqrt{(z^m x_1)} \rvert + \lvert \sqrt{x_2} \rvert +.+ \lvert \sqrt{x_r} \rvert < 1$
A.2.3.14	$u^{f-1} \times F_C^{(r)}(a,\frac{f}{m}+1;c_1,..,c_r; u^m x_1,..,u^m x_r)$ $f > 0$	$\frac{z^f}{f} F_C^{(r)}(a,\frac{f}{m};c_1,..,c_r; z^m x_1,..,z^m x_r)$ $\lvert \sqrt{(z^m x_1)} \rvert +.+ \lvert \sqrt{(z^m x_r)} \rvert < 1$
A.2.3.15	$u^{f-1} \times F_D^{(r)}(a,b_1,..,b_r;\frac{f}{m}; u^m x_1,..,u^m x_r)$ $f > 0$	$\frac{z^f}{f} F_D^{(r)}(a,b_1,..,b_r;\frac{f}{m}+1; z^m x_1,..,z^m x_r)$ $\lvert z^m x_1 \rvert,.., \lvert z^m x_r \rvert < 1$

A.2.4 Definite Integrals of Convolution Type

	$f(v)$	$\int_0^t f(v)\,dv$
A.2.4.1	$v^{f-1}(t-v)^{g-1} \times {}_0F_1(-;f;av) \times {}_0F_1(-;g;b[t-v])$	$\frac{\Gamma(f)\Gamma(g)}{\Gamma(f+g)} t^{f+g-1} \times {}_0F_1(-;f+g;[a+b]t)$ $f,g > 0$
A.2.4.2	$v^{f-1}(t-v)^{g-1} \times {}_1F_1(f+g;c;[x-y]v+yt)$ $f,g > 0$	$\frac{\Gamma(f)\Gamma(g)}{\Gamma(f+g)} t^{f+g-1} \times \Phi_2(f,g;c;xt,yt)$

	$f(v)$	$\int_0^t f(v)\, dv$
A.2.4.3	$v^{f-1}(t-v)^{g-1}$ $\times {}_1F_1(g;f;av)$ $\times {}_1F_1(f;g;b[t-v])$	$\frac{\Gamma(f)\Gamma(g)}{\Gamma(f+g)} t^{f+g-1} e^{(a+b)t}$ $f,g > 0$
A.2.4.4	$v^{c-1}(t-v)^{d-1}$ $\times {}_1F_1(f;c;av)$ $\times {}_1F_1(g;d;b[t-v])$	$\frac{\Gamma(c)\Gamma(d)}{\Gamma(c+d)} t^{c+d-1}$ $\times {}_1F_1(f+g;c+d;[a+b]t)$ $f,g > 0$
A.2.4.5	$v^{f-1}(t-v)^{g-1}$ $\times \Psi_2(f+g;d,d';xv, y[t-v])$ $f,g > 0$	$\frac{\Gamma(f)\Gamma(g)}{\Gamma(f+g)} t^{f+g-1}$ $\times {}_1F_1(f;d;xt)\,{}_1F_1(g;d';yt)$
A.2.4.6	$v^{f-1}(t-v)^{g-1}$ $\times \Psi_2(f+g;f,g;xv, y[t-v])$	$\frac{\Gamma(f)\Gamma(g)}{\Gamma(f+g)} t^{f+g-1} e^{(x+y)t}$ $f,g > 0$
A.2.4.7	$v^{c-1}(t-v)^{d-1}$ $\times {}_FF_G\left(\begin{matrix}(f);\\(g);\end{matrix} av^2\right)$ $\times {}_HF_J\left(\begin{matrix}(h);\\(j);\end{matrix} b[t-v]^2\right)$ $f,g > 0$; series cgt.	$\frac{\Gamma(c)\Gamma(d)}{\Gamma(c+d)} t^{c+d-1}$ $\times F^{0:F+2;G+2}_{2:\ G\ ;\ J}\left[\begin{matrix} - & :(f),\frac{c}{2},\frac{c+1}{2};(h),\frac{d}{2},\frac{d+1}{2}; \\ \frac{c+d}{2},\frac{c+d+1}{2}: & (g)\quad ;\quad (j)\quad ; \end{matrix} at^2, bt^2\right]$
A.2.4.8	$v^{c-1}(t-v)^{d-1}$ $\times {}_FF_G\left(\begin{matrix}(f);\\(g);\end{matrix} av\right)$ $\times {}_HF_J\left(\begin{matrix}(h);\\(j);\end{matrix} b[t-v]\right)$ $f,g > 0$	$\frac{\Gamma(c)\Gamma(d)}{\Gamma(c+d)} t^{c+d-1}$ $\times F^{0:F+1;H+1}_{1:\ G\ ;\ J}\left(\begin{matrix} - & :c,(f);d,(h); \\ c+d: & (g)\ ;\ (j)\ ; \end{matrix} at, bt\right)$ series cgt.
A.2.4.9	$v^{f-1}(t-v)^{g-1}$ $\times F^{A:B;B'}_{C:D;D'}\left(\begin{matrix}(a):(b);(b');\\(c):(d);(d');\end{matrix} xv, y[t-v]\right)$ $f,g > 0$	$\frac{\Gamma(f)\Gamma(g)}{\Gamma(f+g)} t^{c+d-1}$ $\times F^{A\ :B+1;B'+1}_{C+1:\ D\ ;\ D'}\left(\begin{matrix} (a) & :(b),f;(b'),g; \\ (c),f+g: & (d)\ ;\ (d');\end{matrix} xt, yt\right)$ series cgt.

	f(v)	$\int_0^t f(v)\,dv$
A.2.4.10	$v^{c-1}(t-v)^{d-1}$ $\times {}_FF_G\left({(f); \atop (g);}av\right)$ $\times {}_JF_K\left({(j); \atop (k);}b[t-v]\right)$ $\times {}_HF_P\left({(h); \atop (p);}q[t-v]\right)$	$\frac{\Gamma(c)\Gamma(d)}{\Gamma(c+d)}\,t^{c+d-1}$ $\times F^{(3)}\left({-\,::-:d:-;(f),c;(j);(h); \atop c+d::-:-:-;\;(g)\;;(k);(p);}at,bt,qt\right)$ $c,d>0$; series cgt.
A.2.4.11	$v^{d-1}(t-v)^{e-1}$ $\times {}_FF_G\left({(f); \atop (g);}av\right)$ $\times F^{H:J;J'}_{K:P;P'}\left({(h):(j);(j'); \atop (k):(p);(p');}b[t-v],c[t-v]\right)$	$\frac{\Gamma(d)\Gamma(e)}{\Gamma(d+e)}\,t^{d+e-1}$ $\times F^{(3)}\left({-\,::-:(h),e;-;(f),d;(j);(j'); \atop d+e::-:\;(k);-;\;(g)\;;(p);(p');}at,bt,ct\right)$ $d,e>0$; series cgt.
A.2.4.12	$v^{d-1}(t-v)^{e-1}$ $\times {}_FF_G\left({(f); \atop (g);}av\right)$ $\times F^{H:J;J'}_{K:P;P'}\left({(h):(j);(j'); \atop (k):(p);(p');}b[t-v],c\right)$	$\frac{\Gamma(d)\Gamma(e)}{\Gamma(d+e)}\,t^{d+e-1}$ $\times F^{(3)}\left({-::\;-\;:(h):-;(f),d;(j),e;(j'); \atop -::d+e:(k):-;\;(g)\;;\;(p)\;;(p');}at,bt,c\right)$ $d,e>0$; series cgt.

A.3 REPEATED INTEGRALS

A.3.1 Repeated Integrals involving Single and Double Hypergeometric Functions

	f(z)	$\int^z \ldots (r) \ldots \int^z f(z)\,(dz)^r$
A.3.1.1	$z^{f-1}\,{}_AF_B\left({(a); \atop (b);}xz\right)$ series cgt.	$\frac{z^{f+r-1}}{(f,r)}\,{}_{A+1}F_{B+1}\left({(a),\;f\;; \atop (b),f+r;}xz\right)$
A.3.1.2	$z^{f-1}\,\Phi_2(f+r,b;c;xz,y)$	$\frac{z^{f+r-1}}{(f,r)}\,\Phi_2(f,b;c;xz,y)$

	$f(z)$	$\int^z \ldots (r) \ldots \int^z f(z)\ (dz)^r$
A.3.1.3	$z^{f-1}\Phi_2(b,b';f;xz,yz)$	$\frac{z^{f+r-1}}{(f,r)}\Phi_2(b,b';f+r;xz,yz)$
A.3.1.4	$z^{f-1}\Psi_2(a;f,c;xz,y)$	$\frac{z^{f+r-1}}{(f,r)}\Psi_2(a;f+r,c;xz,y)$
A.3.1.5	$z^{f-1}\Psi_2(f+r;c,c';xz,yz)$	$\frac{z^{f+r-1}}{(f,r)}\Psi_2(f;c,c';xz,yz)$
A.3.1.6	$z^{f-1}F_1(a,f+r,b;c;xz,y)$	$\frac{z^{f+r-1}}{(f,r)}F_1(a,f,b;c;xz,y)$ $\lvert xz\rvert,\lvert y\rvert < 1$
A.3.1.7	$z^{f-1}F_1(f+r,b,b';c;xz,yz)$	$\frac{z^{f+r-1}}{(f,r)}F_1(f,b,b';c;xz,yz)$ $\lvert xz\rvert,\lvert yz\rvert < 1$
A.3.1.8	$z^{f-1}F_1(a,b,b';f;xz,yz)$	$\frac{z^{f+r-1}}{(f,r)}F_1(a,b,b';f+r;xz,yz)$ $\lvert xz\rvert,\lvert yz\rvert < 1$
A.3.1.9	$z^{f-1}F_2(a,f+r,b;f,c;xz,y)$	$\frac{z^{f+r-1}}{(f,r)}(1-xz)^{-a}\,{}_2F_1\left({a,b; \atop c\ ;}\frac{y}{1-xz}\right)$ $\lvert xz\rvert+\lvert y\rvert < 1$
A.3.1.10	$z^{f-1}F_2(a,b,b';f,c;xz,y)$	$\frac{z^{f+r-1}}{(f,r)}F_2(a,b,b';f+r,c;xz,y)$ $\lvert xz\rvert+\lvert y\rvert < 1$
A.3.1.11	$z^{f-1}F_2(a,f+r,b;c,c';xz,y)$	$\frac{z^{f+r-1}}{(f,r)}F_2(a,f,b;c,c';xz,y)$ $\lvert xz\rvert+\lvert y\rvert < 1$
A.3.1.12	$z^{f-1}F_2(f+r,b,b';c,c';xz,yz)$	$\frac{z^{f+r-1}}{(f,r)}F_2(f,b,b';c,c';xz,yz)$ $\lvert xz\rvert+\lvert yz\rvert < 1$
A.3.1.13	$z^{f-1}F_3(a,a',b,b';f;xz,yz)$	$\frac{z^{f+r-1}}{(f,r)}F_3(a,a',b,b';f+r;xz,yz)$ $\lvert xz\rvert,\lvert yz\rvert < 1$

	$f(z)$	$\int^{z}..(r)..\int^{z} f(z)\ (dz)^{r}$
A.3.1.14	$z^{f-1}F_3(f+r,a,b,b';c;xz,y)$	$\frac{z^{f+r-1}}{(f,r)}F_3(f,a,b,b';c;xz,y)$ $\lvert xz\rvert,\lvert y\rvert < 1$
A.3.1.15	$z^{f-1}F_4(a,b;f,c;xz,y)$	$\frac{z^{f+r-1}}{(f,r)}F_4(a,b;f+r,c;xz,y)$ $\lvert\sqrt{(xz)}\rvert+\lvert\sqrt{y}\rvert < 1$
A.3.1.16	$z^{f-1}F_4(f+r,b;c,c';xz,yz)$	$\frac{z^{f+r-1}}{(f,r)}F_4(f,b;c,c';xz,yz)$ $\lvert\sqrt{(xz)}\rvert+\lvert\sqrt{(yz)}\rvert < 1$
A.3.1.17	z^{f-1} $\times F^{A:B;B'}_{C:D;D'}\left(\begin{matrix}(a):(b);(b');\\(c):(d);(d');\end{matrix}xz,y\right)$	$\frac{z^{f+r-1}}{(f,r)}F^{A:B+1;B'}_{C:D+1;D'}\left(\begin{matrix}(a):(b),\ f\ ;\ (b');\\(c):(d),f+r;\ (d');\end{matrix}xz,y\right)$ both series cgt.
A.3.1.18	z^{f-1} $\times F^{A:B;B'}_{C:D;D'}\left(\begin{matrix}(a):(b);(b');\\(c):(d);(d');\end{matrix}xz,yz\right)$	$\frac{z^{f+r-1}}{(f,r)}F^{A+1:B;B'}_{C+1:D;D'}\left(\begin{matrix}(a),\ f\ :(b);\ (b');\\(c),f+g:(d);\ (d');\end{matrix}xz,yz\right)$ both series cgt.

A.3.2 Repeated Integrals involving Multiple Hypergeometric Functions

	$f(z)$	$\int^{z}..(r)..\int^{z} f(z)\ (dz)^{r}$
A.3.2.1	z^{f-1} $\times\Phi_2^{(s)}(f+r,b_2,..,b_s;c;x_1z,x_2,..,x_s)$	$\frac{z^{f+r-1}}{(f,r)}\Phi_2^{(s)}(f,b_2,..,b_s;c;x_1z,x_2,..,x_s)$
A.3.2.2	z^{f-1} $\times\Phi_2^{(s)}(b_1,..,b_s;f;x_1z,..,x_sz)$	$\frac{z^{f+r-1}}{(f,r)}\Phi_2^{(s)}(b_1,..,b_s;f+r;x_1z,..,x_sz)$

	$f(z)$	$\int^z..(r)..\int^z f(z)\ (dz)^r$
A.3.2.3	$z^{f-1} \times\Psi_2^{(s)}(a;f,c_2,..,c_s; x_1z,x_2,..,x_s)$	$\frac{z^{f+r-1}}{(f,r)}\Psi_2^{(s)}(a;f+r,c_2,..,c_s; x_1z,x_2,..,x_s)$
A.3.2.4	$z^{f-1} \times\Psi_2^{(s)}(f+r;c_1,..,c_s; x_1z,..,x_sz)$	$\frac{z^{f+r-1}}{(f,r)}\Psi_2^{(s)}(f;c_1,..,c_s; x_1z,..,x_sz)$
A.3.2.5	$z^{f-1} \times F_A^{(s)}(a,f+r,b_2,..,b_s; f,c_2,..,c_s; x_1z,x_2,..,x_s)$	$\frac{z^{f+r-1}}{(f,r)}(1-x_1z)^{-a} \times F_A^{(s-1)}(a,b_2,..,b_s;c_2,..,c_s; x_2/[1-x_1z],..,x_s/[1-x_1z])$ $\lvert x_1z\rvert+\lvert x_2\rvert+.+\lvert x_s\rvert < 1$
A.3.2.6	$z^{f-1} \times F_A^{(s)}(a,b_1,..,b_s; f,c_2,..,c_s; x_1z,x_2,..,x_s)$	$\frac{z^{f+r-1}}{(f,r)}F_A^{(s)}(a,b_1,..,b_s; f+n, c_2,..,c_s;x_1z,x_2,..,x_s)$ $\lvert x_1z\rvert+\lvert x_2\rvert+.+\lvert x_s\rvert < 1$
A.3.2.7	$z^{f-1} \times F_A^{(s)}(f+r,b_1,..,b_s; c_1,..,c_s;x_1z,..,x_sz)$	$\frac{z^{f+r-1}}{(f,r)}F_A^{(s)}(f,b_1,..,b_s; c_1,..,c_s;x_1z,..,x_sz)$ $\lvert x_1z\rvert+.+\lvert x_sz\rvert < 1$
A.3.2.8	$z^{f-1} \times F_B^{(s)}(f+r,a_2,..,a_s, b_1,..,b_s;c; x_1z,x_2,..,x_s)$	$\frac{z^{f+r-1}}{(f,r)}F_B^{(s)}(f,a_2,..,a_s, b_1,..,b_s;c;x_1z,x_2,..,x_s)$ $\lvert x_1z\rvert,\lvert x_2\rvert,..,\lvert x_s\rvert < 1$
A.3.2.9	$z^{f-1} \times F_B^{(s)}(a_1,..,a_s,b_1,..,b_s; f;x_1z,..,x_sz)$	$\frac{z^{f+r-1}}{(f,r)}F_B^{(s)}(a_1,..,a_s,b_1,..,b_s; f+r; x_1z,..,x_sz)$ $\lvert x_1z\rvert,..,\lvert x_sz\rvert < 1$

	$f(z)$	$\int^z..(r)..\int^z f(z)\,(dz)^r$
A.3.2.10	$z^{f-1} \times F_C^{(s)}(f+r,b;c_1,..,c_s;x_1z,..,x_sz)$	$\frac{z^{f+r-1}}{(f,r)} F_C^{(s)}(f,b;c_1,..,c_s;x_1z,..,x_sz)$ $\lvert\sqrt{(x_1z)}\rvert+.+\lvert\sqrt{(x_sz)}\rvert < 1$
A.3.2.11	$z^{f-1} \times F_C^{(s)}(a,b;f,c_2,..,c_s;x_1z,x_2,..,x_s)$	$\frac{z^{f+r-1}}{(f,r)} F_C^{(s)}(a,b;f+r,c_2,..,c_s;x_1z,x_2,..,x_s)$ $\lvert\sqrt{(x_1z)}\rvert+\lvert\sqrt{x_2}\rvert+.+\lvert\sqrt{x_s}\rvert < 1$
A.3.2.12	$z^{f-1} \times F_D^{(s)}(a,f+r,b_2,..,b_s;c;x_1z,x_2,..,x_s)$	$\frac{z^{f+r-1}}{(f,r)} F_D^{(s)}(a,f,b_2,..,b_s;c;x_1z,x_2,..,x_s)$ $\lvert x_1z\rvert,\lvert x_2\rvert,..,\lvert x_s\rvert < 1$
A.3.2.13	$z^{f-1} \times F_D^{(s)}(f+r,b_1,..,b_s;f;x_1z,..,x_sz)$	$\frac{z^{f+r-1}}{(f,r)}(1-x_1z)^{-b_1}..(1-x_sz)^{-b_s}$ $\lvert x_1z\rvert,..,\lvert x_sz\rvert < 1$
A.3.2.14	$z^{f-1} \times F_D^{(s)}(a,b_1,..,b_s;f;x_1z,..,x_sz)$	$\frac{z^{f+r-1}}{(f,r)} F_D^{(s)}(a,b_1,..,b_s;f+r;x_1z,..,x_sz)$ $\lvert x_1s\rvert,..,\lvert x_sz\rvert < 1$
A.3.2.15	$z^{f-1} \times {}^{(k)}_{(1)}E_D^{(s)}(f+r,b_1,..,b_s;c,c';x_1z,..,x_sz)$	$\frac{z^{f+r-1}}{(f,r)}\ {}^{(k)}_{(1)}E_D^{(s)}(f,b_1,..,b_s;c,c';zx_1,..,zx_s)$ both series cgt.
A.3.2.16	$z^{f-1} \times {}^{(k)}_{(1)}E_D^{(s)}(a,a',b_1,..,b_s;f;x_1z,..,x_sz)$	$\frac{z^{f+r-1}}{(f,r)}\ {}^{(k)}_{(2)}E_D^{(s)}(a,a',b_1,..,b_s;f+r;x_1z,..,x_sz)$ both series cgt.
A.3.2.17	$z^{f-1} \times {}^{(k)}_{(1)}E_C^{(s)}(a,a',f+r;b_1,..,b_s;x_1z,..,x_sz)$	$\frac{z^{f+r-1}}{(f,r)}\ {}^{(k)}_{(1)}E_C^{(s)}(a,a',f;b_1,..,b_s;zx_1,..,zx_s)$ both series cgt.

A.4 POCHHAMMER INTEGRALS

The contour of integration of the integrals of this section will be denoted by 'C' and this consists of the Pochhammer double-loop slung about the origin and the point 1.

A.4.1 Pochhammer Integrals involving Single and Double Hypergeometric Functions

	$(-u)^{1-a}(u-1)^{1-b} \times f(u)$	$\frac{\Gamma(1-a)\Gamma(1-b)\Gamma(a+b)}{(2\pi i)^2} \int_C f(u)\, du$
A.4.1.1	${}_1F_1(c;1-a-b;\frac{x}{u}+\frac{y}{1-u})$	$\Psi_2(c;1-a,1-b;x,y)$
A.4.1.2	${}_1F_1(a+b;d;ux+[1-u]y)$	$\Phi_2(a,b;d;x,y)$
A.4.1.3	${}_2F_1\left({c,a+b; \atop d\ ;}ux+[1-u]y\right)$	$F_1(c,a,b;d;x,y)$
A.4.1.4	${}_2F_1\left({c,\ d\ ; \atop 1-a-b;}\frac{x}{u}+\frac{y}{1-u}\right)$	$F_4(c,d;1-a,1-b;x,y)$
A.4.1.5	${}_CF_D\left({(c); \atop (d);}ux\right)$	${}_{C+1}F_{D+1}\left({(c),\ a\ ; \atop (d),a+b;}x\right)$
A.4.1.6	${}_CF_D\left({(c); \atop (d);}u[1-u]x\right)$	${}_{C+2}F_{D+2}\left({(c),\ a\ ,\ b\ ; \atop (d),\frac{a+b}{2},\frac{a+b+1}{2};}\frac{x}{4}\right)$
A.4.1.7	${}_CF_D\left({(c); \atop (d);}\frac{ux}{1-u}\right)$	${}_{C+1}F_{D+1}\left({(c),\ a\ ; \atop (d),1-b;}-x\right)$
A.4.1.8	${}_1F_1(1-a;e;x/u)$ $\times {}_1F_1(1-b;e';y/[1-u])$	$\Psi_2(1-a-b;e,e';x,y)$
A.4.1.9	${}_1F_1(c;a;ux)$ $\times {}_1F_1(c';b;[1-u]y)$	$\Phi_2(c,c';a+b;x,y)$

	$(-u)^{1-a}(u-1)^{1-b}$ $\times f(u)$	$\frac{\Gamma(1-a)\Gamma(1-b)\Gamma(a+b)}{(2\pi i)^2} \int_C f(u)\, du$
A.4.1.10	${}_2F_1\left({c,1-a; \atop e\ ;}x/u\right)$ $\times {}_2F_1\left({c',1-b; \atop e'\ ;}\frac{y}{1-u}\right)$	$F_2(1-a-b,c,c';e,e';x,y)$
A.4.1.11	${}_2F_1\left({c,d; \atop a\ ;}ux\right)$ $\times {}_2F_1\left({c',d'; \atop b\ ;}[1-u]y\right)$	$F_3(c,c',d,d';a+b;x,y)$
A.4.1.12	$\Phi_2(a+b,d;e;ux,y)$	$\Phi_2(a,d;e;x,y)$
A.4.1.13	$\Phi_2(1-a,1-b;1-a-b;$ $x/u,y/[1-u])$	e^{x+y}
A.4.1.14	$\Phi_2(1-a,1-b;d;\frac{x}{u},\frac{y}{1-u})$	${}_1F_1(1-a-b;d;x+y)$
A.4.1.15	$\Phi_2(1-b,d;e;\frac{ux}{1-u},y)$	$\Phi_2(a,d;e;-x,y)$
A.4.1.16	$\Phi_2(d,d';a;ux,uy)$	$\Phi_2(d,d';a+b;\ x,\ y)$
A.4.1.17	$\Phi_2(d,d';1-a-b;$ $x/u,y/[1-u])$	${}_1F_1(d;1-a;x)\,{}_1F_1(d';1-b;y)$
A.4.1.18	$\Psi_2(c;a,e;ux,y)$	$\Psi_2(c;a+b,e;x,y)$
A.4.1.19	$\Psi_2(a+b;a,b;ux,$ $[1-u]y)$	e^{x+y}
A.4.1.20	$\Psi_2(c;a,b;ux,[1-u]y)$	${}_1F_1(c;a+b;x+y)$
A.4.1.21	$\Psi_2(a+b;f,f';ux,$ $[1-u]y)$	${}_1F_1(a;f;x)\,{}_1F_1(b;f';y)$
A.4.1.22	$\Psi_2(c;a,e;\frac{ux}{1-u},y)$	$\Psi_2(c;1-b,e;-x,y)$
A.4.1.23	$\Psi_2(a+b;e,e';ux,uy)$	$\Psi_2(a;e,e';x,y)$
A.4.1.24	$\Psi_2(a+\frac{1}{2};d,d';$ $u[1-u]x,u[1-u]y)$	$\Psi_2(a:d,d';x/4,y/4)$
A.4.1.25	$F_1(c,1-a,d;e;ux,y)$	$F_1(c,1-a-b,d;e;x,y)$

	$(-u)^{1-a}(u-1)^{1-b}$ $\times f(u)$	$\frac{\Gamma(1-a)\Gamma(1-b)\Gamma(a+b)}{(2\pi i)^2}\int_C f(u)\,du$
A.4.1.26	$F_1(c,1-b,d;e;$ $ux/[1-u],y)$	$F_1(c,a,d;e;-x,y)$
A.4.1.27	$F_1(c,1-a,1-b;1-a-b;$ $x/u,y/[1-u])$	$(1-x-y)^{-c}$
A.4.1.28	$F_1(c,1-a,1-b;d;$ $x/u,y/[1-u])$	${}_2F_1(c,1-a-b;d;x+y)$
A.4.1.29	$F_1(\frac{a+b}{2},c,c';a;$ $u[1-u]x,u[1-u]y)$	$F_1(b,c,c';\frac{a+b+1}{2};x/4,y/4)$
A.4.1.30	$F_1(a+b,c,c';d;$ $ux,[1-u]y)$	$F_3(a,c,b,c';d;x,y)$
A.4.1.31	$F_2(c,a+b,d;e,e';$ $ux,y)$	$F_2(c,a,d;e,e';x,y)$
A.4.1.32	$F_2(c,a+b,d;a,e;$ $ux,y)$	$(1-x)^{-c}{}_2F_1(c,d;e;y/[1-x])$
A.4.1.33	$F_2(c,1-b,d;a,e;$ $ux/[1-u],y)$	$(1+x)^{-c}{}_2F_1(c,d;e;y/[1+x])$
A.4.1.34	$F_2(c,1-b,d;e,e';$ $ux/[1-u],y)$	$F_2(c,a,d;e,e';-x,y)$
A.4.1.35	$F_2(c,d,d';a,e;$ $ux/[1-u],y)$	$F_2(c,d,d';1-b,e;-x,y)$
A.4.1.36	$F_2(c,\frac{a+b}{2},e;a,f;$ $u[1-u]x,y)$	$F_2(c,b,e;\frac{a+b+1}{2},f;x/4,y)$
A.4.1.37	$F_2(c,a+\frac{1}{2},e;a,f;$ $u[1-u]x,y)$	$(1-\frac{x}{4})^{-c}{}_2F_1(c,e;f;\frac{4y}{4-x})$
A.4.1.38	$F_2(c,1-a,1-b;d,d';$ $x/u,y/[1-u])$	$F_4(c,1-a-b;d,d';x,y)$
A.4.1.39	$F_2(a+b,d,d';a,b;$ $ux,[1-u]y)$	$(1-x)^{-d}(1-y)^{-d'}$
A.4.1.40	$F_2(a+b,d,d';f,f';$ $ux,[1-u]y)$	${}_2F_1\left(\begin{matrix}d,a;\\f\ ;\end{matrix}x\right){}_2F_1\left(\begin{matrix}d',b;\\f'\ ;\end{matrix}y\right)$

	$(-u)^{1-a}(u-1)^{1-b} \times f(u)$	$\frac{\Gamma(1-a)\Gamma(1-b)\Gamma(a+b)}{(2\pi i)^2} \int_C f(u)\, du$
A.4.1.41	$F_2(c,d,d';a,b;ux,[1-u]y)$	$F_1(c,d,d';a+b;x,y)$
A.4.1.42	$F_3(c,c',a+b,d;e;ux,y)$	$F_3(c,c',a,d;e;x,y)$
A.4.1.43	$F_3(\frac{a+b}{2},c,\frac{a+b+1}{2},d;f;ux[1-u],y)$	$F_3(a,c,b,d;f;x/4,y)$
A.4.1.44	$F_3(d,d',e,e';a;ux,uy)$	$F_3(d,d',e,e';a+b;x,y)$
A.4.1.45	$F_3(c,c',1-a,1-b;1-a-b;x/u,y/[1-u])$	$(1-x)^{-c}(1-y)^{-c'}$
A.4.1.46	$F_3(c,c',d,d';1-a-b;x/u,y/[1-u])$	${}_2F_1(\begin{matrix}c,d;\\1-a;\end{matrix}x)\,{}_2F_1(\begin{matrix}c',d';\\1-b\ ;\end{matrix}y)$
A.4.1.47	$F_3(c,c',1-a,1-b;d\ ;x/u,y/[1-u])$	$F_1(1-a-b,c,c';d;x,y)$
A.4.1.48	$F_3(c,c',1-b,d;e;ux/[1-u],y)$	$F_3(c,c',a,d;e;-x,y)$
A.4.1.49	$F_3(c,c',d,d';a;ux/[1-u],uy/[1-u])$	$F_3(c,c',d,d';1-b;-x,-y)$
A.4.1.50	$F_4(c,d;a,e;ux,y)$	$F_4(c,d;a+b,e;x,y)$
A.4.1.51	$F_4(c,d;a,e;\frac{ux}{1-u},y)$	$F_4(c,d;1-b,e;-x,y)$
A.4.1.52	$F_4(a+b,c;d,d';ux,uy)$	$F_4(a,c;d,d';x,y)$
A.4.1.53	$F_4(\frac{a+b}{2},\frac{a+b+1}{2};f,g;ux[1-u],uy[1-u])$	$F_4(a,b;f,g;x/4,y/4)$
A.4.1.54	$F_4(1-b,c;d,d';ux/[1-u],uy/[1-u])$	$F_4(a,c;d,d';-x,-y)$
A.4.1.55	$F_4(a+b,d;f,f';ux,[1-u]y)$	$F_2(d,a,b;f,f';x,y)$

	$(-u)^{1-a}(u-1)^{1-b} \times f(u)$	$\frac{\Gamma(1-a)\Gamma(1-b)\Gamma(a+b)}{(2\pi i)^2} \int_C f(u)\, du$
A.4.1.56	$F_4(c,d;a,b;ux,[1-u]y)$	${}_2F_1(c,d;a+b;x+y)$
A.4.1.57	$F^{C:E;E'}_{D:F;F'}\left(\begin{matrix}(c):(e);\\(d):(f);\end{matrix}\begin{matrix}(e');\\(f');\end{matrix}ux,y\right)$	$F^{C:E+1;E'}_{D:F+1;F'}\left(\begin{matrix}(c):(e),\ a\ ;\\(d):(f),a+b;\end{matrix}\begin{matrix}(e');\\(f');\end{matrix}x,y\right)$
A.4.1.58	$F^{C:E;E'}_{D:F;F'}\left(\begin{matrix}(c):(e);\\(d):(f);\end{matrix}\begin{matrix}(e');\\(f');\end{matrix}ux[1-u],y\right)$	$F^{C:E+2;E'}_{D:F+2;F'}\left(\begin{matrix}(c):(e),\ a\ ,\ b\ ;\\(d):(f),\frac{a+b}{2},\frac{a+b+1}{2};\end{matrix}\begin{matrix}(e');\\(f');\end{matrix}x/4,y\right)$
A.4.1.59	$F^{C:E;E'}_{D:F;F'}\left(\begin{matrix}(c):(e);\\(d):(f);\end{matrix}\begin{matrix}(e');\\(f');\end{matrix}ux/[1-u],y\right)$	$F^{C:E+1;E'}_{D:F+1;F'}\left(\begin{matrix}(c):(e)\ \ a\ ;\\(d):(f),1-b;\end{matrix}\begin{matrix}(e');\\(f');\end{matrix}-x,y\right)$
A.4.1.60	$F^{C:E;E'}_{D:F;F'}\left(\begin{matrix}(c):(e);\\(d):(f);\end{matrix}\begin{matrix}(e');\\(f');\end{matrix}ux,uy\right)$	$F^{C+1:E;E'}_{D+1:F;F'}\left(\begin{matrix}(c),\ a\ :(e);\\(d),a+b:(f);\end{matrix}\begin{matrix}(e');\\(f');\end{matrix}x,y\right)$
A.4.1.61	$F^{C:E;E'}_{D:F;F'}\left(\begin{matrix}(c):(e);\\(d):(f);\end{matrix}\begin{matrix}(e');\\(f');\end{matrix}x/u,y/[1-u]\right)$	$F^{C+1:\ E\ ;\ E'}_{D\ :F+1;F'+1}\left(\begin{matrix}(c),1-a-b:\\(d)\ :\end{matrix}\begin{matrix}(e)\ ;\\(f),1-a;\end{matrix}\begin{matrix}(e')\ ;\\(f'),1-b;\end{matrix}x,y\right)$
A.4.1.62	$F^{C:E;E'}_{D:F;F'}\left(\begin{matrix}(c):(e);\\(d):(f);\end{matrix}\begin{matrix}(e');\\(f');\end{matrix}ux,[1-u]y\right)$	$F^{C\ :E+1;E'+1}_{D+1:\ F\ ;\ F'}\left(\begin{matrix}(c)\ :(e),a;\\(d),a+b:\ (f)\ ;\end{matrix}\begin{matrix}(e'),b;\\(f');\end{matrix}x,y\right)$
A.4.1.63	$F^{C:E;E'}_{D:F;F'}\left(\begin{matrix}(c):(e);\\(d):(f);\end{matrix}\begin{matrix}(e');\\(f');\end{matrix}ux[1-u],uy[1-u]\right)$	$F^{C+2:E;E'}_{D+2:F;F'}\left(\begin{matrix}(c),\ a\ ,\ b\ :(e);\\(d),\frac{a+b}{2},\frac{a+b+1}{2}:(f);\end{matrix}\begin{matrix}(e');\\(f');\end{matrix}x/4,y/4\right)$

A.4.2 Pochhammer Integrals involving Multiple Hypergeometric Functions

	$(-u)^{1-a}(u-1)^{1-b}$ $\times f(u)$	$\dfrac{\Gamma(1-a)\Gamma(1-b)\Gamma(a+b)}{(2\pi i)^2}\int_C f(u)\,du$
A.4.2.1	$\Phi_2^{(s)}(a+b,d_2,..,d_s;f;$ $ux_1,x_2,..,x_s)$	$\Phi_2^{(s)}(a,d_2,..,d_s;f;x_1,..,x_s)$
A.4.2.2	$\Phi_2^{(s)}(1-b,d_2,..,d_s;f;$ $ux_1/[1-u],x_2,..,x_s)$	$\Phi_2^{(s)}(a,d_2,..,d_s;f;-x_1,x_2,..,x_s)$
A.4.2.3	$\Phi_2^{(s)}(c_1,..,c_s;a;$ $ux_1,..,ux_s)$	$\Phi_2^{(s)}(c_1,..,c_s;a+b;x_1,..,x_s)$
A.4.2.4	$\Phi_2^{(s)}(c_1,..,c_s;a;$ $\frac{ux_1}{1-u},..,\frac{ux_s}{1-u})$	$\Phi_2^{(s)}(c_1,..,c_s;1-b;-x_1,..,-x_s)$
A.4.2.5	$\Psi_2^{(s)}(c;a,e_2,..,e_s;$ $ux_1,x_2,..,x_s)$	$\Psi_2^{(s)}(c;a+b,e_2,..,e_s;x_1,..,x_s)$
A.4.2.6	$\Psi_2^{(s)}(c;a,e_2,..,e_s;$ $ux_1/[1-u],x_2,..,x_s)$	$\Psi_2^{(s)}(c;1-b,e_2,..,e_s;$ $-x_1,x_2,..,x_s)$
A.4.2.7	$\Psi_2^{(s)}(a+b;e_1,..,e_s;$ $ux_1,..,ux_s)$	$\Psi_2^{(s)}(a;e_1,..,e_s;x_1,..,x_s)$
A.4.2.8	$\Psi_2^{(s)}(1-b;d_1,..,d_s;$ $\frac{ux_1}{1-u},..,\frac{ux_s}{1-u})$	$\Psi_2^{(s)}(a;d_1,..,d_s;-x_1,..,-x_s)$
A.4.2.9	$F_A^{(s)}(c,a+b,d_2,..,d_s;$ $a,e_2,..,e_s;$ $ux_1,x_2,..,x_s)$	$(1-x_1)^{-c}\,F_A^{(s-1)}(c,d_2,..,d_s;$ $e_2,..,e_s;\frac{x_2}{1-x_1},..,\frac{x_s}{1-x_1})$

	$(-u)^{1-a}(u-1)^{1-b} \times f(u)$	$\frac{\Gamma(1-a)\Gamma(1-b)\Gamma(a+b)}{(2\pi i)^2} \int_C f(u)\, du$
A.4.2.10	$F_A^{(s)}(c,d_1,..,d_s;\ a,e_2,..,e_s;\ ux_1,x_2,..,x_s)$	$F_A^{(s)}(c,d_1,..,d_s;a+b,e_2,..,e_s;\ x_1,..,x_s)$
A.4.2.11	$F_A^{(s)}(c,a+b,d_2,..,d_s;\ e_1,..,e_s;\ ux_1,x_2,..,x_s)$	$F_A^{(s)}(c,a,d_2,..,d_s;e_1,..,e_s;\ x_1,..,x_s)$
A.4.2.12	$F_A^{(s)}(c,\frac{a+b}{2},e_2,..,e_s;\ a,f_2,..,f_s;\ u[1-u]x_1,x_2,..,x_s)$	$F_A^{(s)}(c,b,e_2,..,e_s;\frac{a+b+1}{2},f_2,..,f_s;\ x_1/4,x_2,..,x_s)$
A.4.2.13	$F_A^{(s)}(c,1-b,d_2,..,d_s;\ a,e_2,..,e_s;\ \frac{ux_1}{1-u},x_2,..,x_s)$	$(1+x_1)^{-c}\ F_A^{(s-1)}(c,d_2,..,d_s;\ e_2,..,e_s;\frac{x_2}{1+x_1},..,\frac{x_s}{1+x_1})$
A.4.2.14	$F_A^{(s)}(c,d_1,..,d_s;\ a,e_2,..,e_s;\ \frac{ux_1}{1-u},x_2,..,x_s)$	$F_A^{(s)}(c,d_1,..,d_s;1-b,e_2,..,e_s;\ -x_1,x_2,..,x_s)$
A.4.2.15	$F_A^{(s)}(a+b,d_1,..,d_s;\ e_1,..,e_s;\ ux_1,..,ux_s)$	$F_A^{(s)}(a,d_1,..,d_s;e_1,..,e_s;\ x_1,..,x_s)$
A.4.2.16	$F_A^{(s)}(1-b,c_1,..,c_s;\ d_1,..,d_s;\ \frac{ux_1}{1-u},..,\frac{ux_s}{1-u})$	$F_A^{(s)}(a,c_1,..,c_s;d_1,..,d_s;\ -x_1,..,-x_s)$
A.4.2.17	$F_B^{(s)}(a+b,d_2,..,d_s,\ e_1,..,e_s;f;\ ux_1,x_2,..,x_s)$	$F_B^{(s)}(a,d_2,..,d_s,e_1,..,e_s;f;\ x_1,..,x_s)$

	$(-u)^{1-a}(u-1)^{1-b}$ $\times f(u)$	$\frac{\Gamma(1-a)\Gamma(1-b)\Gamma(a+b)}{(2\pi i)^2}\int_C f(u)\,du$
A.4.2.18	$F_B^{(s)}(\frac{a+b}{2},c_2,..,c_s,$ $\frac{a+b+1}{2},d_2,..,d_s;$ $f;ux_1[1-u],x_2,..,x_s)$	$F_B^{(s)}(a,c_2,..,c_s,b,d_2,..,d_s;f;$ $x_1/4,x_2,..,x_s)$
A.4.2.19	$F_B^{(s)}(1-b,d_2,..,d_s,$ $e_1,..,e_s;f;$ $\frac{ux_1}{1-u},x_2,..,x_s)$	$F_B^{(s)}(a,d_2,..,d_s,e_1,..,e_s;f;$ $-x_1,x_2,..,x_s)$
A.4.2.20	$F_B^{(s)}(c_1,..,c_s,d_1,..,d_s;$ $a;ux_1,..,ux_s)$	$F_B^{(s)}(c_1,..,c_s,d_1,..,d_s;a+b;$ $x_1,..,x_s)$
A.4.2.21	$F_C^{(s)}(c,d;a,e_2,..,e_s;$ $\frac{ux_1}{1-u},x_2,..,x_s)$	$F_C^{(s)}(c,d;1-b,e_2,..,e_s;$ $-x_1,x_2,..,x_s)$
A.4.2.22	$F_C^{(s)}(a+b,c;d_1,..,d_s;$ $ux_1,..,ux_s)$	$F_C^{(s)}(a,c;d_1,..,d_s;x_1,..,x_s)$
A.4.2.23	$F_B^{(s)}(c_1,..,c_s,d_1,..,d_s;$ $a;\frac{ux_1}{1-u},..,\frac{ux_s}{1-u})$	$F_B^{(s)}(c_1,..,c_s,d_1,..,d_s;1-b;$ $-x_1,..,-x_s)$
A.4.2.24	$F_C^{(s)}(c,d;a,e_2,..,e_s;$ $ux_1,x_2,..,x_s)$	$F_C^{(s)}(c,d;a+b,e_2,..,e_s;x_1,..,x_s)$
A.4.2.25	$F_C^{(s)}(\frac{a+b}{2},\frac{a+b+1}{2};$ $c_1,..,c_s;$ $ux_1[1-u],..,ux_s[1-u])$	$F_C^{(s)}(a,b;c_1,..,c_s;$ $x_1/4,..,x_s/4)$

	$(-u)^{1-a}(u-1)^{1-b} \times f(u)$	$\frac{\Gamma(1-a)\Gamma(1-b)\Gamma(a+b)}{(2\pi i)^2} \int_C f(u)\, du$
A.4.2.26	$F_C^{(s)}(1-b,c;d_1,..,d_s;\frac{ux_1}{1-u},..,\frac{ux_s}{1-u})$	$F_C^{(s)}(a,c;d_1,..,d_s;-x_1,..,-x_s)$
A.4.2.27	$F_D^{(s)}(c,a+b,d_2,..,d_s;e;ux_1,x_2,..,x_s)$	$F_D^{(s)}(c,a,d_2,..,d_s;e;x_1,..,x_s)$
A.4.2.28	$F_D^{(s)}(c,1-b,d_2,..,d_s;e;\frac{ux_1}{1-u},x_2,..,x_s)$	$F_D^{(s)}(c,a,d_2,..,d_s;e;-x_1,x_2,..,x_s)$
A.4.2.29	$F_D^{(s)}(a+b,c_1,..,c_s;d;ux_1,..,ux_s)$	$F_D^{(s)}(a,c_1,..,c_s;d;x_1,..,x_s)$
A.4.2.30	$F_D^{(s)}(\frac{a+b+1}{2},c_1,..,c_s;a;u[1-u]x_1,..,u[1-u]x_s)$	$F_D^{(s)}(b,c_1,..,c_s;\frac{a+b}{2};x_1/4,..,x_s/4)$
A.4.2.31	$F_D^{(s)}(\frac{a+b}{2},c_1,..,c_s;a;u[1-u]x_1,..,u[1-u]x_s)$	$F_D^{(s)}(b,c_1,..,c_s;\frac{a+b+1}{2};x_1/4,..,x_s/4)$
A.4.2.32	$F_D^{(s)}(c,d_1,..,d_s;a;\frac{ux_1}{1-u},..,\frac{ux_s}{1-u})$	$F_D^{(s)}(c,d_1,..,d_s;1-b;-x_1,..,-x_s)$
A.4.2.33	$F^{C:E';.;E^{(s)}}_{D:F';.;F^{(s)}}\left[\begin{matrix}(c): (e');.;(e^{(s)}); \\ (d): (f');.;(f^{(s)});\end{matrix} u[1-u]x_1,x_2,..,x_s\right]$	$F^{C:E'+2;E'';.;E^{(s)}}_{D:F'+2;F'';.;F^{(s)}}\left[\begin{matrix}(c):(e'), a, b :(e'');.;(e^{(s)}); \\ (d):(f'), \frac{a+b}{2},\frac{a+b+1}{2};(f'');.;(f^{(s)});\end{matrix} x_1/4,x_2,..,x_s\right]$

	$(-u)^{1-a}(u-1)^{1-b} \times f(u)$	$\frac{\Gamma(1-a)\Gamma(1-b)\Gamma(a+b)}{(2\pi i)^2} \int_C f(u)\, du$
A.4.2.34	$F^{C:E';.;E^{(s)}}_{D:F';.;F^{(s)}}\left[\begin{matrix}(c): (e');.;(e^{(s)});\\ (d): (f');.;(f^{(s)});\end{matrix} \frac{ux_1}{1-u}, x_2,.,x_s\right]$	$F^{C:E'+1;E'';.;E^{(s)}}_{D:F'+1;F'';.;F^{(s)}}\left[\begin{matrix}(c):(e'), a\ ;(e'');.;(e^{(s)});\\ (d):(f'), 1-b;(f'');.;(f^{(s)});\end{matrix} -x_1,x_2,.,x_s\right]$
A.4.2.35	$F^{C:E';.;E^{(s)}}_{D:F';.;F^{(s)}}\left[\begin{matrix}(c): (e');.;(e^{(s)});\\ (d): (f');.;(f^{(s)});\end{matrix} ux_1,x_2,.,x_s\right]$	$F^{C:E'+1;E'';.;E^{(s)}}_{D:F'+1;F'';.;F^{(s)}}\left[\begin{matrix}(c):(e'), a\ ;(e'');.;(e^{(s)});\\ (d):(f'), a+b;(f'');.;(f^{(s)});\end{matrix} x_1,.,x_s\right]$
A.4.2.36	$F^{C:E';.;E^{(s)}}_{D:F';.;F^{(s)}}\left[\begin{matrix}(c): (e');.;(e^{(s)});\\ (d): (f');.;(f^{(s)});\end{matrix} ux_1,.,ux_s\right]$	$F^{C+1:E';.;E^{(s)}}_{D+1:F';.;F^{(s)}}\left[\begin{matrix}(c),\ a\ : (e');.;(e^{(s)});\\ (d),a+b: (f');.;(f^{(s)});\end{matrix} x_1,.,x_s\right]$
A.4.2.37	$F^{C:E';.;E^{(s)}}_{D:F';.;F^{(s)}}\left[\begin{matrix}(c): (e');.;(e^{(s)});\\ (d): (f');.;(f^{(s)});\end{matrix} ux_1[1-u],.,ux_s[1-u]\right]$	$F^{C+2:E';.;E^{(s)}}_{D+2:F';.;F^{(s)}}\left[\begin{matrix}(c),\ a\ ,\ b\ : (e');.;(e^{(s)});\\ (d),\frac{a+b}{2},\frac{a+b+1}{2}: (f');.;(f^{(s)});\end{matrix} x_1/4,.,x_s/4\right]$

	$(-u)^{1-a}(u-1)^{1-b}$ $\times f(u)$	$\frac{\Gamma(1-a)\Gamma(1-b)\Gamma(a+b)}{(2\pi i)^2}\int_C f(u)\,du$
A.4.2.38	$F^{C:E';.;E^{(s)}}_{D:F';.;F^{(s)}}\left[\begin{matrix}(c): (e');.;(e^{(s)});\\ (d): (f');.;(f^{(s)});\end{matrix}\ \frac{ux_1}{1-u},..,\frac{ux_s}{1-u}\right]$	$F^{C+1:E';.;E^{(s)}}_{D+1:F';.;F^{(s)}}\left[\begin{matrix}(c),\ a: (e');.;(e^{(s)});\\ (d),1-b: (f');.;(f^{(s)});\end{matrix}\ -x_1,..,-x_s\right]$

	$(-u)^{1-a}(u-1)^{-a}$ $\times f(u)$	$\frac{\Gamma(1-a)\Gamma(-a)\Gamma(2a+1)}{(2\pi i)^2}\int_C f(u)\,du$
A.4.2.39	$\Phi_2^{(s)}(a+\frac{1}{2},e_2,..,e_s;d;$ $ux_1[1-u],x_2,..,x_s)$	$\Phi_2^{(s)}(a,e_2,..,e_s;d;\frac{x_1}{4},x_2,..,x_s)$
A.4.2.40	$\Psi_2^{(s)}(c;a,f_2,..,f_s;$ $ux_1[1-u],x_2,..,x_s)$	$\Psi_2^{(s)}(c;a+\frac{1}{2},f_2,..,f_s;x_1/4,x_2,..,x_s)$
A.4.2.41	$F_A^{(s)}(c,a+\frac{1}{2},e_2,..,e_s;$ $a,f_2,..,f_s;$ $ux_1[1-u],x_2,..,x_s)$	$(1-x_1/4)^{-c}$ $\times F_A^{(s-1)}(c,e_2,..,e_s;f_2,..,f_s;$ $\frac{4x_2}{4-x_1},..,\frac{4x_s}{4-x_1})$
A.4.2.42	$F_A^{(s)}(a+\frac{1}{2},c_1,..,c_s;$ $d_1,..,d_s;$ $ux_1[1-u],..,ux_s[1-u])$	$F_A^{(s)}(a,c_1,..,c_s;d_1,..,d_s;$ $x_1/4,..,x_s/4)$
A.4.2.43	$F_C^{(s)}(c,d;a,f_2,..,f_s;$ $ux_1[1-u],x_2,..,x_s)$	$F_C^{(s)}(c,d;a+1/2,f_2,..,f_s;$ $x_1/4,x_2,..,x_s)$
A.4.2.44	$F_D^{(s)}(c,a+\frac{1}{2},e_2,..,e_s;$ $d;ux_1[1-u],x_2,..,x_s)$	$F_D^{(s)}(c,a,e_2,..,e_s;d;$ $x_1/4,x_2,..,x_s)$

	$(-u)^{1-a}(u-1)^{1-a}$ $\times f(u)$	$\frac{[\Gamma(1-a)]^2\Gamma(2a)}{(2\pi i)^2}\int_C f(u)\,du$
A.4.2.45	$\Phi_2^{(s)}(c_1,..,c_s;a;$ $ux_1[1-u],..,ux_s[1-u])$	$\Phi_2^{(s)}(c_1,..,c_s;a+1/2;$ $x_1/4,..,x_s/4)$
A.4.2.46	$\Psi_2^{(s)}(c;a,f_2,..,f_s;$ $ux_1[1-u],x_2,..,x_s)$	$\Psi_2^{(s)}(c;a+1/2,f_2,..,f_s;$ $x_1/4,x_2,..,x_s)$
A.4.2.47	$\Psi_2^{(s)}(a+\frac{1}{2};d_1,..,d_s;$ $ux_1[1-u],..,ux_s[1-u])$	$\Psi_2^{(s)}(a;d_1,..,d_s;$ $x_1/4,..,x_s/4)$
A.4.2.48	$F_B^{(s)}(c_1,..,c_s,d_1,..,d_s;$ $a;ux_1[1-u],..,ux_s[1-u])$	$F_B^{(s)}(c_1,..,c_s,d_1,..,d_s;a+1/2;$ $x_1/4,..,x_s/4)$
A.4.2.49	$F_D^{(s)}(a+\frac{1}{2},c_1,..,c_s;a;$ $ux_1[1-u],..,ux_s[1-u])$	$(1-x_1/4)^{-c_1}...(1-x_s/4)^{-c_s}$

A.5 INTEGRALS WITH RESPECT TO PARAMETERS

A.5.1 Integrals with respect to Parameters of Barnes Type

	$f(u)$	$\int_{\gamma-i\infty}^{\gamma+i\infty} f(u)\,du$
A.5.1.1	$\Gamma(a_1+u)\Gamma(c-u)$ $\times {}_AF_B\binom{a_1+u,a_2,..,a_A;x}{(b)\qquad;}$	$\frac{\Gamma(a_1+c)}{2^{a_1+c}}\,{}_AF_B\binom{a_1+c,a_2,..,a_A;x/2}{(b)\qquad;}$ $a_1+c>0$
A.5.1.2	$\Gamma(a_1+u)\Gamma(a_2-u)$ $\times {}_AF_B\binom{a_1+u,a_2-u,\ a_3,..,a_A;x}{(b)\quad;}$	$\frac{\Gamma(a_1+a_2)}{2^{a_1+a_2}}\,{}_AF_B\binom{\frac{a_1+a_2}{2},\frac{a_1+a_2+1}{2},\ a_3,..,a_A;x}{(b)\quad;}$ $a_1+a_2>0$
A.5.1.3	$\Gamma(b+u)\Gamma(b'-u)$ $\times {}_1F_1(b+u;c;x)$ $\times {}_1F_1(b'-u;c';y)$	$\frac{\Gamma(b+b')}{2^{b+b'}}\Psi_2(b+b';c,c';x/2,y/2)$ $b+b'>0$

	$f(u)$	$\int_{\gamma-i\infty}^{\gamma+i\infty} f(u)\,du$
A.5.1.4	$\Gamma(b+u)\Gamma(b'-u)$ $\times {}_2F_1(a,b+u;c;x)$ $\times {}_2F_1(a',b'-u;c';y)$	$\frac{\Gamma(b+b')}{2^{b+b'}} F_2(b+b',a,a';c,c';\frac{x}{2},\frac{y}{2})$ $b+b' > 0$
A.5.1.5	$\Gamma(b+u)\Gamma(b'-u)$ $\times\Phi_2(b+u,b'-u;c;x,y)$	$\frac{\Gamma(b+b')}{2^{b+b'}} {}_1F_1(b+b';c;[x+y]/2)$ $b+b' > 0$
A.5.1.6	$\Gamma(b+u)\Gamma(f-u)$ $\times\Phi_2(b+u,b';c;x,y)$	$\frac{\Gamma(b+f)}{2^{b+f}}\Phi_2(b+f,b';c;x/2,y)$ $b+f > 0$
A.5.1.7	$\Gamma(a+u)\Gamma(f-u)$ $\times\Psi_2(a+u;c,c';x,y)$	$\frac{\Gamma(a+f)}{2^{a+f}}\Psi_2(a+f;c,c';x/2,y/2)$ $a+f > 0$
A.5.1.8	$\Gamma(a+u)\Gamma(f-u)$ $\times F_1(a+u,b,b';c;x,y)$	$\frac{\Gamma(a+f)}{2^{a+f}}F_1(a+f,b,b';c;x/2,y/2)$ $a+f > 0$
A.5.1.9	$\Gamma(b+u)\Gamma(f-u)$ $\times F_1(a,b+u,b';c;x,y)$	$\frac{\Gamma(b+f)}{2^{b+f}}F_1(a,b+f,b';c;x,y)$ $b+f > 0$
A.5.1.10	$\Gamma(b+u)\Gamma(b'-u)$ $\times F_1(a,b+u,b'-u;c;x,y)$	$\frac{\Gamma(b+b')}{2^{b+b'}}{}_2F_1(a,b+b';c;[x+y]/2)$ $b+b' > 0$
A.5.1.11	$\Gamma(b+u)\Gamma(f-u)$ $\times F_2(a,b+u,b';c,c';x,y)$	$\frac{\Gamma(b+f)}{2^{b+f}}F_2(a,b+f,b';c,c';x/2,y)$ $b+f > 0$
A.5.1.12	$\Gamma(a+u)\Gamma(f-u)$ $\times F_2(a+u,b,b';c,c';x,y)$	$\frac{\Gamma(a+f)}{2^{a+f}}F_2(a+f,b,b';c,c';x/2,y/2)$ $a+f > 0$
A.5.1.13	$\Gamma(b+u)\Gamma(b'-u)$ $\times F_2(a,b+u,b'-u;c,c';x,y)$	$\frac{\Gamma(b+b')}{2^{b+b'}}F_4(a,b+b';c,c';x/2,y/2)$ $b+b' > 0$
A.5.1.14	$\Gamma(a+u)\Gamma(f-u)$ $\times F_3(a+u,a',b,b';c;x,y)$	$\frac{\Gamma(a+f)}{2^{a+f}}F_3(a+f,a',b,b';c;x/2,y)$ $a+f > 0$

	$f(u)$	$\int_{\gamma-i\infty}^{\gamma+i\infty} f(u)\,du$
A.5.1.15	$\Gamma(a+u)\Gamma(b-u) \times F_3(a+u,a',b+u,b';c;x,y)$	$\frac{\Gamma(a+b)}{2^{a+b}} F_3(\frac{a+b}{2},a',\frac{a+b+1}{2},b';c;x,y)$ $a+b > 0$
A.5.1.16	$\Gamma(b+u)\Gamma(b'-u) \times F_3(a,a',b+u,b'-u;c;x,y)$	$\frac{\Gamma(b+b')}{2^{b+b'}} F\ (b+b',a,a';c;x/2,y/2)$ $b+b' > 0$
A.5.1.17	$\Gamma(a+u)\Gamma(f-u) \times F_4(a+u,b;c,c';x,y)$	$\frac{\Gamma(a+f)}{2^{a+f}} F_4(a+f,b;c,c';x/2,y/2)$ $a+f > 0$
A.5.1.18	$\Gamma(a+u)\Gamma(b-u) \times F_4(a+u,b-u;c,c';x,y)$	$\frac{\Gamma(a+b)}{2^{a+b}} F_4(\frac{a+b}{2},\frac{a+b+1}{2};c,c';x,y)$ $a+b > 0$
A.5.1.19	$\Gamma(a_1+u)\Gamma(f-u) \times F^{A:B;B'}_{C:D;D'}\left[\begin{matrix} a_1+u,a_2,..,a_A : (b);(b'); \\ (c) : (d);(d'); \end{matrix} x,y\right]$	$\frac{\Gamma(a_1+f)}{2^{a_1+f}} F^{A:B;B'}_{C:D;D'}\left[\begin{matrix} a_1+f,a_2,..,a_A : (b);(b'); \\ (c) : (d);(d'); \end{matrix} x/2,y/2\right]$ $a_1+f > 0$
A.5.1.20	$\Gamma(a_1+u)\Gamma(a_2-u) \times F^{A:B;B'}_{C:D;D'}\left[\begin{matrix} a_1+u,a_2-u,a_3,..,a_A : (b);(b'); \\ (c) : (d);(d'); \end{matrix} x,y\right]$	$\frac{\Gamma(a_1+a_2)}{2^{a_1+a_2}} F^{A:B;B'}_{C:D;D'}\left[\begin{matrix} \frac{a_1+a_2}{2},\frac{a_1+a_2+1}{2},a_3,..,a_A : (b);(b'); \\ (c) : (d);(d'); \end{matrix} x,y\right]$ $a_1+a_2 > 0$
A.5.1.21	$\Gamma(b_1+u)\Gamma(f-u) \times F^{A:B;B'}_{C:D;D'}\left[\begin{matrix} (a):b_1+u,b_2,..,b_B;(b'); \\ (c): (d)\ ;(d'); \end{matrix} x,y\right]$	$\frac{\Gamma(b_1+f)}{2^{b_1+f}} F^{A:B;B'}_{C:D;D'}\left[\begin{matrix} (a):b_1+f,b_2,..,b_B;(b'); \\ (c): (d);(d'); \end{matrix} x/2,y\right]$ $b_1+f > 0$

	$f(u)$	$\int_{\gamma-i\infty}^{\gamma+i\infty} f(u)\,du$
A.5.1.22	$\Gamma(b_1+u)\Gamma(f-u)$ $\times\Phi_2^{(s)}(b_1+u,b_2,..,b_s;c;x_1,..,x_s)$	$\frac{\Gamma(b_1+f)}{2^{b_1+f}}\Phi_2^{(s)}(b_1+f,b_2,..,b_s;c;x_1/2,..,x_s)$ $b_1+f>0$
A.5.1.23	$\Gamma(a+u)\Gamma(f-u)$ $\times\Psi_2^{(s)}(a+u;c_1,..,c_s;x_1,..,x_s)$	$\frac{\Gamma(a+f)}{2^{a+f}}\Psi_2^{(s)}(a+f;c_1,..,c_s;x_1/2,..,x_s/2)$ $a+f>0$
A.5.1.24	$\Gamma(b_1+u)\Gamma(f-u)$ $\times F_A^{(s)}(a,b_1+u,b_2,..,b_s;c_1,..,c_s;x_1,..,x_s)$	$\frac{\Gamma(b_1+f)}{2^{b_1+f}}F_A^{(s)}(a,b_1+f,b_2,..,b_s;c_1,..,c_s;\frac{x_1}{2},x_2,..,x_s)$ $b_1+f>0$
A.5.1.25	$\Gamma(a_1+u)\Gamma(b_1-u)$ $\times F_B^{(s)}(a_1+u,a_2,..,a_s,b_1-u,b_2,..,b_s;c;x_1,..,x_s)$	$\frac{\Gamma(a_1+b_1)}{2^{a_1+b_1}}F_B^{(s)}(\frac{a_1+b_1}{2},a_2,..,a_s,\frac{a_1+b_1+1}{2},b_2,..,b_s;c;x_1,..,x_s)$ $a_1+b_1>0$
A.5.1.26	$\Gamma(a+u)\Gamma(f-u)$ $\times F_C^{(s)}(a+u,b;c_1,..,c_s;x_1,..,x_s)$	$\frac{\Gamma(a+f)}{2^{a+f}}F_C^{(s)}(a+f,b;c_1,..,c_s;x_1/2,..,x_s/2)$ $a+f>0$
A.5.1.27	$\Gamma(a+u)\Gamma(b-u)$ $\times F_C^{(s)}(a+u,b-u;c_1,..,c_s;x_1,..,x_s)$	$\frac{\Gamma(a+b)}{2^{a+b}}F_C^{(s)}(\frac{a+b}{2},\frac{a+b+1}{2};c_1,..,c_s;x_1,..,x_s)$ $a+b>0$
A.5.1.28	$\Gamma(b_1+u)\Gamma(f-u)$ $\times F_D^{(s)}(a,b_1+u,b_2,..,b_s;c;x_1,..,x_s)$	$\frac{\Gamma(b_1+f)}{2^{b_1+f}}F_D^{(s)}(a,b_1+f,b_2,..,b_s;c;x_1/2,x_2,..,x_s)$ $b_1+f>0$

A.5.2 Infinite Integrals along the Real Axis

	$f(u)$	$\int_{-\infty}^{\infty} f(u)\, du$
A.5.2.1	$[\Gamma(b_1+u)\Gamma(f-u)]^{-1} \times {}_AF_B\left[\begin{matrix}(a) & ; x\\ b_1+u, b_2, .., b_B; & \end{matrix}\right]$	$\frac{2^{b_1+f-2}}{\Gamma(b_1+f-1)}\, {}_AF_B\left(\begin{matrix}(a)\\ b_1+f-1, b_2, .., b_B\end{matrix}; 2x\right)$ $b_1+f > 1$
A.5.2.2	$[\Gamma(b_1+u)\Gamma(b_2-u)]^{-1} \times {}_AF_B\left[\begin{matrix}(a) & ; x\\ b_1+u, b_2-u, b_3, .., b_B; & \end{matrix}\right]$ $b_1+b_2 > 1$	$\frac{2^{b_1+b_2-2}}{\Gamma(b_1+b_2-1)} \times {}_AF_B\left[\begin{matrix}(a) & ; x\\ \frac{b_1+b_2}{2}, \frac{b_1+b_2+1}{2}, b_3, .., b_B; & \end{matrix}\right]$
A.5.2.3	$[\Gamma(c+u)\Gamma(f-u)]^{-1} \times \Phi_2(b,b';c+u;x,y)$	$\frac{2^{c+f-2}}{\Gamma(c+f-1)}\Phi_2(b,b';c+f-1;2x,2y)$ $c+f > 1$
A.5.2.4	$[\Gamma(c+u)\Gamma(f-u)]^{-1} \times \Psi_2(a;c+u,c';x,y)$	$\frac{2^{c+f-2}}{\Gamma(c+f-1)}\Psi_2(a;c+f-1,c';2x,y)$ $c+f > 1$
A.5.2.5	$[\Gamma(c+u)\Gamma(f-u)]^{-1} \times F_1(a,b,b';c+u;x,y)$	$\frac{2^{c+f-2}}{\Gamma(c+f-1)}F_1(a,b,b';c+f-1;2x,2y)$ $c+f > 1$
A.5.2.6	$[\Gamma(c+u)\Gamma(f-u)]^{-1} \times F_2(a,b,b';c+u,c';x,y)$	$\frac{2^{c+f-2}}{\Gamma(c+f-1)}F_2(a,b,b';c+f-1;2x,y)$ $c+f > 1$
A.5.2.7	$[\Gamma(c+u)\Gamma(c'-u)]^{-1} \times F_2(a,b,b';c+u,c'-u;x,y)$	$\frac{2^{c+c'-2}}{\Gamma(c+c'-1)}F_1(a,b,b';c+c'-1;2x,2y)$ $c+c' > 1$
A.5.2.8	$[\Gamma(c+u)\Gamma(f-u)]^{-1} \times F_3(a,a',b,b';c+u;x,y)$	$\frac{2^{c+f-2}}{\Gamma(c+f-1)}F_3(a,a',b,b';c+f-1;2x,2y)$ $c+f > 1$

	$f(u)$	$\int_{-\infty}^{\infty} f(u)\,du$
A.5.2.9	$[\Gamma(c+u)\Gamma(c'-u)]^{-1}$ $\times F_4(a,b;c+u,c'-u;x,y)$	$\frac{2^{c+c'-2}}{\Gamma(c+c'-1)}\ {}_2F_1\left({a,\ b\ ; \atop c+c'-1;}2x+2y\right)$ $c+c' > 1$
A.5.2.10	$[\Gamma(c+u)\Gamma(f-u)]^{-1}$ $\times F_4(a,b;c+u,c';x,y)$	$\frac{2^{c+f-2}}{\Gamma(c+f-1)}\ F_4(a,b;c+f-1,c';2x,y)$ $c+f > 1$
A.5.2.11	$[\Gamma(c_1+u)\Gamma(f-u)]^{-1}$ $\times F^{A:B;B'}_{C:D;D'}\left[{(a) \atop c_1+u,c_2,..,c_C}{:(b);(b'); \atop :(d);(d');}x,y\right]$	$\frac{2^{c_1+f-2}}{\Gamma(c_1+f-1)} F^{A:B;B'}_{C:D;D'}\left[{(a) \atop c_1+f-1,c_2,..,c_C}{:(b);(b'); \atop :(d);(d');}2x,2y\right]$ $c_1+f > 1$
A.5.2.12	$[\Gamma(c+u)\Gamma(c'-u)]^{-1}$ $\times\Psi_2(a;c+u,c'-u;x,y)$	$\frac{2^{c+c'-2}}{\Gamma(c+c'-1)}\ {}_1F_1(a;c+c';2x+2y)$ $c+c' > 1$
A.5.2.13	$[\Gamma(c+u)\Gamma(f-u)]^{-1}$ $\times\Phi_2^{(s)}(b_1,..,b_s;c+u;x_1,..,x_s)$	$\frac{2^{c+f-2}}{\Gamma(c+f-1)}\Phi_2^{(s)}(b_1,..,b_s;c+f-1;2x_1,..,2x_s)$ $c+f > 1$
A.5.2.14	$[\Gamma(c_1+u)\Gamma(f-u)]^{-1}$ $\times\Psi_2^{(s)}(a;c_1+u,c_2,..,c_s;x_1,..,x_s)$	$\frac{2^{c_1+f-2}}{\Gamma(c_1+f-1)}\Psi_2^{(s)}(a;c_1+f-1,c_2,..,c_s;2x_1,x_2,..,x_s)$ $c_1+f > 1$
A.5.2.15	$[\Gamma(c_1+u)\Gamma(f-u)]^{-1}$ $\times F_A^{(s)}(a,b_1,..,b_s;c_1+u,c_2,..,c_s;x_1,..,x_s)$	$\frac{2^{c_1+f-2}}{\Gamma(c_1+f-1)}F_A^{(s)}(a,b_1,..,b_s;c_1+f-1,c_2,..,c_s;2x_1,x_2,..,x_s)$ $c_1+f > 1$
A.5.2.16	$[\Gamma(d+u)\Gamma(f-u)]^{-1}$ $\times F_B^{(s)}(b_1,..,b_s,c_1,..,c_s;d+u;x_1,..,x_s)$	$\frac{2^{d+f-2}}{\Gamma(d+f-1)}F_B^{(s)}(b_1,..,b_s,c_1,..,c_s;d+f-1;2x_1,..,2x_s)$ $d+f > 1$

	$f(u)$	$\int_{-\infty}^{\infty} f(u)\,du$
A.5.2.17	$[\Gamma(c_1+u)\Gamma(f-u)]^{-1}$ $\times F_C^{(s)}(a,b;c_1+u,c_2,..,c_s;$ $x_1,..,x_s)$	$\frac{2^{c_1+f-2}}{\Gamma(c_1+f-1)} F_C^{(s)}(a,b;c_1+f-1,c_2,..,$ $c_s;2x_1,x_2,..,x_s)$ $c_1+f > 1$
A.5.2.18	$[\Gamma(c+u)\Gamma(f-u)]^{-1}$ $\times F_D^{(s)}(c+f-1,b_1,..,b_s;$ $c+u;x_1,..,x_s)$	$\frac{2^{c+f-2}}{\Gamma(c+f-1)}(1-2x_1)^{-b_1}..(1-2x_s)^{-b_s}$ $c+f > 1$
A.5.2.19	$[\Gamma(c+u)\Gamma(f-u)]^{-1}$ $\times F_D^{(s)}(a,b_1,..,b_s;c+u;$ $x_1,..,x_s)$	$\frac{2^{c+f-2}}{\Gamma(c+f-1)} F_D^{(s)}(a,b_1,..,b_s;c+f-1;$ $2x_1,..,2x_s)$ $c+f > 1$
A.5.2.20	$[\Gamma(c+u)\Gamma(f-u)]^{-1}$ $\times {}^{(k)}_{(1)}E_D^{(s)}(a,b_1,..,b_s;$ $c+u,c';x_1,..,x_s)$	$\frac{2^{c+f-2}}{\Gamma(c+f-1)}\, {}^{(k)}_{(1)}E_D^{(s)}(a,b_1,..,b_s;$ $c+f-1,c';2x_1,..,2x_k,$ $x_{k+1},..,x_s)$ $c+f > 1$
A.5.2.21	$[\Gamma(c+u)\Gamma(c'-u)]^{-1}$ $\times {}^{(k)}_{(1)}E_D^{(s)}(c+c'-1,b_1,..,$ $b_s;c+u,c'-u;$ $x_1,..,x_s)$	$\frac{2^{c+c'-2}}{\Gamma(c+c'-1)}(1-2x_1)^{-b_1}..$ $..(1-2x_s)^{-b_s}$ $c+c' > 1$
A.5.2.22	$[\Gamma(c+u)\Gamma(c'-u)]^{-1}$ $\times {}^{(k)}_{(1)}E_D^{(s)}(a,b_1,..,b_s;$ $c+u,c'-u;x_1,..,x_s)$	$\frac{2^{c+c'-2}}{\Gamma(c+c'-1)} F_D^{(s)}(a,b_1,..,b_s;$ $c+c'-1;2x_1,..,2x_s)$ $c+c' > 1$
A.5.2.23	$[\Gamma(c+u)\Gamma(f-u)]^{-1}$ $\times {}^{(k)}_{(2)}E_D^{(s)}(a,a',b_1,..,b_s;$ $c+u;x_1,..,x_s)$	$\frac{2^{c+f-2}}{\Gamma(c+f-1)}\, {}^{(k)}_{(2)}E_D^{(s)}(a,a';b_1,..,b_s;$ $c+f-1;2x_1,..,2x_s)$ $c+f > 1$
A.5.2.24	$[\Gamma(c_1+u)\Gamma(f-u)\ ]^{-1}$ $\times {}^{(k)}_{(1)}E_C^{(s)}(a,a',b;c_1+u,$ $c_2,..,c_s;x_1,..,x_s)$	$\frac{2^{c_1+f-2}}{\Gamma(c_1+f-1)}\, {}^{(k)}_{(1)}E_C^{(s)}(a,a',b;$ $c_1+f-1,c_2,..,c_s;2x_1,x_2,..,x_s)$ $c_1+f>1$

A.6 LAPLACE INTEGRALS

A.6.1 Laplace Integrals involving Single and Double Hypergeometric Functions

	$s^p f(t)/\Gamma(p)$	$\int_0^\infty e^{-st}\, t^{p-1}\, f(t)\, dt;\ p,s > 0$
A.6.1.1	${}_0F_1(-;c;xt^2)$	${}_2F_1(\frac{p}{2},\frac{p+1}{2};c;4x/s^2)$
A.6.1.2	${}_0F_1(-;p/2;xt^2)$	$(1-4x/s^2)^{-(p+1)/2}$
A.6.1.3	${}_0F_1(-;\frac{p+1}{2};xt^2)$	$(1-4x/s^2)^{-p/2}$
A.6.1.4	${}_0F_1(-;c;xt)$	${}_1F_1(p;c;x/s)$
A.6.1.5	${}_0F_1(-;p;xt)$	$\exp(x/s)$
A.6.1.6	${}_1F_1(a;c;xt)$	${}_2F_1(p,a;c;x/s)$
A.6.1.7	${}_1F_1(a;p;xt)$	$(1-x/s)^{-a}$
A.6.1.8	${}_0F_1(-;c;xt+y)$	$\Phi_3(p;c;x/s,y)$
A.6.1.9	${}_1F_1(a;c;xt+y)$	$\Phi_1(a,p;c;y,x/s)$
A.6.1.10	${}_AF_B\left(\begin{matrix}(a);\\(b);\end{matrix}xt^m\right)$	${}_{A+m}F_B\left(\begin{matrix}(a),\frac{p}{m},..,\frac{p+m-1}{m};\\(b)\quad;\end{matrix}m^m x/s^m\right)$
A.6.1.11	$G^{m,n}_{r,q}\left(xt\,\middle\|\,\begin{matrix}(a)\\(b)\end{matrix}\right)$	$1/\Gamma(p)\ G^{m\ ,n+1}_{r+1,\ q}\left(\frac{x}{s}\,\middle\|\,\begin{matrix}p,(a)\\(b)\end{matrix}\right)$
A.6.1.12	${}_0F_1(-;c;xt)$ $\times{}_0F_1(-;c';yt)$	$\Psi_2(p;c,c';x/s,y/s)$
A.6.1.13	${}_0F_1(-;c;xt^2)$ $\times{}_0F_1(-;c';yt^2)$	$F_4(\frac{p}{2},\frac{p+1}{2};c,c';4x/s^2,4y/s^2)$

	$s^p f(t)/\Gamma(p)$	$\int_0^\infty e^{-st}\, t^{p-1}\, f(t)\, dt;\ p,s > 0$
A.6.1.14	${}_1F_1(a;c;xt)$ $\times {}_1F_1(a';c';yt)$	$F_2(p,a,a';c,c';x/s,y/s)$
A.6.1.15	$\Phi_2(b,b';c;xt,yt)$	$F_1(p,b,b';c;x/s,y/s)$
A.6.1.16	$\Phi_2(b,b';p;xt,yt)$	$(1-x/s)^{-b}(1-y/s)^{-b'}$
A.6.1.17	$\Psi_2(a;p,c;xt,y)$	$(1-x/s)^{-a}\,{}_1F_1(a;c;ys/[s-x])$
A.6.1.18	$\Psi_2(a;c,c';xt,y)$	$\Psi_1(a,p;c,c';x/s,y)$
A.6.1.19	$\Psi_2(a;c,c';xt,yt)$	$F_4(p,a;c,c';x/s,y/s)$
A.6.1.20	${}_1F_1(-N;b;st)$ $\times {}_FF_G\left({(f);\atop (g);}xt\right)$	$(b-p,N)$ $\times {}_{F+2}F_{G+1}\left({p,1+p-b,(f);\atop 1+p-b-N,(g);}x/s\right)$

A.6.2 Laplace Integrals involving Multiple Hypergeometric Functions

	$s^p f(t)/\Gamma(p)$	$\int_0^\infty e^{-st}\, t^{p-1}\, f(t)\, dt;\ p,s > 0$
A.6.2.1	${}_0F_1(-;c_1;x_1t)\ldots$ $\times {}_0F_1(-;c_r;x_rt)$	$\Psi_2^{(r)}(p;c_1,..,c_r;x_1/s,..,x_r/s)$
A.6.2.2	${}_0F_1(-;c_1;x_1t^2)$ $\times {}_0F_1(-;c_r;x_rt^2)$	$F_C^{(r)}(\frac{p}{2},\frac{p+1}{2};c_1,..,c_r;\frac{4x_1}{s},..,\frac{4x_r}{s})$
A.6.2.3	${}_1F_1(a_1;c_1;x_1t)\ldots$ $\times {}_1F_1(a_r;c_r;x_rt)$	$F_A^{(r)}(p,a_1,..,a_r;c_1,..,c_r;$ $x_1/s,..,x_r/s)$

	$s^p f(t)/\Gamma(p)$	$\int_0^\infty e^{-st} t^{p-1} f(t)\, dt;\ p,s > 0$
A.6.2.4	$\Phi_2^{(r)}(b_1,..,b_r;c;x_1t,..,x_rt)$	$F_D^{(r)}(p,b_1,..,b_r;c;x_1/s,..,x_r/s)$
A.6.2.5	$\Phi_2^{(r)}(b_1,..,b_r;p;x_1t,..,x_rt)$	$(1-x_1/s)^{-b_1}\ldots(1-x_r/s)^{-b_r}$
A.6.2.6	$\Psi_2^{(r)}(a;c_1,..,c_r;x_1t,..,x_rt)$	$F_C^{(r)}(a,p;c_1,..,c_r;x_1/s,..,x_r/s)$
A.6.2.7	$\Phi_2^{(q)}(b_1,..,b_q;c;x_1t,..,x_qt)$ $\times\Phi_2^{(r)}(d_1,..,d_r;f;y_1t,..,y_rt)$	${}^{(q)}_{(1)}E_D^{(q+r)}(p,b_1,..,b_q,d_1,..,d_r;c,f;x_1/s,..,x_q/s,y_1/s,..,y_r/s)$
A.6.2.8	$\Psi_2^{(q)}(c;b_1,..,b_q;x_1t,..,x_qt)$ $\times\Psi_2^{(r)}(f;d_1,..,d_r;y_1t,..,y_rt)$	${}^{(q)}_{(1)}E_C^{(q+r)}(p,c,f;b_1,..,b_q,d_1,..,d_r;x_1/s,..,x_q/s,y_1/s,..,y_r/s)$

A.7 HANKEL LOOP INTEGRALS

A.7.1 Hankel Loop Integrals involving Single and Double Hypergeometric Functions

	$s^{1-p}\Gamma(p)\ f(t)$	$1/2\pi i\int_{-\infty}^{(0+)} e^{st}\ t^{-p}\ f(t)\, dt;\ s>0$
A.7.1.1	${}_0F_1(-;c;xt)$	${}_1F_1(1-p;c;-x/s)$
A.7.1.2	${}_0F_1(-;1-p;xt)$	$\exp(-x/s)$
A.7.1.3	${}_0F_1(-;c;xt^2)$	${}_2F_1(\frac{1-p}{2},1-\frac{p}{2};c;-4x/s^2)$
A.7.1.4	${}_0F_1(-;\frac{1-p}{2};xt^2)$	$(1+4x/s^2)^{p/2-1}$

	$s^{1-p}\Gamma(p)\ f(t)$	$1/2\pi i\int_{-\infty}^{(0+)} e^{st}\ t^{-p}\ f(t)\ dt;\ s > 0$
A.7.1.5	${}_0F_1(-;1-p/2;xt^2)$	$(1+4x/s^2)^{[p-1]/2}$
A.7.1.6	${}_1F_1(a;b;xt)$	${}_2F_1(a,1-p;b;-x/s)$
A.7.1.7	${}_1F_1(a;1-p;xt)$	$(1+x/s)^{-a}$
A.7.1.8	${}_2F_1(-n,a;b;x/t)$	${}_2F_2\left(\begin{matrix}-n,a;\\ b,p;\end{matrix}sx\right)$
A.7.1.9	${}_2F_1(-n,p;b;x/t)$	${}_1F_1(-n;b;sx)$
A.7.1.10	${}_2F_0(-n,a;-;x/t)$	${}_2F_1(-n,a;p;sx)$
A.7.1.11	${}_2F_0(-n,p;-;x/t)$	$(1-sx)^n$
A.7.1.12	${}_2F_0(-n,p/2;-;x/t^2)$	${}_1F_1(-n;\frac{p+1}{2};s^2x/4)$
A.7.1.13	${}_2F_0(-n,\frac{p+1}{2};-;x/t^2)$	${}_1F_1(-n;p/2;s^2x/4)$
A.7.1.14	${}_2F_0(-n,a;-;x/t^2)$	${}_2F_2\left(\begin{matrix}-n, & a & ;\\ p/2, & [p+1]/2 & ;\end{matrix}s^2x/4\right)$
A.7.1.15	${}_AF_B\left(\begin{matrix}(a);\\ (b);\end{matrix}xt^m\right)$	${}_{A+m}F_B\left(\begin{matrix}(a),\frac{1-p}{m},..,\frac{m-p}{m}; & \frac{(-m)^m x}{s^m}\\ (b) \qquad\qquad ; & \end{matrix}\right)$
A.7.1.16	${}_AF_B\left(\begin{matrix}(a);\\ (b);\end{matrix}x/t^m\right)$	${}_AF_{B+m}\left(\begin{matrix}(a) \qquad\qquad ; & \frac{s^m x}{m^m}\\ (b),\frac{p}{m},..,\frac{p+m-1}{m}; & \end{matrix}\right)$
A.7.1.17	${}_0F_1(-;c;xt^2)$ $\times{}_0F_1(-;c';yt^2)$	$F_4(\frac{1-p}{2},1-\frac{p}{2};c,c';4x/s^2,4y/s^2)$
A.7.1.18	${}_1F_1(a;c;xt)$ $\times{}_1F_1(a';c';yt)$	$F_2(1-p,a,a';c,c';x/s,y/s)$
A.7.1.19	${}_2F_0(-m,a;-;x/t)$ $\times{}_2F_0(-n,b;-;y/t)$	$F_3(-m,-n,a,b;p;xs,ys)$
A.7.1.20	$\Phi_1(a,b;c;xt,y)$	$F_1(a,1-p,b;c;-x/s,y)$

	$s^{1-p}\Gamma(p)\ f(t)$	$\frac{1}{2\pi i}\int_{-\infty}^{(0+)} e^{st}\ t^{-p}\ f(t)\ dt; s > 0$
A.7.1.21	$\Phi_1(a,p;c;y,x/t)$	${}_1F_1(a;c;xs+y)$
A.7.1.22	$\Psi_1(a,b;c',c;y,xt)$	$F_2(a,1-p,b;c,c';-x/s,y)$
A.7.1.23	$\Phi_2(p,b;c;x/t,y)$	$\Phi_3(b;c;y,xs)$
A.7.1.24	$\Phi_2(b,b';c;xt,yt)$	$F_1(1-p,b,b';c;-x/s,-y/s)$
A.7.1.25	$\Phi_2(b,b';1-p;xt,yt)$	$(1+x/s)^{-b}(1+y/s)^{-b'}$
A.7.1.26	$\Psi_2(a;c,c';xt,yt)$	$F_4(a,1-p;c,c';-x/s,-y/s)$
A.7.1.27	$\Psi_2(p;c,c';x/t,y/t)$	${}_0F_1(-;c;xs)\,{}_0F_1(-;c';ys)$
A.7.1.28	$F_1(a,p,b;c;x/t,y)$	$\Phi_1(a,b;c;xs,y)$
A.7.1.29	$F_1(p,b,b';c;x/t,y/t)$	$\Phi_2(b,b';c;sx,ys)$
A.7.1.31	$F_2(a,p,b;c,c';\frac{x}{t},y)$	$\Psi_1(a,b;c,c';y,xs)$
A.7.1.32	$F_2(p,a,a';b,b';x/t,y/t)$	${}_1F_1(a;b;xs)\,{}_1F_1(a';b';ys)$
A.7.1.33	$F_3(-m,-n,b,b';1-p;xt,yt)$	${}_2F_0(-m,b;-;-x/s)$ $\times {}_2F_0(-n,b';-;-y/s)$
A.7.1.34	$F_4(p,b;c,c';\frac{x}{t},\frac{y}{t})$	$\Psi_2(b;c,c';xs,ys)$
A.7.1.35	$F_4(\frac{p}{2},\frac{p+1}{2};c,c';x/t^2,y/t^2)$	${}_0F_1(-;c;xs^2/4)\,{}_0F_1(-;c';ys^2/4)$
A.7.1.36	$\Psi_1(a,b;1-p,c;y,xt)$	$(1+x/s)^{-a}\,{}_2F_1(a,b;c;\frac{ys}{s+x})$
A.7.1.37	$F_4(p,p;p,p;x/t,y/t)$	$e^{xs+ys}\ {}_0F_1(-;p;xys^2)$

A.2.7 Hankel Loop Integrals involving Multiple Hypergeometric Functions

	$s^{1-p}\Gamma(p)\ f(t)$	$\frac{1}{2\pi i}\int_{-\infty}^{(0+)} e^{st}\ t^{-p}\ f(t)\ dt;\ s > 0$
A.7.2.1	$\Phi_2^{(r)}(b_1,..,b_r;1-p;x_1t,..,x_rt)$	$(1+x_1/s)^{-b_1}...(1+x_r/s)^{-b_r}$
A.7.2.2	$\Phi_2^{(r)}(b_1,..,b_r;c;x_1t,..,x_rt)$	$F_D^{(r)}(1-p,b_1,..,b_r;c;-\frac{x_1}{s},..,-\frac{x_r}{s})$
A.7.2.3	$\Psi_2^{(r)}(p;c_1,..,c_r;x_1/t,..,x_r/t)$	${}_0F_1(-;c_1;x_1s)..{}_0F_1(-;c_r;x_rs)$
A.7.2.4	$\Psi_2^{(r)}(a;c_1,..,c_r;x_1t,..,x_rt)$	$F_C^{(r)}(a,1-p;c_1,..,c_r;-\frac{x_1}{s},..,-\frac{x_r}{s})$
A.7.2.5	$F_A^{(r)}(p,b_1,..,b_r;c_1,..,c_r;x_1/t,..,x_r/t)$	${}_1F_1(b_1;c_1;x_1s)..{}_1F_1(b_r;c_r;x_rs)$
A.7.2.6	$F_B^{(r)}(-n_1,..,-n_r,b_1,..,b_r;1-p;x_1t,..,x_rt)$	${}_2F_0(-n_1,b_1;-;-x_1/s)$ $\times{}_2F_0(-n_r;b_r;-;-x_r/s)$
A.7.2.7	$F_C^{(r)}(p,b;c_1,..,c_r;x_1/t,..,x_r/t)$	$\Psi_2^{(r)}(b;c_1,..,c_r;x_1s,..,x_rs)$
A.7.2.8	$F_D^{(r)}(p,b_1,..,b_r;c;x_1/t,..,x_r/t)$	$\Phi_2^{(r)}(b_1,..,b_r;c;x_1s,..,x_rs)$
A.7.2.9	${}_0F_1(-;c_1;x_1t)...$ $\times{}_0F_1(-;c_r;x_rt)$	$\Psi_2(1-p;c_1,..,c_r;-\frac{x_1}{s},..,-\frac{x_r}{s})$
A.7.2.10	${}_1F_1(a_1;c_1;x_1t)...$ $\times{}_1F_1(a_r;c_r;x_rt)$	$F_A^{(r)}(1-p,a_1,..,a_r;c_1,..,c_r;-x_1/s,..,-x_r/s)$

A.8 MELLIN INTEGRALS

A.8.1 Mellin Integrals involving Single and Double Hypergeometric Functions

	$f(x)$	$\int_0^\infty x^{s-1} f(x)\,dx$
A.8.1.1	${}_AF_B\left({(a);\atop (b);}-kx\right)$ $k > 0$	$k^{-s}\,\Gamma\left[{(b),(a)-s,s \atop (a),(b)-s}\right]$ $0 < s < \mathrm{Min}(a)$
A.8.1.2	$G^{m,n}_{p,q}\left(\omega x\,\Big\vert\,{(a) \atop (b)}\right)$ $\lvert\arg\omega\rvert<\frac{\pi}{2}(2m+2n-p-q)$ $-\min(b_r)<s<1-\max(a_k)$ $1\le r\le m;\ 1\le k\le n$	$\dfrac{\omega^{-s}\prod_{j=1}^{m}\Gamma(b_j+s)\prod_{j=1}^{n}\Gamma(1-a_j-s)}{\prod_{j=m+1}^{q}\Gamma(1-b_j-s)\prod_{j=n+1}^{p}\Gamma(a_j+s)}$
A.8.1.3	$K_n(yx)\,{}_FF_G\left({(f);\atop (g);}zx^2\right)$ $s+n > 0$	$\dfrac{2^{s-2}}{y^s}\Gamma\left(\frac{s+n}{2}\right)\Gamma\left(\frac{s-n}{2}\right)$ $\times{}_{F+2}F_G\left({(f),\frac{s+n}{2},\frac{s-n}{2}; \atop (g)\qquad ;}4z\,y^{-2}\right)$
A.8.1.4	$\Psi_2(a;c,c';-hx,k)$ $0 < s < a$	$h^{-s}\,\Gamma\left[{c,a-s,s \atop c-s,a}\right]{}_1F_1(a-s;c';k)$ $h > 0$
A.8.1.5	$\Phi_2(b,b';c;-hx,-kx)$ $0 < s < b$	$h^{-s}\Gamma\left[{c \atop b,b',c-s}\right]G^{2,2}_{2,2}\left(\frac{h}{k}\,\Big\vert\,{1-b+s,1 \atop b',s}\right)$ $h,k > 0$
A.8.1.6	$\Psi_2(a;c,c';-hx,-kx)$ $0 < s < a$	$h^{-s}\Gamma\left[{c,c',a-s \atop a}\right]G^{1,1}_{2,2}\left(\frac{h}{k}\,\Big\vert\,{1,c \atop s,1-c'-s}\right)$ $h,k > 0$
A.8.1.7	$\Phi_2(b,b';c;-hx,k)$ $0 < s < b$	$h^{-s}\Gamma\left[{c,b-s,s \atop c-s,b}\right]{}_1F_1(b';c-s;k)$ $h > 0$
A.8.1.8	${}_0F_1(-;c_1;-h_1x^2)$ $\times{}_0F_1(-;c_2;-h_2x^2)$ $c_1+c_2-s < 1;\ s,h_1,h_2 > 0$	$(h_2/4)^{-s/2}\Gamma\left[{s,c_2,\frac{1}{2} \atop \frac{s+1}{2},c_2-\frac{s}{2}}\right]$ ${}_2F_1(s/2,1-c_2+s/2;c_1;h_1/h_2)$

A.8.2 Mellin Integrals involving Multiple Hypergeometric Functions

	$f(x)$	$\int_0^\infty x^{s-1} f(x)\,dx$
A.8.2.1	$\Psi_2^{(r)}(a;c_1,.,c_r;$ $-h_1x,h_2,.,h_r)$ $0 < s < a;\ h_1 > 0$	$h_1^{-s}\,\Gamma\left[{a-s,c_1,s \atop c_1-s,\ a}\right]$ $\times\Psi_2^{(r-1)}(a-s;c_2,.,c_r;h_2,.,h_r)$
A.8.2.2	$\Phi_2^{(r)}(b_1,.,b_r;c;$ $-h_1x,h_2,.,h_r)$ $0 < s < b_1;\ h_1 > 0$	$h_1^{-s}\Gamma\left[{c,b_1-s,s \atop b_1,\ c-s}\right]$ $\times\Phi_2^{(r-1)}(b_2,.,b_r;c-s;h_2,.,h_r)$
A.8.2.3	${}_1F_1(c_1+N;c_1;-h_1x)$ $\times{}_1F_1(a_2;c_2;-h_2x)..$ $\times{}_1F_1(a_r;c_r;-h_rx)$ $s,h_1,.,h_r > 0$	$\dfrac{\Gamma(s)}{(c_1,N)h^s}F^{2:1}_{1:1}\left({s,1+c_1-s\ :a_2;..; \atop 1+c_1-s-N:c_2;..;}\right.$ $\left.{a_r; \atop c_r;}-h_2/h_1,.,-h_r/h_1\right)$ $c_1+N+a_2+.+a_r+s < 0$
A.8.2.4	${}_1F_1(a_1;c_1;-h_1x)..$ $\times{}_1F_1(a_r;c_r;-h_rx)$ $s,h_1 > 0$ $c_1+a_1+.+a_r+s < 0$	$\Gamma(s)(h_1+h_2)^{-s}$ $\times F_A^{(r)}(s,c_1-a_1,c_2-a_2,a_3,.,a_r;$ $c_1,.,c_r;$ $h_1/[h_1+h_2],h_2/[h_1+h_2],$ $-h_3/[h_1+h_2],.,-h_r/[h_1+h_2])$
A.8.2.5	$\Psi_2^{(r)}(a;c_1,.,c_r;$ $-h_1x,.,-h_rx)$ $0 < s < a$ $h_1,.,h_r > 0$	$\Gamma\left[{s,c_r,a-s \atop a,\ c_r-s}\right]h_r^{-s}$ $\times F_C^{(r-1)}(s,1+s-c_r;c_1,.,c_{r-1};$ $h_1/h_r,.,h_{r-1}/h_r)$
A.8.2.6	$\Phi_2^{(r)}(b_1,.,b_r;c;$ $-h_1x,.,-h_rx)$ $0 < s < b_r$ $h_1,.,h_r > 0$	$\Gamma\left[{s,c,b_r-s \atop c-s,\ b_r}\right]h_r^{-s}$ $\times F_D^{(r-1)}(s,b_1,.,b_{r-1};1-b_r+s;$ $h_1/h_r,.,h_{r-1}/h_r)$

	$f(x)$	$\int_0^\infty x^{s-1}\, f(x)\, dx$
A.8.2.7	${}_0F_1(-;c_1;-h_1x^2)\ldots$ $\times {}_0F_1(-;c_r;-h_rx^2)$ $s,h_1,..,h_r > 0$ $c_1+.+c_r-s < r/2$	$\Gamma\left[{c_r,s,1/2 \atop s/2+1/2,c_r-s/2}\right](h_r/4)^{-s/2}$ $\times F_C^{(r-1)}(s/2,1-c_r+s/2;c_1,..,c_{r-1};$ $h_1/h_r,..,h_{r-1}/h_r)$
A.8.2.8	${}_0F_1(-;c_1;h_1x^2)\ldots$ $\times {}_0F_1(-;c_r;h_rx^2)$ $\times K_a(yx)$ $s+a > 0$	$\Gamma(\frac{s+a}{2})\Gamma(\frac{s-a}{2})y^{-s}\, 2^{s-2}$ $\times F_C^{(r)}(\frac{s+a}{2},\frac{s-a}{2};c_1,..,c_r;$ $h_1/[4y^2],..,h_r/[4y^2])$

A.9 MULTIPLE INTEGRALS

A.9.1 Multiple Integrals of Euler Type

	$\prod_{j=1}^{r}[u_j^{a_j-1}(1-u_j)^{b_j-1}]^{-1}$ $\times f(u_1,..,u_r)$	$\Gamma\left[{a_1+b_1,..,a_r+b_r \atop a_1,b_1,..,a_r,b_r}\right]$ $a_j,b_j > 0$ $\times\int_0^1.(r).\int_0^1 f(u_1,..,u_r)du_1..du_r$
A.9.1.1	${}_CF_D\left({(c); \atop (d);}xu_1..u_r\right)$	${}_{C+r}F_{D+r}\left({(c),a_1,......,a_r; \atop (d),a_1+b_1,..,a_r+b_r;}x\right)$
A.9.1.2	${}_CF_D\left({(c); \atop (d);}u_1x_1+.+u_rx_r\right)$	$F^{C:1}_{D:1}\left({(c):\ a_1\ ;.;\ a_r\ ; \atop (d):b_1+a_1;.;b_r+a_r;}x_1,..,x_r\right)$
A.9.1.3	${}_CF_D\left({(c); \atop (d);}u_1[1-u_1]x_1+.\right.$ $\left..+u_r[1-u_r]x_r\right)$	$F^{C:2}_{D:2}\left[{(c):\ a_1\ ,\ b_1\ ; \atop (d):[a_1+b_1]/2,[a_1+b_1+1]/2;}\right.$ $..;\ a_r\ ;\ b_r\ ;$ $..;[a_r+b_r]/2,[a_r+b_r+1]/2;$ $\left.x_1/4,..,x_r/4\right]$
A.9.1.4	$\Phi_2^{(r)}(a_1+b_1,..,a_r+b_r;d;$ $u_1x_1,..,u_rx_r)$	$\Phi_2^{(r)}(a_1,..,a_r;d;x_1,..,x_r)$

	$\prod_{j=1}^{r}[u_j^{a_j-1}(1-u_j)^{b_j-1}]^{-1}$ $\times f(u_1,..,u_r)$	$\Gamma\left[{a_1+b_1,..,a_r+b_r \atop a_1,b_1,..,a_r,b_r}\right]$ $a_j,b_j > 0$ $\times\int_0^1 .(r). \int_0^1 f(u_1,..,u_r)du_1..du_r$
A.9.1.5	$\Psi_2^{(r)}(c;a_1,..,a_r;$ $u_1x_1,..,u_rx_r)$	$\Psi_2^{(r)}(c;a_1+b_1,..,a_r+b_r;x_1,..,x_r)$
A.9.1.6	$F_A^{(r)}(c,d_1,..,d_r;$ $a_1,..,a_r;u_1x_1,..,u_rx_r)$	$F_A^{(r)}(c,d_1,..,d_r;a_1+b_1,..,a_r+b_r;$ $x_1,..,x_r)$ $\Sigma\lvert x_i\rvert<1$
A.9.1.7	$F_A^{(r)}(c,a_1+b_1,..,a_r+b_r;$ $f_1,..,f_r;u_1x_1,..,u_rx_r)$	$F_A^{(r)}(c,a_1,..,a_r;f_1,..,f_r;$ $x_1,..,x_r)$ $\Sigma\lvert x_i\rvert<1$
A.9.1.8	$F_A^{(r)}(c,\frac{a_1+b_1+1}{2},..,$ $\frac{a_r+b_r+1}{2};a_1,..,a_r;$ $u_1[1-u_1]x_1,..,$ $u_r[1-u_r]x_r)$	$F_A^{(r)}(c,b_1,..,b_r;\frac{a_1+b_1}{2},..,\frac{a_r+b_r}{2};$ $x_1/4,..,x_r/4)$ $\Sigma\lvert x_i\rvert<4$
A.9.1.9	$F_B^{(r)}(a_1+b_1,..,a_r+b_r,$ $c_1,..,c_r;d;$ $u_1x_1,..,u_rx_r)$	$F_B^{(r)}(a_1,..,a_r,c_1,..,c_r;d;$ $x_1,..,x_r)$ $\Sigma\lvert x_i\rvert<1$
A.9.1.10	$F_C^{(r)}(c,d;a_1,..,a_r;$ $u_1x_1,..,u_rx_r)$	$F_C^{(r)}(c,d;a_1+b_1,..,a_r+b_r;$ $x_1,..,x_r)$ $\Sigma\lvert\sqrt{x_i}\rvert<1$
A.9.1.11	$F_D^{(r)}(c,a_1+b_1,..,a_r+b_r;$ $d;u_1x_1,..,u_rx_r)$	$F_D^{(r)}(c,a_1,..,a_r;d;x_1,..,x_r)$ $\Sigma\lvert x_i\rvert<1$
A.9.1.12	$F_D^{(r)}(c,a_1+b_1,..,a_r+b_r;$ $d;u_1x,..,u_kx,$ $u_{k+1}x_{k+1},..,u_rx_r)$	$F_D^{(r-k)}(c,a_1+b_1+.+a_k+b_k,$ $a_{k+1}+b_{k+1},..,a_r+b_r;d;$ $x,x_{k+1},..,x_r)$ $\lvert x\rvert,\lvert x_i\rvert<1,\ k+1\le i\le r$

A.9.2 Multiple Integrals of Dirichlet Type

$b,a_j>0$ $u_j \geqq 0$ $\Sigma u_j \leqq 1$	$\prod_{j=1}^{r} [u_j^{a_j-1}]^{-1}$ $\times(1-u_1-.-u_r)^{1-b}$ $\times\ f(u_1,.,u_r)$	$\Gamma\left[{b+a_1+.+a_r \atop b,a_1,.,a_r}\right]$ $\times \int.(r).\int f(u_1,.,u_r)du_1..du_r$
A.9.2.1	${}_2F_1(c,b+a_1+.+a_r;d;$ $x_1u_1+.+x_ru_r)$	$F_D^{(r)}(c,a_1,.,a_r;d;x_1,.,x_r)$
A.9.2.2	${}_CF_D\left({(c); \atop (d);}xu_1..u_r\right)$	${}_{C+r}F_{D+r}\left[{(c),\ a_1,.....,a_r\ ; \atop (d),\frac{b+a_1+.+a_r}{r},...,\frac{b+a_1+.+a_r+r-1}{r};}\ x/r^r\right]$
A.9.2.3	${}_CF_D\left({(c); \atop (d);}u_1x_1+.+u_rx_r\right)$	$F^{C\ :1}_{D+1;0}\left({(c) \atop (d),b+a_1+.+a_r}:{a_1,.,a_r; \atop -\ ;}\ x_1,.,x_r\right)$
A.9.2.4	${}_1F_1(d_1;a_1;u_1x_1)...$ $\times{}_1F_1(d_r;a_r;u_rx_r)$	$\Phi_2^{(r)}(d_1,.,d_r;b+a_1+.+a_r;$ $x_1,.,x_r)$
A.9.2.5	${}_2F_1(c_1,d_1;a_1;u_1x_1)..$ $\times{}_2F_1(c_r,d_r;a_r;u_rx_r)$	$F_B^{(r)}(c_1,.,c_r,d_1,.,d_r;$ $b+a_1+.+a_r;x_1,.,x_r)\ \lvert x_i\rvert<1$
A.9.2.6	${}_1F_1(b+a_1+.+a_r;d;$ $x_1u_1+.+x_ru_r)$	$\Phi_2^{(r)}(a_1,.,a_r;d;x_1,.,x_r)$
A.9.2.7	$\Psi_2^{(r)}(b+a_1+.+a_r;e_1,..e_r;$ $u_1x_1,.,u_rx_r)$	${}_1F_1(a_1;e_1;x_1)...$ $\times{}_1F_1(a_r;e_r;x_r)$
A.9.2.8	$\Psi_2^{(r)}(b+a_1+.+a_r;a_1,.,a_r;$ $u_1x_1,.,u_rx_r)$	$\exp(x_1+.+x_r)$
A.9.2.9	$\Psi_2^{(r)}(c;a_1,.,a_r;$ $u_1x_1,.,u_rx_r)$	${}_1F_1(c;b+a_1+.+a_r;x_1+.+x_r)$

$b, a_j > 0$ $u_j \geq 0$ $\Sigma u_j \leq 1$	$\prod_{j=1}^{r} [u_j^{a_j - 1}]^{-1}$ $\times (1-u_1-.-u_r)^{1-b}$ $\times f(u_1,.,u_r)$	$\Gamma\left[{b+a_1+.+a_r \atop b,a_1,.,a_r}\right]$ $\times \int.(r).\int f(u_1,.,u_r)du_1..du_r$
A.9.2.10	$F_A^{(r)}(b+a_1+.+a_r,d_1,.,d_r;$ $e_1,.,e_r;u_1x_1,.,u_rx_r)$	${}_2F_1(a_1,d_1;e_1;x_1)\ldots$ $\times{}_2F_1(a_r,d_r;e_r;x_r)\quad \lvert x_i\rvert<1$
A.9.2.11	$F_A^{(r)}(b+a_1+.+a_r,d_1,.,d_r;$ $a_1,.,a_r;u_1x_1,.,u_rx_r)$	$(1-x_1)^{-d_1}\ldots(1-x_r)^{-d_r}$
A.9.2.12	$F_A^{(r)}(c,d_1,.,d_r;$ $a_1,.,a_r;u_1x_1,.,u_rx_r)$	$F_D^{(r)}(c,d_1,.,d_r;b+a_1+.+a_r;$ $x_1,.,x_r)\quad \lvert x_i\rvert<1$
A.9.2.13	$F_C^{(r)}(c,d;a_1,.,a_r;$ $u_1x_1,.,u_rx_r)$	${}_2F_1(c,d;b+a_1+.+a_r;x_1+.+x_r)$ $\Sigma\lvert x_i\rvert<1$
A.9.2.14	$F_C^{(r)}(c,d;a_1,.,a_r;$ $u_1[1-x_2-.-x_r],$ $u_2x_2,.,u_rx_r)$	$\frac{\Gamma(b+a_1+.+a_r)\Gamma(b+a_1+.+a_r-c-d)}{\Gamma(b+a_1+.+a_r-c)\Gamma(b+a_1+.+a_r-d)}$ $b-c-d+\Sigma a_i>0$
A.9.2.15	$F_C^{(r)}(c,b+a_1+.+a_r;$ $d_1,.,d_r;u_1x_1,.,u_rx_r)$	$F_A^{(r)}(c,a_1,.,a_r;d_1,.,d_r;$ $x_1,.,x_r)\quad \Sigma\lvert x_i\rvert<1$
A.9.2.16	$F_C^{(r)}(c,b+a_1+.+a_r;$ $a_1,.,a_r;u_1x_1,.,u_rx_r)$	$(1-x_1-.-x_r)^{-c}\quad \Sigma\lvert x_i\rvert<1$
A.9.2.17	$F_D^{(r)}(b+a_1+.+a_r,d_1,.,d_r;$ $e;u_1x_1,.,u_rx_r)$	$F_B^{(r)}(d_1,.,d_r,a_1,.,a_r;e;$ $x_1,.,x_r)\quad \lvert x_i\rvert<1$
A.9.2.18	${}_0F_1(-;a_1;u_1x_1)\ldots$ $\times{}_0F_1(-;a_r;u_rx_r)$	${}_0F_1(-;b+a_1+.+a_r;x_1+.+x_r)$

A.9.3 Multiple Laplace Integrals

	$\exp(p_1t_1+.+p_rt_r)$ $\times t_1^{1-a_1}..t_r^{1-a_r}$ $\times f(t_1,.,t_r)$	$\frac{p_1^{a_1}.....p_r^{a_r}}{\Gamma(a_1)..\Gamma(a_r)}$ $\times\int_0^\infty.(r).\int_0^\infty f(t_1,.,t_r)\,dt_1..dt_r$ $p_1,.,p_r,\ a_1,.,a_r > 0$
A.9.3.1	${}_0F_1(-;d;t_1x_1+.+t_rx_r)$	$\Phi_2^{(r)}(a_1,.,a_r;d;x_1/p_1,.,x_r/p_r)$
A.9.3.2	${}_1F_1(c;d;t_1x_1+.+t_rx_r)$	$F_D^{(r)}(c,a_1,.,a_r;d;x_1/p_1,.,x_r/p_r)$
A.9.3.3	${}_CF_D\left({(c);\atop (d);}xt_1..t_r\right)$	${}_{C+r}F_D\left({(c),a_1,.,a_r;\atop (d)\quad ;}\frac{x}{p_1..p_r}\right)$
A.9.3.4	${}_CF_D\left({(c);\atop (d);}t_1x_1+.+t_rx_r\right)$	$F_{D:0}^{C:1}\left({(c):a_1;.;a_r;\atop (d):\quad -\quad ;}\frac{x_1}{p_1},.,\frac{x_r}{p_r}\right)$
A.9.3.5	$\Phi_2^{(r)}(b_1,.,b_r;c;$ $x_1t_1,.,x_rt_r)$	$F_B^{(r)}(a_1,.,a_r,b_1,.,b_r;c;$ $x_1/p_1,.,x_r/p_r)$ $\lvert x_i/p_i\rvert<1$
A.9.3.6	$\Psi_2^{(r)}(b;c_1,.,c_r;$ $t_1x_1,.,t_rx_r)$	$F_A^{(r)}(b,a_1,.,a_r;c_1,.,c_r;$ $x_1/p_1,.,x_r/p_r)$ $\Sigma\lvert x_i/p_i\rvert<1$
A.9.3.7	$\Psi_2^{(r)}(b;a_1,.,a_r;$ $x_1t_1,.,x_rt_r)$	$(1-x_1/p_1-.-x_r/p_r)^{-b}$

A.9.4 Multiple Definite Integrals

	$u_1^{1-a_1}..u_r^{1-a_r}\ f(u_j)$ $a_1,.,a_r > 0$	$z_1^{-a_1}..z_r^{-a_r}\ (a_1..a_r)$ $\times\int^{z_1}.(r)\int^{z_r} f(u_j)du_1..du_r$
A.9.4.1	${}_CF_D\left({(c);\atop (d);}u_1x_1+.+u_rx_r\right)$	$F_{D:1}^{C:1}\left({(c):\ a_1\ ;.;\ a_r\ ;\atop (d):1+a_1;.;1+a_r;}x_1z_1,.,x_rz_r\right)$
A.9.4.2	$F_A^{(r)}(b;c_1,.,c_r;a_1,.,a_r;$ $u_1x_1,.,u_rx_r)$	$F_A^{(r)}(b,c_1,.,c_r;1+a_1,.,1+a_r;$ $u_1z_1,.,u_rz_r)$

A.9.5 Multiple Mellin Integrals

	$x_1^{1-s_1}..x_r^{1-s_r} \times f(x_1,..,x_r)$	$p_1^{s_1}..p_r^{s_r}\int_0^\infty.(r).\int_0^\infty f(x_1,..,x_r) \times dx_1...dx_r$
A.9.5.1	${}_0F_1(-;c;-p_1x_1-.-p_rx_r)$ $p_1,s_1,..,p_r,s_r > 0$	$\frac{\Gamma(c)\Gamma(s_1)..\Gamma(s_r)}{\Gamma(c-s_1-.-s_r)}$ $c-s_1-.-s_r > -1/2$
A.9.5.2	${}_1F_1(a;c;-p_1x_1-.-p_rx_r)$ $p_1,s_1,..,p_r,s_r > 0$	$\frac{\Gamma(c)\Gamma(a-s_1-.-s_r)\Gamma(s_1)\Gamma(s_r)}{\Gamma(a)\Gamma(c-s_1-.-s_r)}$ $a-s_1-.-s_r > 0$
A.9.5.3	$\Psi_2^{(r)}(a;c_1,..,c_r;-p_1x_1,..,-p_rx_r)$ $p_1,s_1,..,p_r,s_r > 0$	$\frac{\Gamma(c_1)..\Gamma(c_r)\Gamma(a-s_1-.-s_r)}{\Gamma(a)\Gamma(c_1-s_1)..\Gamma(c_r-s_r)}$ $\times\Gamma(s_1)..\Gamma(s_r)$; $a-s_1-.-s_r > 0$
A.9.5.4	$\Phi_2^{(r)}(a_1,..,a_r;c;-p_1x_1-.-p_rx_r)$ $a_1-s_1,..,a_r-s_r > 0$	$\frac{\Gamma(c)\Gamma(a_1-s_1)..\Gamma(a_r-s_r)}{\Gamma(c-s_1-.-s_r)\Gamma(a_1)..\Gamma(a_r)}$ $\times\Gamma(s_1)..\Gamma(s_r)$; $a_1,s_1,..,a_r,s_r>0$

A.9.6 Multiple Integrals with respect to Parameters

	$f(u_1,..,u_r)$	$(2\pi i)^{-r}\int_{\gamma_1-i\infty}^{\gamma_1+i\infty}.(r).\int_{\gamma_r-i\infty}^{\gamma_r+i\infty} \times f(u_1,..,u_r)du_1..du_r$
A.9.6.1	$\Gamma(b_1+u_1)\Gamma(f_1-u_1)..$ $\times\Gamma(b_r+u_r)\Gamma(f_r-u_r)$ $\times F_D^{(r)}(a,b_1+u_1,..,b_r+u_r;c;x_1,..,x_r)$	$\frac{\Gamma(b_1+f_1)..\Gamma(b_r+f_r)}{2^{b_1+f_1+.+b_r+f_r}}$ $\times F_D^{(r)}(a,b_1+f_1,..,b_r+f_r;c;x_1/2,..,x_r/2)$ $b_1+f_1,..,b_r+f_r > 0$, $\lvert x_i\rvert<1$

	$f(u_1,..,u_r)$	$(2\pi i)^{-r}\int_{\gamma_1-i\infty}^{\gamma_1+i\infty}.(r).\int_{\gamma_r-i\infty}^{\gamma_r+i\infty}$ $\times f(u_1,..,u_r)du_1..du_r$
A.9.6.2	$\Gamma(a_1+u_1)\Gamma(f_1-u_1)..$ $\times\Gamma(a_r+u_r)\Gamma(f_r-u_r)$ $\times\Phi_2^{(r)}(a_1+u_1,..,a_r+u_r;$ $b;x_1,..,x_r)$ $a_1+f_1,..,a_r+f_r > 0$	$\dfrac{\Gamma(a_1+f_1)..\Gamma(a_r+f_r)}{2^{a_1+f_1+.+a_r+f_r}}$ $\times\Phi_2^{(r)}(a_1+f_1,..,a_r+f_r;b;$ $x_1/2,..,x_r/2)$
A.9.6.3	$\Gamma(b_1+u_1)\Gamma(f_1-u_1)..$ $\times\Gamma(b_r+u_r)\Gamma(f_r-u_r)$ $\times F_B^{(r)}(a_1,..,a_r,$ $b_1+u_1,..,b_r+u_r;c;$ $x_1,..,x_r)$	$\dfrac{\Gamma(b_1+f_1)..\Gamma(b_r+f_r)}{2^{b_1+f_1+.+b_r+f_r}}$ $\times F_B^{(r)}(a_1,..,a_r,b_1+f_1,..,b_r+f_r;c;$ $x_1/2,..,x_r/2)$ $b_1+f_1,..,b_r+f_r > 0, \lvert x_i\rvert<1$
A.9.6.4	$\Gamma(b_1+u_1)\Gamma(f_1-u_1)..$ $\times\Gamma(b_r+u_r)\Gamma(f_r-u_r)$ $\times F_A^{(r)}(a,b_1+u_1,..,$ $b_r+u_r;c_1,..,c_r;$ $x_1,..,x_r)$	$\dfrac{\Gamma(b_1+f_1)..\Gamma(b_r+f_r)}{2^{b_1+f_1+.+b_r+f_r}}$ $\times F_A^{(r)}(a,b_1+f_1,..,b_r+f_r;$ $c_1,..,c_r;x_1/2,..,x_r/2)$ $b_1+f_1,..,b_r+f_r > 0, \Sigma\lvert x_i\rvert<1$
A.9.6.5	$\Gamma(b_1+u_1)\Gamma(f_1-u_1)..$ $\times\Gamma(b_r+u_r)\Gamma(f_r-u_r)$ $\times F_A^{(r)}(a,b_1+u_1,..,$ $b_r+u_r;b_1+f_1,..,$ $b_r+f_r;x_1,..,x_r)$	$\dfrac{\Gamma(b_1+f_1)..\Gamma(b_r+f_r)}{2^{b_1+f_1+.+b_r+f_r}}$ $\times\ (1-x_1/2-.-x_r/2)^{-a}$ $b_1+f_1,..,b_r+f_r > 0$

	$f(u_1,..,u_r)$	$(2\pi i)^{-r}\int_{\gamma_1-i\infty}^{\gamma_1+i\infty}.(r).\int_{\gamma_r-i\infty}^{\gamma_r+i\infty}$ $\times f(u_1,..,u_r)du_1..du_r$
A.9.6.6	$\Gamma(a_1+u_1)\Gamma(b_1-u_1)..$ $\times\Gamma(a_r+u_r)\Gamma(b_r-u_r)$ $\times F_B^{(r)}(a_1+u_1,..,a_r+u_r,$ $b_1-u_1,..,b_r-u_r;c;$ $x_1,..,x_r)$ $a_1+b_1,..,a_r+b_r > 0$	$\dfrac{\Gamma(a_1+b_1)..\Gamma(a_r+b_r)}{2^{a_1+b_1+.+a_r+b_r}}$ $\times F_B^{(r)}(\dfrac{a_1+b_1}{2},..,\dfrac{a_r+b_r}{2},$ $\dfrac{a_1+b_1+1}{2},..,\dfrac{a_r+b_r+1}{2};c;$ $x_1,..,x_r)$ $\lvert x_i\rvert<1$

	$f(u_1,..,u_r)$ $\times\Gamma(c_1+u_1)\Gamma(f_1-u_1)..$ $\times\Gamma(c_r+u_r)\Gamma(f_r-u_r)$ $c_1+f_1,..,c_r+f_r > 1$	$\dfrac{\Gamma(c_1+f_1-1)..\Gamma(c_r+f_r-1)}{2^{c_1+f_1+.+c_r+f_r-2r}}$ $\times\int_{-\infty}^{\infty}.(r).\int_{-\infty}^{\infty}f(u_1,..,u_r)du_1..du_r$
A.9.6.7	$F_C^{(r)}(a,b;c_1+u_1,..,$ $c_r+u_r;x_1,..,x_r)$	$F_C^{(r)}(a,b;c_1+f_1-1,..,c_r+f_r-1;$ $2x_1,..,2x_r)$ $\Sigma\lvert\sqrt{(2x_i)}\rvert<1$
A.9.6.8	$F_A^{(r)}(a,b_1,..,b_r;$ $c_1+u_1,..,c_r+u_r;$ $x_1,..,x_r)$	$F_A^{(r)}(a,b_1,..,b_r;$ $c_1+f_1-1,..,c_r+f_r-1;$ $2x_1,..,2x_r)$ $\Sigma\lvert x_i\rvert<1/2$
A.9.6.9	$F_A^{(r)}(a,c_1+f_1-1,..,$ $c_r+f_r-1;c_1+u_1,..,$ $c_r+u_r;x_1,..,x_r)$	$(1-2x_1-..-2x_r)^{-a}$
A.9.6.10	$\Psi_2^{(r)}(a;c_1+u_1,..,$ $c_r+u_r;x_1,..,x_r)$	$\Psi_2^{(r)}(a;c_1+f_1-1,..,c_r+f_r-1;$ $2x_1,..,2x_r)$

B
Computer Programs

Introduction. The list of fifty computer programs which follows consists of representative examples appropriate to the evaluation of the hypergeometric integrals tabulated above. It is taken that all quantities are real unless otherwise indicated. For obvious reasons, the number of parameters and summations has been limited, but the form of the programs is such that they may easily be extended to cover even the most complicated cases of the integrals under consideration. The international computer language FORTRAN IV is employed.

It is suggested that these programs are run for a low value of M, say M=5, and then re-run for M=6. By this means, a practical indication of the speed of convergence of the summations will be obtained and M may be increased until the desired degree of accuracy is acheived. It must always be ascertained that the series, single or multiple, being investigated is either convergent or a suitable asymptotic series.

Note. The symbols 'm' and 'n' in the READ and WRITE orders should be replaced by the appropriate numbers for the input and output channels on the equipment being used.

B.1 Programs for the Evaluation of Euler Integrals

B.1.1 $\frac{\Gamma(a+b)}{\Gamma(a)\Gamma(b)} \int_0^1 u^{a-1} (1-u)^{b-1} {}_2F_2(c_1,c_2;d_1,d_2;ux)\, du,\ a,b > 0.$

```
1 READ(m,6)A,B,C1,C2,D1,D2,X
  IF(A)8,8,9
6 FØRMAT(7F4.2)
9 READ(m,7)M
7 FØRMAT(I2)
  CALL F(A,B,C1,C2,D1,D2,X,M,S)
  WRITE(n,5)A,B,C1,C2,D1,D2,X,M,S
5 FØRMAT(13H PARAMETERS =, 7F5.2/3H M=,I2,3H F=,1PE14.6)
  GØTØ 1
8 STØP
  END

  SUBRØUTINE F(A,B,C1,C2,D1,D2,X,M,S)
  S=0.0
  T=1.0
  DØ 1 N=1,M
```

```
      AN=FLØAT(N)-1.0
      S=S+T
      T=T*((A+AN)/(A+B+AN))*((C1+AN)/(D1+AN))*((C2+AN)/(D2+AN))
      T=T*(X/(1.0+AN))
    1 CØNTINUE
      RETURN
      END
```

B.1.2 $$\frac{\Gamma(a+b)}{\Gamma(a)\Gamma(b)}\int_0^1 u^{a-1}\,(1-u)^{b-1} \times F^{1:1;1}_{0:2;2}\left[\begin{matrix} c: d_1 \;\; ; d_2 \;\; ; \\ -:g_1,h_1;g_2,h_2; \end{matrix} ux,(1-u)y\right] du,$$

$$a,b > 0.$$

```
    1 READ(m,6)A,B,C,D1,D2,G1,G2,H1,H2,X,Y
      IF(A)8,8,9
    6 FØRMAT(11F4.2)
    9 READ(m,7)M
    7 FØRMAT(I2)
      CALL F(A,B,C,D1,D2,G1,G2,H1,H2,X,Y,M,S)
      WRITE(n,5)A,B,C,D1,D2,G1,G2,H1,H2,X,Y,S
    5 FØRMAT(13H PARAMETERS =, 11F5.2/3H M=,I2,3H F=,1PE14.6)
      GØTØ 1
    8 STØP
      END

      SUBRØUTINE F(A,B,C,D1,D2,G1,G2,H1,H2,X,Y,M,S)
      S=0.0
      T1=1.0
      DØ 1 N1=1,M
      AN1=FLØAT(N1)
      C1=C+AN1-1.0
      A1=A+B+AN1-1.0
      T2=T1
      DØ 2 N2=1,M
      AN2=FLØAT(N2)
      C2=C1+AN2-1.0
      A2=A1+AN2-1.0
      S=S+T2
      T2=T2*(C2/A2)*((D2+AN2-1.0)/(G2+AN2-1.0))
      T2=T2*((B+AN2-1.0)/(H2+AN2-1.0))*(Y/AN2)
    2 CØNTINUE
      T1=T1*(C1/A1)*((D1+AN1-1.0)/(G1+AN1-1.0))
      T1=T1*((A+AN1-1.0)/(H1+AN1-1.0))*(X/AN1)
    1 CØNTINUE
      RETURN
      END
```

B.1.3 $$\frac{\Gamma(a+b)}{\Gamma(a)\Gamma(b)}\int_0^1 u^{a-1}(1-u)^{b-1} \times F^{1:1}_{1:1}\left[\begin{matrix}c:d_1;d_2;d_3;\\ g:h_1;h_2;h_3;\end{matrix} ux,uy,uz\right]du, \quad a,b>0.$$

```
1 READ(m,6)A,B,C,D1,D2,D3,G,H1,H2,H3,X,Y,Z
  IF(A)8,8,9
6 FØRMAT(13F4.2)
9 READ(m,7)M
7 FØRMAT(I2)
  CALL F(A,B,C,D1,D2,D3,G,H1,H2,H3,X,Y,Z,M,S)
  WRITE(n,5)A,B,C,D1,D2,D3,G,H1,H2,H3,X,Y,Z,M,S
5 FØRMAT(13H PARAMETERS =, 13F5.2/3H M=,I2,3H F=,1PE14.6)
  GØTØ 1
8 STØP
  END

  SUBRØUTINE F(A,B,C,D1,D2,D3,G,H1,H2,H3,X,Y,Z,M,S)
  S=0.0
  T1=1.0
  DØ 1 N1=1,M
  AN1=FLØAT(N1)
  A1=A+AN1-1.0
  B1=A+B+AN1-1.0
  C1=C+AN1-1.0
  G1=G+AN1-1.0
  T2=T1
  DØ 2 N2=1,M
  AN2=FLØAT(N2)
  A2=A1+AN2-1.0
  B2=B1+AN2-1.0
  C2=C1+AN2-1.0
  G2=G1+AN2-1.0
  T3=T2
  DØ 3 N3=1,M
  AN3=FLØAT(N3)
  A3=A2+AN3-1.0
  B3=B2+AN3-1.0
  C3=C2+AN3-1.0
  G3=G2+AN3-1.0
  S=S+T3
  T3=T3*(A3/B3)*(C3/G3)*((D3+AN3-1.0)/(H3+AN3-1.0))*(Z/AN3)
3 CØNTINUE
  T2=T2*(A2/B2)*(C2/G2)*((D2+AN2-1.0)/(H2+AN2-1.0))*(Y/AN2)
2 CØNTINUE
  T1=T1*(A1/B1)*(C1/G1)*((D1+AN1-1.0)/(H1+AN1-1.0))*(X/AN1)
1 CØNTINUE
  RETURN
  END
```

B.1.4 $$\frac{\Gamma(a+b)}{\Gamma(a)\Gamma(b)} \int_0^1 u^{a-1} (1-u)^{b-1} {}_1F_1(c;d;uy) \times {}_2F_2(g_1,g_2;h_1,h_2;ux)\, du, \quad a,b > 0.$$

```
1 READ(m,6)A,B,C,D,G1,G2,H1,H2,X,Y
  IF(A)8,8,9
6 FØRMAT(10F4.2)
9 READ(m,7)M
7 FØRMAT(I2)
  CALL F(A,B,C,D,G1,G2,H1,H2,X,Y,M,S)
  WRITE(n,5)A,B,C,D,G1,G2,H1,H2,X,Y,M,S
5 FØRMAT(13H PARAMETERS =, 10F5.2/3H M=,I2,3H F=,1PE14.6)
  GØTØ 1
8 STØP
  END

  SUBRØUTINE F(A,B,C,D,G1,G2,H1,H2,X,Y,M,S)
  S=0.0
  T1=1.0
  DØ 1 N1=1,M
  AN1=FLØAT(N1)
  A1=A+AN1-1.0
  B1=A+B+AN1-1.0
  T2=T1
  DØ 2 N2=1,M
  AN2=FLØAT(N2)
  A2=A1+AN2-1.0
  B2=B1+AN2-1.0
  S=S+T2
  T2=T2*A2*(A2/B2)*((C+AN2-1.0)/(D+AN2-1.0))*(Y/AN2)
2 CØNTINUE
  T1=T1*A1*(A1/B1)*((G1+AN1-1.0)/(H1+AN1-1.0))
  T1=T1*((G2+AN1-1.0)/(H2+AN1-1.0))*(X/AN1)
1 CØNTINUE
  RETURN
  END
```

B.1.5 $$\frac{\Gamma(a+b)}{\Gamma(a)\Gamma(b)} \int_0^1 u^{a-1} (1-u)^{b-1} {}_1F_1(c;d;uz) \times F^{0:2;2}_{1:1;1}\left[\begin{matrix} -:g_1,h_1;g_2;h_2; \\ p:\ q_1\ ;\ q_2\ ; \end{matrix} ux,uy\right] du, \quad a,b > 0.$$

```
1 READ(m,6)A,B,C,D,G1,G2,H1,H2,P,Q1,Q2,X,Y,Z
  IF(A)8,8,9
6 FØRMAT(14F4.2)
9 READ(m,7)M
7 FØRMAT(I2)
```

```
  CALL F(A,B,C,D,G1,G2,H1,H2,P,Q1,Q2,X,Y,Z,M,S)
  WRITE(n,5)A,B,C,D,G1,G2,H1,H2,P,Q1,Q2,X,Y,Z,M,S
5 FØRMAT(13H PARAMETERS =, 14F 5.2/3H M=,I2,3H F=,1PE14.6)
  GØTØ 1
8 STØP
  END

  SUBRØUTINE F(A,B,C,D,G1,G2,H1,H2,P,Q1,Q2,X,Y,Z,M,S)
  S=0.0
  T1=1.0
  DØ 1 N1=1,M
  AN1=FLØAT(N1)
  A1=A+AN1-1.0
  B1=A+B+AN1-1.0
  P1=P+AN1-1.0
  T2=T1
  DØ 2 N2=1,M
  AN2=FLØAT(N2)
  A2=A1+AN2-1.0
  B2=B1+AN2-1.0
  P2=P1+AN2-1.0
  T3=T2
  DØ 3 N3=1,M
  AN3=FLØAT(N3)
  A3=A2+AN3-1.0
  B3=B2+AN3-1.0
  S=S+T3
  T3=T3*(A3/B3)*((C+AN3-1.0)/(D+AN3-1.0))*(Z/AN3)
3 CØNTINUE
  T2=T2*(A2/B2)*((G2+AN2-1.0)/P2)*((H2+AN2-1.0)/(Q2+AN2-1.0))
  T2=T2*(Y/AN2)
2 CØNTINUE
  T1=T1*(A1/B1)*((G1+AN1-1.0)/P1)*((H1+AN1-1.0)/(Q1+AN1-1.0))
  T1=T1*(X/AN1)
1 CØNTINUE
  RETURN
  END
```

B.1.6 $$\frac{\Gamma(a+b)}{\Gamma(a)\Gamma(b)}\int_0^1 u^{a-1}(1-u)^{b-1}\,{}_1F_1(c;d;uw) \times F^{1:1}_{1:1}\left[\begin{matrix} g:h_1;h_2;h_3; \\ p:q_1;q_2;q_3; \end{matrix} ux,uy,uz\right] du,\quad a,b > 0.$$

```
1 READ(m,6)A,B,C,D,G,H1,H2,H3,P,Q1,Q2,Q3,X,Y,Z,W
  IF(A)8,8,9
6 FØRMAT(16F4.2)
9 READ(m,7)M
```

```
7 FØRMAT(I2)
  CALL F(A,B,C,D,G,H1,H2,H3,P,Q1,Q2,Q3,X,Y,Z,W,M,S)
  WRITE(n,5)A,B,C,D,G,H1,H2,H3,P,Q1,Q2,Q3,X,Y,Z,W,M,S
5 FØRMAT(13H PARAMETERS =, 8F5.2/2X,8F5.2/3H M=,I2,3H F=,1PE14.6)
  GØTØ 1
8 STØP
  END

  SUBRØUTINE F(A,B,C,D,G,H1,H2,H3,P,Q1,Q2,Q3,X,Y,Z,W,M,S)
  S=0.0
  T1=1.0
  DØ 1 N1=1,M
  AN1=FLØAT(N1)
  A1=A+AN1=1.0
  B1=A+B+AN1-1.0
  G1=G+AN1-1.0
  P1=P+AN1-1.0
  T2=T1
  DØ 2 N2=1,M
  AN2=FLØAT(N2)
  A2=A1+AN2-1.0
  B2=B1+AN2-1.0
  G2=G1+AN2-1.0
  P2=P1+AN2-1.0
  T3=T2
  DØ 3 N3=1,M
  AN3=FLØAT(N3)
  A3=A2+AN3-1.0
  B3=B2+AN3-1.0
  G3=G2+AN3-1.0
  P3=P2+AN3-1.0
  T4=T3
  DØ 4 N4=1,M
  AN4=FLØAT(N4)
  A4=A3+AN4-1.0
  B4=B3+AN4-1.0
  S=S+T4
  T4=T4*(A4/B4)*((C+AN4-1.0)/(D+AN4-1.0))*(W/AN4)
4 CØNTINUE
  T3=T3*(A3/B3)*(G3/P3)*((H3+AN3-1.0)/(Q3+AN3-1.0))*(Z/AN3)
3 CØNTINUE
  T2=T2*(A2/B2)*(G2/P2)*((H2+AN2-1.0)/(Q2+AN2-1.0))*(Y/AN2)
2 CØNTINUE
  T1=T1*(A1/B1)*(G1/P1)*((H1+AN1-1.0)/(Q1+AN1-1.0))*(X/AN1)
1 CØNTINUE
  RETURN
  END
```

B.1.7 $\dfrac{\Gamma(a+b)}{\Gamma(a)\Gamma(b)} \int_0^1 u^{a-1} (1-u)^{b-1} {}_2F_1(c_1,c_2;d;uy)$

$$\times {}_3F_2(g_1,g_2,g_3;h_1,h_2;[1-u]x)\ du, \quad a,b > 0, \quad -1 < x,y < 1.$$

```
1 READ(m,6)A,B,C1,C2,D,G1,G2,G3,H1,H2,X,Y
  IF(A)8,8,9
6 FØRMAT(12F4.2)
9 READ(m,7)M
7 FØRMAT(I2)
  CALL F(A,B,C1,C2,D,G1,G2,G3,H1,H2,X,Y,M,S)
  WRITE(n,5)A,B,C1,C2,D,G1,G2,G3,H1,H2,X,Y,M,S
5 FØRMAT(13H PARAMETERS =, 12F5.2/3H M=, I2,3H F=,1PE14.6)
  GØTØ 1
8 STØP
  END

  SUBRØUTINE F(A,B,C1,C2,D,G1,G2,G3,H1,H2,X,Y,M,S)
  S=0.0
  T1=1.0
  DØ 1 N1=1,M
  AN1=FLØAT(N1)
  B1=A+B+AN1-1.0
  T2=T1
  DØ 2 N2=1,M
  AN2=FLØAT(N2)
  B2=B1+AN2-1.0
  S=S+T2
  T2=T2*((C1+AN2-1.0)/B2)*((C2+AN2-1.0)/(D+AN2-1.0))
  T2=T2*((A+AN2-1.0)/AN2)*Y
2 CØNTINUE
  T1=T1*((G1+AN1-1.0)/B1)*((G2+AN1-1.0)/(H1+AN1-1.0))
  T1=T1*((G3+AN1-1.0)/(H2+AN1-1.0))*((B+AN1-1.0)/AN1)*X
1 CØNTINUE
  RETURN
  END
```

B.1.8 $\dfrac{\Gamma(a+b)}{\Gamma(a)\Gamma(b)} \int_0^1 u^{a-1} (1-u)^{b-1} {}_2F_1(c_1,c_2;d;uz)$

$$\times F^{1:2;2}_{0:2;2}\left[\begin{matrix} e:g_1,h_1;g_2,h_2; \\ -:p_1,q_1:p_2,q_2; \end{matrix}\ ux,uy\right] du, \quad a,b > 0, \quad -1 < x+y,z < 1.$$

```
1 READ(m,6)A,B,C1,C2,D,E,G1,G2,H1,H2,P1,P2,Q1,Q2,X,Y,Z
  IF(A)8,8,9
6 FØRMAT(17F4.2)
9 READ(m,7)M
```

```
7 FØRMAT(I2)
  CALL F(A,B,C1,C2,D,E,G1,G2,H1,H2,P1,P2,Q1,Q2,X,Y,Z,M,S)
  WRITE(n,5)A,B,C1,C2,D,E,G1,G2,H1,H2,P1,P2,Q1,Q2,X,Y,Z,M,S
5 FØRMAT(13H PARAMETERS =, 9F5.2/2X,8F5.2/3H M=,I2,3H F=,1PE14.6)
  GØTØ 1
8 STØP
  END

  SUBRØUTINE F(A,B,C1,C2,D,E,G1,G2,H1,H2,P1,P2,Q1,Q2,X,Y,Z,M,S)
  S=0.0
  T1=1.0
  DØ 1 N1=1,M
  AN1=FLØAT(N1)
  A1=A+AN1-1.0
  B1=A+B+AN1-1.0
  E1=E+AN1-1.0
  T2=T1
  DØ 2 N2=1,M
  AN2=FLØAT(N2)
  A2=A1+AN2-1.0
  B2=B1+AN2-1.0
  E2=E1+AN2-1.0
  T3=T2
  DØ 3 N3=1,M
  AN3=FLØAT(N3)
  A3=A2+AN3-1.0
  B3=B2+AN3-1.0
  S=S+T3
  T3=T3*(A3/B3)*((C1+AN3-1.0)/(D+AN3-1.0))*((C2+AN3-1.0)/AN3)*Z
3 CØNTINUE
  T2=T2*(A2/B2)*(E2/(P2+AN2-1.0))*((G2+AN2-1.0)/(Q2+AN2-1.0))
  T2=T2*((H2+AN2-1.0)/AN2)*Y
2 CØNTINUE
  T1=T1*(A1/B1)*(E1/(P1+AN1-1.0))*((G1+AN1-1.0)/(Q1+AN1-1.0))
  T1=T1*((H1+AN1-1.0)/AN1)*X
1 CØNTINUE
  RETURN
  END
```

B.1.9 $$\frac{\Gamma(a+b)}{\Gamma(a)\Gamma(b)}\int_0^1 u^{a-1}(1-u)^{b-1}\,{}_2F_1(c_1,c_2;d;uw) \times F^{1:2}_{2:0}\left[\begin{matrix} e\ :g_1,h_1;g_2,h_2;g_3,h_3; \\ p,q:\qquad - \qquad ; \end{matrix}\ ux,y,z\right] du,$$

$$a,b > 0,\ -1 < x+y+z,w < 1.$$

```
1 READ(m,6)A,B,C1,C2,D,E,G1,G2,G3,H1,H2,H3,P,Q,X,Y,Z,W
  IF(A)8,8,9
```

```
6 FØRMAT(18F4.2)
9 READ(m,7)M
7 FØRMAT(I2)
  CALL F(A,B,C1,C2,D,E,G1,G2,G3,H1,H2,H3,P,Q,X,Y,Z,W,M,S)
  WRITE(n,5)A,B,C1,C2,E,G1,G2,G3,H1,H2,H3,P,Q,X,Y,Z,W,M,S
5 FØRMAT(13H PARAMETERS =, 9F .2/2X,9F5.2/3H M=,I2,3H F=,1PE14.6)
  GØTØ 1
8 STØP
  END

  SUBRØUTINE F(A,B,C1,C2,D,E,G1,G2,G3,H1,H2,H3,P,Q,X,Y,Z,W,M,S)
  S=0.0
  T1=1.0
  DØ 1 N1=1,M
  AN1=FLØAT(N1)
  A1=A+AN1-1.0
  B1=A+B+AN1-1.0
  E1=E+AN1-1.0
  P1=P+AN1-1.0
  Q1=Q+AN1-1.0
  T2=T1
  DØ 2 N2=1,M
  AN2=FLØAT(N2)
  E2=E1+AN2-1.0
  P2=P1+AN2-1.0
  Q2=Q1+AN2-1.0
  T3=T2
  DØ 3 N3=1,M
  AN3=FLØAT(N3)
  E3=E2+AN3-1.0
  P3=P2+AN3-1.0
  Q3=Q2+AN3-1.0
  T4=T3
  DØ 4 N4=1,M
  AN4=FLØAT(N4)
  A4=A1+AN4-1.0
  B4=B1+AN4-1.0
  S=S+T4
  T4=T4*(A4/B4)*((C1+AN4-1.0)/(D+AN4-1.0))*((C2+AN4-1.0)/AN4)*W
4 CØNTINUE
  T3=T3*(E3/P3)*((G3+AN3-1.0)/Q3)*((H3+AN3-1.0)/AN3)*Z
3 CØNTINUE
  T2=T2*(E2/P2)*((G2+AN2-1.0)/Q2)*((H2+AN2-1.0)/AN2)*Y
2 CØNTINUE
  T1=T1*(A1/B1)*(E1/P1)*((G1+AN1-1.0)/Q1)*((H1+AN1-1.0)/AN1)*X
1 CØNTINUE
  RETURN
  END
```

B.2 Programs for the Evaluation of Definite Integrals

B.2.1 $$a\, z^{-a} \int_0^z z^{a-1}\; {}_2F_2(c_1,c_2;d_1,d_2;uz)\, dz, \quad a > 0.$$

```
1 READ (m,6)A,C1,C2,D1,D2,U,Z
  IF(A)8,8,9
6 FØRMAT(7F4.2)
9 READ(m,7)M
7 FØRMAT(I2)
  CALL F(A,C1,C2,D1,D2,U,Z,M,S)
  WRITE(n,5)A,C1,C2,D1,D2,U,Z,M,S
5 FØRMAT(13H PARAMETERS =, 7F5.2/3H M=,I2,3H F=,1PE14.6)
  GØTØ 1
8 STØP
  END

  SUBRØUTINE F(A,C1,C2,D1,D2,U,Z,M,S)
  S=0.0
  T=1.0
  DØ 1 N=1,M
  AN=FLØAT(N)-1.0
  S=S+T
  T=T*((A+AN)/(A+AN+1.0))*((C1+AN)/(D1+AN))*((C2+AN)/(D2+AN))
  T=T*(U/(1.0+AN))*Z
1 CØNTINUE
  RETURN
  END
```

B.2.2 $$a\, z^{-a} \int_0^z z^{a-1}\; F^{1:1;1}_{0:2;2}\left[\begin{matrix} g: h_1 \;\; ; h_2 \;\; ; \\ -: p_1, q_1; p_2, q_2; \end{matrix}\; uz, vz\right] dz, \quad a > 0.$$

```
1 READ(m,6)A,G,H1,H2,P1,P2,Q1,Q2,U,V,Z
  IF(A)8,8,9
6 FØRMAT(11F4.2)
9 READ(m,7)M
7 FØRMAT(I2)
  CALL F(A,G,H1,H2,P1,P2,Q1,Q2,U,V,Z,M,S)
  WRITE(n,5)A,G,H1,H2,P1,P2,Q1,Q2,U,V,Z,M,S
5 FØRMAT(13H PARAMETERS =, 11F5.2/3H M=,I2,3H F=,1PE14.6)
  GØTØ 1
8 STØP
  END

  SUBRØUTINE F(A,G,H1,H2,P1,P2,Q1,Q2,U,V,Z,M,S)
  S=0.0
  T1=1.0
```

```
  DØ 1 N1=1,M
  AN1=FLØAT(N1)
  A1=A+AN1-1.0
  B1=A+AN1
  G1=G+AN1-1.0
  T2=T1
  DØ 2 N2=1,M
  AN2=FLØAT(N2)
  A2=A1+AN2-1.0
  B2=B1+AN2-1.0
  G2=G1+AN2-1.0
  S=S+T2
  T2=T2*(A2/B2)*(G2/(P2+AN2-1.0))*((H2+AN2-1.0)/(Q2+AN2-1.0))
  T2=T2*(V/AN2)*Z
2 CØNTINUE
  T1=T1*(A1/B1)*(G1/(P1+AN1-1.0))*((H1+AN1-1.0)/(Q1+AN1-1.0))
  T1=T1*(U/AN1)*Z
1 CØNTINUE
  RETURN
  END
```

B.2.3 $$a\, z^{-a} \int_0^z z^{a-1}\, F^{1:1}_{2:0}\left[\begin{matrix} g & :h_1;h_2;h_3; \\ p,q: & - \quad ; \end{matrix}\; uz,vz,wz\right] dz, \quad a > 0.$$

```
1 READ(m,6)A,G,H1,H2,H3,P,Q,U,V,W,Z
  IF(A)8,8,9
6 FØRMAT(11F4.2)
9 READ(m,7)M
7 FØRMAT(I2)
  CALL F(A,G,H1,H2,H3,P,Q,U,V,W,Z,M,S)
  WRITE(n,5)A,G,H1,H2,H3,P,Q,U,V,W,Z,M,S
5 FØRMAT(13H PARAMETERS =, 11F5.2/3H M=,I2,3H F=,1PE14.6)
  GØTØ 1
8 STØP
  END

  SUBRØUTINE F(A,G,H1,H2,H3,P,Q,U,V,W,Z,M,S)
  S=0.0
  T1=1.0
  DØ 1 N1=1,M
  AN1=FLØAT(N1)
  A1=A+AN1-1.0
  B1=A+AN1
  G1=G+AN1-1.0
  P1=P+AN1-1.0
  Q1=Q+AN1-1.0
  T2=T1
  DØ 2 N2=1,M
```

```
      AN2=FLØAT(N2)
      A2=A1+AN2-1.0
      B2=B1+AN2-1.0
      G2=G1+AN2-1.0
      P2=P1+AN2-1.0
      Q2=Q1+AN2-1.0
      T3=T2
      DØ 3 N3=1,M
      AN3=FLØAT(N3)
      A3=A2+AN3-1.0
      B3=B2+AN3-1.0
      G3=G2+AN3-1.0
      P3=P2+AN3-1.0
      Q3=Q2+AN3-1.0
      S=S+T3
      T3=T3*(A3/B3)*(G3/P3)*((H3+AN3-1.0)/Q3)*(W/AN3)*Z
    3 CØNTINUE
      T2=T2*(A2/B2)*(G2/P2)*((H2+AN2-1.0)/Q2)*(V/AN2)*Z
    2 CØNTINUE
      T1=T1*(A1/B1)*(G1/P1)*((H1+AN1-1.0)/Q1)*(U/AN1)*Z
    1 CØNTINUE
      RETURN
      END
```

B.2.4 $\quad a\, z^{-a} \int_0^z z^{a-1}\, {}_1F_1(c;d;uz)\, {}_2F_2(g_1,g_2;h_1,h_2;vz)\, dz, \quad a > 0.$

```
    1 READ(m,6)A,C,D,G1,G2,H1,H2,U,V,Z
      IF(A)8,8,9
    6 FØRMAT(10F4.2)
    9 READ(m,7)M
    7 FØRMAT(I2)
      CALL F(A,C,D,G1,G2,H1,H2,U,V,Z,M,S)
      WRITE(n,5)A,C,D,G1,G2,H1,H2,U,V,Z,M,S
    5 FØRMAT(13H PARAMETERS =,10F5.2/3H M=,I2,3H F=,1PE14.6)
      GØTØ 1
    8 STØP
      END

      SUBRØUTINE F(A,C,D,G1,G2,H1,H2,U,V,Z,M,S)
      S=0.0
      T1=1.0
      DØ 1 N1=1,M
      AN1=FLØAT(N1)
      A1=A+AN1-1.0
      B1=A+AN1
      T2=T1
      DØ 2 N2=1,M
      AN2=FLØAT(N2)
```

```
  A2=A1+AN2-1.0
  B2=B1+AN2-1.0
  S=S+T2
  T2=T2*(A2/B2)*((G1+AN2-1.0)/(H1+AN2-1.0))
  T2=T2*((G2+AN2-1.0)/(H2+AN2-1.0))*(V/AN2)*Z
2 CØNTINUE
  T1=T1*(A1/B1)*((C+AN1-1.0)/(D+AN1-1.0))*(U/AN1)*Z
1 CØNTINUE
  RETURN
  END
```

B.2.5 $$a\, z^{-a} \int_0^z z^{a-1}\ {}_1F_1(c;d;uz)\, F^{1:1;1}_{0:2;2}\left[\begin{matrix} e:\ g\ ;\ h\ ; \\ -:p_1,q_1;p_2,q_2; \end{matrix}\ vz,wz\right] dz,$$

$a > 0.$

```
1 READ(m,6)A,C,D,E,G,H,P1,P2,Q1,Q2,U,V,W,Z
  IF(A)8,8,9
6 FØRMAT(14F4.2)
9 READ(m,7)M
7 FØRMAT(I2)
  CALL F(A,C,D,E,G,H,P1,P2,Q1,Q2,U,V,W,Z,M,S)
  WRITE(n,5)A,C,D,E,G,H,P1,P2,Q1,Q2,U,V,W,Z,M,S
5 FØRMAT(13H PARAMETERS =, 14F5.2/3H M=,I2,3H F=,1PE14.6)
  GØTØ 1
8 STØP
  END

  SUBRØUTINE F(A,C,D,E,G,H,P1,P2,Q1,Q2,U,V,W,Z,M,S)
  S=0.0
  T1=1.0
  DØ 1 N1=1,M
  AN1=FLØAT(N1)
  A1=A+AN1-1.0
  B1=A+AN1
  T2=T1
  DØ 2 N2=1,M
  AN2=FLØAT(N2)
  A2=A1+AN2-1.0
  B2=B1+AN2-1.0
  E2=E +AN2-1.0
  T3=T2
  DØ 3 N3=1,M
  AN3=FLØAT(N3)
  A3=A2+AN3-1.0
  B3=B2+AN3-1.0
  E3=E2+AN3-1.0
  S=S+T3
```

```
  T3=T3*(A3/B3)*(E3/(P2+AN3-1.0))*((H+AN3-1.0)/(Q2+AN3-1.0))
  T3=T3*Z*(W/AN3)
3 CØNTINUE
  T2=T2*(A2/B2)*(E2/(P1+AN2-1.0))*((G+AN2-1.0)/(Q1+AN2-1.0))
  T2=T2*Z*(V/AN2)
2 CØNTINUE
  T1=T1*(A1/B1)*((C+AN1-1.0)/(D+AN1-1.0))*(U/AN1)*Z
1 CØNTINUE
  RETURN
  END
```

B.2.6 $$a\, z^{-a} \int_0^z z^{a-1}\, {}_1F_1(c;d;uz) \times F^{0:2}_{1:1}\left[\begin{matrix} :e_1,g_1;e_2,g_2;e_3,g_3; \\ p:\ q_1\ \ ;\ q_2\ \ ;\ q_3\ \ ; \end{matrix} vz,wz,x\right] dz,$$

$$a > 0.$$

```
1 READ(m,6)A,C,D,E1,E2,E3,G1,G2,G3,P,Q1,Q2,Q3,U,V,W,X,Z
  IF(A)8,8,9
6 FØRMAT(18F4.2)
9 READ(m,7)M
7 FØRMAT(I2)
  CALL F(A,C,D,E1,E2,E3,G1,G2,G3,P,Q1,Q2,Q3,U,V,W,X,Z,M,S)
  WRITE(n,5)A,C,D,E1,E2,E3,G1,G2,G3,P,Q1,Q2,Q3,U,V,W,X,Z,M,S
5 FØRMAT(13H PARAMETERS =, 9F5.2/2X,9F5.2/3H M=,I2,3H F=,1PE14.6)
  GØTØ 1
8 STØP
  END

  SUBRØUTINE F(A,C,D,E1,E2,E3,G1,G2,G3,P,Q1,Q2,Q3,U,V,W,X,Z,M,S)
  S=0.0
  T1=1.0
  DØ 1 N1=1,M
  AN1=FLØAT(N1)
  A1=A+AN1-1.0
  B1=A+AN1
  T2=T1
  DØ 2 N2=1,M
  AN2=FLØAT(N2)
  A2=A1+AN2-1.0
  B2=B1+AN2-1.0
  P2=P+AN2-1.0
  T3=T2
  DØ 3 N3=1,M
  AN3=FLØAT(N3)
  A3=A2+AN3-1.0
  B3=B2+AN3-1.0
```

```
      P3=P2+AN3-1.0
      T4=T3
      DØ 4 N4=1,M
      AN4=FLØAT(N4)
      P4=P3+AN4-1.0
      S=S+T4
      T4=T4*((E3+AN4-1.0)/P4)*((G3+AN4-1.0)/(Q3+AN4-1.0))*(X/AN4)
    4 CØNTINUE
      T3=T3*(A3/B3)*((E2+AN3-1.0)/P3)*((G2+AN3-1.0)/(Q2+AN3-1.0))
      T3=T3*(W/AN3)*Z
    3 CØNTINUE
      T2=T2*(A2/B2)*((E1+AN2-1.0)/P2)*((G1+AN2-1.0)/(Q1+AN2-1.0))
      T2=T2*(V/AN2)*Z
    2 CØNTINUE
      T1=T1*(A1/B1)*((C+AN1-1.0)/(D+AN1-1.0))*(U/AN1)*Z
    1 CØNTINUE
      RETURN
      END
```

B.2.7 $a\, z^{-a} \int_0^z z^{a-1}\, {}_2F_1(c_1,c_2;d;uz)\, {}_3F_2(e_1,e_2,e_3;g_1,g_2;vz)\, dz,$

$$a > 0, \quad -1 < uz, vz < 1.$$

```
    1 READ(m,6)A,C1,C2,D,E1,E2,E3,G1,G2,U,V,Z
      IF(A)8,8,9
    6 FØRMAT(12F4.2)
    9 READ(m,7)M
    7 FØRMAT(I2)
      CALL F(A,C1,C2,D,E1,E2,E3,G1,G2,U,V,Z,M,S)
      WRITE(n,5)A,C1,C2,D,E1,E2,E3,G1,G2,U,V,Z,M,S
    5 FØRMAT(13H PARAMETERS =,12F 5.2/3H M=,I2,3H F=,1PE14.6)
      GØTØ 1
    8 STØP
      END

      SUBRØUTINE F(A,C1,C2,D,E1,E2,E3,G1,G2,U,V,Z,M,S)
      S=0.0
      T1=1.0
      DØ 1 N1=1,M
      AN1=FLØAT(N1)
      A1=A+AN1=1.0
      B1=A+AN1
      T2=T1
      DØ 2 N2=1,M
      AN2=FLØAT(N2)
      A2=A1+AN2-1.0
      B2=B1+AN2-1.0
      S=S+T2
```

```
  T2=T2*(A2/B2)*((E1+AN2-1.0)/(G1+AN2-1.0))
  T2=T2*((E2+AN2-1.0)/(G2+AN2-1.0))*((E3+AN2-1.0)/AN2)*V*Z
2 CØNTINUE
  T1=T1*(A1/B1)*((C1+AN1-1.0)/(D+AN1-1.0))
  T1=T1*((C2+AN1-1.0)/AN1)*U*Z
1 CØNTINUE
  RETURN
  END
```

B.2.8
$$a\, z^{-a} \int_0^z z^{a-1}\; {}_2F_1(c_1,c_2;d;uz) \times F^{1:2;2}_{0:2;2}\left[\begin{matrix} e:g_1,h_1;g_2,h_2; \\ -:p_1,q_1;p_2,q_2; \end{matrix} vz,wz\right] dz,$$
$$a > 0, \quad -1 < uz, \quad (v+w)z < 1.$$

```
1 READ(m,6)A,C1,C2,D,E,G1,G2,H1,H2,P1,P2,Q1,Q2,U,V,W,Z
  IF(A)8,8,9
6 FØRMAT(17F4.2)
9 READ(m,7)M
7 FØRMAT(I2)
  CALL F(A,C1,C2,D,E,G1,G2,H1,H2,P1,P2,Q1,Q2,U,V,W,Z,M,S)
  WRITE(n,5)A,C1,C2,D,E,G1,G2,H1,H2,P1,P2,Q1,Q2,U,V,W,Z,M,S
5 FØRMAT(13H PARAMETERS =, 9F5.2/2X,8F5.2/3H M=,I2,3H F=,1PE14.6)
  GØTØ 1
8 STØP
  END

  SUBRØUTINE F(A,C1,C2,D,E,G1,G2,H1,H2,P1,P2,Q1,Q2,U,V,W,Z,M,S)
  S=0.0
  T1=1.0
  DØ 1 N1=1,M
  AN1=FLØAT(N1)
  A1=A+AN1-1.0
  B1=A+AN1
  T2=T1
  DØ 2 N2=1,M
  AN2=FLØAT(N2)
  A2=A1+AN1-1.0
  B2=B1+AN2-1.0
  E2=E+AN2-1.0
  T3=T2
  DØ 3 N3=1,M
  AN3=FLØAT(N3)
  A3=A2+AN3-1.0
  B3=B2+AN3-1.0
  E3=E2+AN3-1.0
```

```
      S=S+T3
      T3=T3*(A3/B3)*(E3/(P2+AN3-1.0))*((G2+AN3-1.0)/(Q2+AN3-1.0))
      T3=T3*((H2+AN3-1.0)/AN3)*W*Z
    3 CØNTINUE
      T2=T2*(A2/B2)*(E2/(P2+AN2-1.0))*((G1+AN2-1.0)/(Q1+AN2-1.0))
      T2=T2*((H1+AN2-1.0)/AN2)*V*Z
    2 CØNTINUE
      T1=T1*(A1/B1)*((C1+AN1-1.0)/(D+AN1-1.0))*((C2+AN1-1.0)/AN1)
      T1=T1*U*Z
    1 CØNTINUE
      RETURN
      END
```

B.2.9 $$a\, z^{-a} \int_0^z z^{a-1}\; {}_2F_1(c_1,c_2;d;uz) \times F^{3:0}_{0:2}\left[\begin{matrix} p,q,r: & - & ; \\ - & :g_1,h_1;g_2,h_2;g_3,h_3; & \end{matrix}\; vz,wz,xz\right] dz,$$

$$a > 0, \quad -1,< uz, \ (vz)^{1/3}+(wz)^{1/3}+(xz)^{1/3} < 1.$$

```
    1 READ(m,6)A,C1,C2,D,G1,G2,G3,H1,H2,H3,P,Q,R,U,V,W,X,Z
      IF(A)8,8,9
    6 FØRMAT(18F4.2)
    9 READ(m,7)M
    7 FØRMAT(I2)
      CALL F(A,C1,C2,D,G1,G2,G3,H1,H2,H3,P,Q,R,U,V,W,X,Z,M,S)
      WRITE(n,5)A,C1,C2,D,G1,G2,G3,H1,H2,H3,P,Q,R,U,V,W,X,Z,M,S
    5 FØRMAT(13H PARAMETERS =, 9F 5.2/2X,9F5.2/3H M=,I2,3H F=,1PE14.6)
      GØTØ 1
    8 STØP
      END

      SUBRØUTINE F(A,C1,C2,D,G1,G2,G3,H1,H2,H3,P,Q,R,U,V,W,X,Z,M,S)
      S=0.0
      T1=1.0
      DØ 1 N1=1,M
      AN1=FLØAT(N1)
      A1=A+AN1-1.0
      B1=A+AN1
      T2=T1
      DØ 2 N2=1,M
      AN2=FLØAT(N2)
      A2=A1+AN2-1.0
      B2=B1+AN2-1.0
      P2=P+AN2-1.0
      Q2=Q+AN2-1.0
      R2=R+AN2-1.0
```

```
      T3=T2
      DØ 3 N3=1,M
      AN3=FLØAT(N3)
      A3=A2+AN3-1.0
      B3=B2+AN3-1.0
      P3=P2+AN3-1.0
      Q3=Q2+AN3-1.0
      R3=R2+AN3-1.0
      T4=T3
      DØ 4 N4=1,M
      AN4=FLØAT(N4)
      A4=A3+AN4-1.0
      B4=B3+AN4-1.0
      P4=P3+AN4-1.0
      Q4=Q3+AN4-1.0
      R4=R3+AN4-1.0
      S=S+T4
      T4=T4*(A4/B4)*(P4/(G3+AN4-1.0))*(Q4/(H3+AN4-1.0))*(R4/AN4)*X*Z
    4 CØNTINUE
      T3=T3*(A3/B3)*(P3/(G2+AN3-1.0))*(Q3/(H2+AN3-1.0))*(R3/AN3)*W*Z
    3 CØNTINUE
      T2=T2*(A2/B2)*(P2/(G1+AN2-1.0))*(Q2/(H1+AN2-1.0))*(R2/AN2)*V*Z
    2 CØNTINUE
      T1=T1*(A1/B1)*((C1+AN1-1.0)/(D+AN1-1.0))*((C2+AN1-1.0)/AN1)*U*Z
    1 CØNTINUE
      RETURN
      END
```

B.3 Programs for the Evaluation of Repeated Integrals

B.3.1 $(a,n)\ z^{1-a-n} \int_0^z \ldots(n)\ldots \int_0^z z^{a-1}\ {}_2F_2(c_1,c_2;d_1,d_2;uz)\ dz,$

$a > 0.$

```
    1 READ(m,6)A,C1,C2,D1,D2,U,Z,N
      IF(A)8,8,9
    6 FØRMAT(7F4.2,I2)
    9 READ(m,7)M
    7 FØRMAT(I2)
      CALL F(A,C1,C2,D1,D2,U,Z,N,M,S)
      WRITE(n,5)A,C1,C2,D1,D2,U,Z,N,M,S
    5 FØRMAT(13H PARAMETERS =, 7F5.2/3H N=,I2,3H M=,I2,3H F=,1PE14.6)
      GØTØ 1
    8 STØP
      END

      SUBRØUTINE F(A,C1,C2,D1,D2,U,Z,N,M,S)
      S=0.0
      T=1.0
```

```
  DØ 1 N1=1,M
  AN=FLØAT(N1)-1.0
  BN=FLØAT(N)
  S=S+T
  T=T*((A+AN)/(A+BN+AN))*((C1+AN)/(D1+AN))*((C2+AN)/(D2+AN))
  T=T*(U/(AN+1.0))*Z
1 CØNTINUE
  RETURN
  END
```

B.3.2 $$(a,n)\ z^{1-a-n}\int_0^z..(n)..\int_0^z z^{a-1}F^{1:1;1}_{1:1;1}\left[\begin{matrix}g:h_1;h_2;\\p:q_1;q_2;\end{matrix}uz,vz\right]dz,$$

$$a > 0.$$

```
1 READ(m,6)A,G,H1,H2,P,Q1,Q2,U,V,Z,N
  IF(A)8,8,9
6 FØRMAT(10F4.2,I2)
9 READ(m,7)M
7 FØRMAT(I2)
  CALL F(A,G,H1,H2,P,Q1,Q2,U,V,Z,N,M,S)
  WRITE(n,5)A,G,H1,H2,P,Q1,Q2,U,V,Z,N,M,S
5 FØRMAT(13H PARAMETERS =,10F5.2/3H N=,I2,3H M=,I2,3H F=,1PE14.6)
  GØTØ 1
8 STØP
  END

  SUBRØUTINE F(A,G,H1,H2,P,Q1,Q2,U,V,Z,N,M,S)
  S=0.0
  T1=1.0
  DØ 1 N1=1,M
  AN1=FLØAT(N1)
  BN=FLØAT(N)
  A1=A+AN1-1.0
  B1=BN+AN1-1.0
  G1=G+AN1-1.0
  P1=P+AN1-1.0
  T2=T1
  DØ 2 N2=1,M
  AN2=FLØAT(N2)
  A2=A1+AN2-1.0
  B2=B1+AN2-1.0
  G2=G1+AN2-1.0
  P2=P1+AN2-1.0
  S=S+T2
  T2=T2*(A2/B2)*(G2/P2)*((H2+AN2-1.0)/(Q2+AN2-1.0))*(V/AN2)*Z
2 CØNTINUE
  T1=T1*(A1/B1)*(G1/P1)*((H1+AN1-1.0)/(Q1+AN1-1.0))*(U/AN1)*Z
```

```
1 CØNTINUE
  RETURN
  END
```

B.3.3 $$(a,n)z^{1-a-n}\int_0^z .(n). \int_0^z z^{a-1} F^{1:1}_{0:2}\left[\begin{matrix} g: & h_1 & ; h_2 & ; h_3 & ; \\ -: & p_1, q_1; & p_2, q_2; & p_3, q_3; \end{matrix} uz, vz, wz\right] dz,$$

$a > 0.$

```
1 READ(m,6)A,G,H1,H2,H3,P1,P2,P3,Q1,Q2,Q3,U,V,W,Z,N
  IF(A)8,8,9
6 FØRMAT(15F4.2,I2)
9 READ(m,7)M
7 FØRMAT(I2)
  CALL F(A,G,H1,H2,H3,P1,P2,P3,Q1,Q2,Q3,U,V,W,Z,N,M,S)
  WRITE(n,5)A,G,H1,H2,H3,P1,P2,P3,Q1,Q2,Q3,U,V,W,Z,N,M,S
5 FØRMAT(13H PARAMETERS =, 8F5.2/2X,7F5.2/3H N=,I2,3H M=,I2,
 13H F=,1PE14.6)
  GØTØ 1
8 STØP
  END

  SUBRØUTINE F(A,G,H1,H2,H3,P1,P2,P3,Q1,Q2,Q3,U,V,W,Z,N,M,S)
  S=0.0
  T1=1.0
  DØ 1 N1=1,M
  AN1=FLØAT(N1)
  BN=FLØAT(N)
  A1=A+AN1-1.0
  B1=BN+AN1-1.0
  G1=G+AN1-1.0
  T2=T1
  DØ 2 N2=1,M
  AN2=FLØAT(N2)
  A2=A1+AN2-1.0
  B2=B1+AN2-1.0
  G2=G1+AN2-1.0
  T3=T2
  DØ 3 N3=1,M
  AN3=FLØAT(N3)
  A3=A2+AN3-1.0
  B3=B2+AN3-1.0
  G3=G2+AN3-1.0
  S=S+T3
  T3=T3*(A3/B3)*(G3/(P3+AN3-1.0))*((H3+AN3-1.0)/(Q3+AN3-1.0))
  T3=T3*(W/AN3)*Z
3 CØNTINUE
  T2=T2*(A2/B2)*(G2/(P2+AN2-1.0))*((H2+AN2-1.0)/(Q2+AN2-1.0))
```

```
  T2=T2*(V/AN2)*Z
2 CØNTINUE
  T1=T1*(A1/B1)*(G1/(P1+AN1-1.0))*((H1+AN1-1.0)/(Q1+AN1-1.0))
  T1=T1*(U/AN1)*Z
1 CØNTINUE
  RETURN
  END
```

B.4 Programs for the Evaluation of Barnes Integrals and Related Integrals

B.4.1 $$\frac{2^{c_1+c_2-1}}{\pi i\Gamma(c_1+c_2)}\int_{-i\infty}^{i\infty}\Gamma(c_1+u)\Gamma(c_2-u)\,{}_3F_2(c_1+u,c_2-u,c_3;d_1,d_2;x)\,du,$$

$$c_1+c_2 > 0,\quad -1 < x < 1.$$

```
1 READ(m,6)C1,C2,C3,D1,D2,X
  IF(C1+C2)8,8,9
6 FØRMAT(6F4.2)
9 READ(m,7)M
7 FØRMAT(I2)
  CALL F(C1,C2,C3,D1,D2,X,M,S)
  WRITE(n,5)C1,C2,C3,D1,D2,X,M,S
5 FØRMAT(13H PARAMETERS =, 6F5.2/3H M=,I2,3H F=,1PE14.6)
  GØTØ 1
8 STØP
  END

  SUBRØUTINE F(C1,C2,C3,D1,D2,X,M,S)
  S=0.0
  T=1.0
  DØ 1 N1=1,M
  AN1=FLØAT(N1)
  E1=(C1+C2)/2.0
  E2=E1+.5
  S=S+T
  T=T*((E1+AN1-1.0)/(D1+AN1-1.0))*((E2+AN1-1.0)/(D2+AN2-1.0))
  T=T*((C3+AN1-1.0)/AN1)*X
1 CØNTINUE
  RETURN
  END
```

B.4.2 $$\frac{2^{c_1+g_1-1}}{\pi i\Gamma(c_1+g_1)}\int_{-i\infty}^{i\infty}\Gamma(c_1+u)\Gamma(g_1-u)\,{}_2F_2(c_1+u,c_2;d_1,d_2;x)$$

$$\times{}_2F_2(g_1-u,g_2;h_1,h_2;y)\,du,\quad c_1+g_1 > 0.$$

```
1 READ(m,6)C1,C2,D1,D2,G1,G2,H1,H2,X,Y
  IF(C1+C2)8,8,9
```

```
6 FØRMAT(10F4.2)
9 READ(m,7)M
7 FØRMAT(I2)
  CALL F(C1,C2,D1,D2,G1,G2,H1,H2,X,Y,M,S)
  WRITE(n,5)C1,C2,D1,D2,G1,G2,H1,H2,X,Y,M,S
5 FØRMAT(13H PARAMETERS =,10F5.2/3H M=,I2,3H F=,1PE14.6)
  GØTØ 1
8 STØP
  END

  SUBRØUTINE F(C1,C2,D1,D2,G1,G2,H1,H2,X,Y,M,S)
  S=0.0
  T1=1.0
  DØ 1 N1=1,M
  AN1=FLØAT(N1)
  A1=C1+G1+AN1-1.0
  T2=T1
  DØ 2 N2=1,M
  AN2=FLØAT(N2)
  A2=A1+AN2-1.0
  S=S+T2
  T2=T2*(A2/2.0)*((G2+AN2-1.0)/(H1+AN2-1.0))*(Y/(H2+AN2-1.0))/AN2
2 CØNTINUE
  T1=T1*(A1/2.0)*((C2+AN1-1.0)/(D1+AN1-1.0))*(X/(D2+AN1-1.0))/AN1
1 CØNTINUE
  RETURN
  END
```

B.4.3 $$\frac{2^{c_1+c_2-1}}{\pi i \Gamma(c_1+c_2)} \int_{-i\infty}^{i\infty} \Gamma(c_1+u)\Gamma(c_2-u) \times F^{2:0}_{0:2}\left[\begin{matrix} c_1+u, c_2-u: & ; \\ - & :p_1, q_1; p_2, q_2; p_3, q_3; \end{matrix}\ x, y, z\right] du,$$

$$c_1+c_2 > 0.$$

```
1 READ(m,6)C1,C2,P1,P2,P3,Q1,Q2,Q3,X,Y,Z
  IF(C1+C2)8,8,9
6 FØRMAT(11F4.2)
9 READ(m,7)M
7 FØRMAT(I2)
  CALL F(C1,C2,P1,P2,P3,Q1,Q2,Q3,X,Y,Z,M,S)
  WRITE(n,5)C1,C2,P1,P2,P3,Q1,Q2,Q3,X,Y,Z,M,S
5 FØRMAT(13H PARAMETERS =,11F5.2/3H M=,I2,3H F=,1PE14.6)
  GØTØ 1
8 STØP
  END

SUBRØUTINE F(C1,C2,P1,P2,P3,Q1,Q2,Q3,X,Y,Z,M,S)
```

```
  S=0.0
  T1=1.0
  DØ 1 N1=1,M
  AN1=FLØAT(N1)
  E1=(C1+C2)/2.0
  E2=E1+.5
  A1=E1+AN1-1.0
  B1=E2+AN1=1.0
  T2=T1
  DØ 2 N2=1,M
  AN2=FLØAT(N2)
  A2=A1+AN2-1.0
  B2=B1+AN2-1.0
  T3=T2
  DØ 3 N3=1.M
  AN3=FLØAT(N3)
  A3=A2+AN3-1.0
  B3=B2+AN3-1.0
  S=S+T3
  T3=T3*(A3/(P3+AN3-1.0))*(B3/(Q3+AN3-1.0))*(Z/AN3)
3 CØNTINUE
  T2=T2*(A2/(P2+AN2-1.0))*(B2/(Q2+AN2-1.0))*(Y/AN2)
2 CØNTINUE
  T1=T1*(A1/(P1+AN1-1.0))*(B1/(Q1+AN1-1.0))*(X/AN1)
1 CØNTINUE
  RETURN
  END
```

B.4.4 $$\frac{\Gamma(d_1+d_2-1)}{2^{d_1+d_2-1}}\int_{-\infty}^{\infty} {}_2F_2(c_1,c_2;d_1+u,d_2-u;x)\,\frac{du}{\Gamma(d_1+u)\Gamma(d_2-u)},$$

$$d_1+d_2 > 1.$$

```
1 READ(m,6)C1,C2,D1,D2,X
  IF(D1+D2-1.0)8,8,9
6 FØRMAT(5F4.2)
9 READ(m,7)M
7 FØRMAT(I2)
  CALL F(C1,C2,D1,D2,X,M,S)
  WRITE(n,5)C1,C2,D1,D2,X,M,S
5 FØRMAT(13H PARAMETERS =, 5F5.2/3H M=,I2,3H F=,1PE14.6)
  GØTØ 1
8 STØP
  END

  SUBRØUTINE F(C1,C2,D1,D2,X,M,S)
  S=0.0
  T=1.0
  DØ 1 N1=1,M
```

```
  AN1=FLØAT(N1)
  E1=(D1+D2-1.0)/2.0
  E2=E1+.5
  S=S+T
  T=T*((C1+AN1-1.0)/(E1+AN1-1.0))*((C2+AN1-1.0)/(E2+AN1-1.0))
  T=T*(X/AN1)
1 CØNTINUE
  RETURN
  END
```

B.4.5
$$\frac{\Gamma(h_1+h_2-1)}{2^{h_1+h_2-1}} \int_{-\infty}^{\infty} F^{0:2;2}_{1:1;1}\left[\begin{matrix} -:b_1,c_1;b_2,c_2; \\ g:h_1+u\ ;h_2-u\ ; \end{matrix} x,y\right] \frac{du}{\Gamma(h_1+u)\Gamma(h_2-u)},$$
$$h_1+h_2 > 1.$$

```
1 READ(m,6)B1,B2,C1,C2,G,H1,H2,X,Y
  IF(H1+H2-1)8,8,9
6 FØRMAT(9F4.2)
9 READ(m,7)
7 FØRMAT(I2)
  CALL F(B1,B2,C1,C2,G,H1,H2,X,Y,M,S)
  WRITE(n,5)B1,B2,C1,C2,G,H1,H2,X,Y,M,S
5 FØRMAT(13H PARAMETERS =, 9F5.2/3H M=,I2,3H F=,1PE14.6)
  GØTØ 1
8 STØP
  END

  SUBRØUTINE F(B1,B2,C1,C2,G,H1,H2,X,Y,M,S)
  S=0.0
  T1=1.0
  DØ 1 N1=1,M
  AN1=FLØAT(N1)
  E1=H1+H2-2.0+AN1
  G1=G+AN1-1.0
  T2=T1
  DØ 2 N2=1,M
  AN2=FLØAT(N2)
  E2=E1+AN2-1.0
  G2=G1+AN2-1.0
  S=S+T2
  T2=T2*((B2+AN2-1.0)/E2)*((C2+AN2-1.0)/G2)*(Y/AN2)*2
2 CØNTINUE
  T1=T1*((B1+AN1-1.0)/E1)*((C1+AN1-1.0)/G1)*(X/AN1)*2
1 CØNTINUE
  RETURN
  END
```

B.4.6 $$\frac{\Gamma(g_1+h_1-1)}{2^{g_1+h_1-1}}\int_{-\infty}^{\infty} F^{1:1}_{0:2}\left[\begin{matrix} a: & b_1 & ; & b_2 & ; & b_3 & ; \\ -: & g_1+u,h_1-u & ; & g_2,h_2 & ; & g_3,h_3 & ; \end{matrix} x,y,z\right] \times \frac{du}{\Gamma(g_1+u)\Gamma(h_1-u)}, \quad g_1+h_1 > $$

```
1 READ(m,6)A,B1,B2,B3,G1,G2,G3,H1,H2,H3,X,Y,Z
  IF(G1+H1-1.0)8,8,9
6 FØRMAT(13F4.2)
9 READ(m,7)M
7 FØRMAT(I2)
  CALL F(A,B1,B2,B3,G1,G2,G3,H1,H2,H3,X,Y,Z,M,S)
  WRITE(n,5)A,B1,B2,B3,G1,G2,G3,H1,H2,H3,X,Y,Z,M,S
5 FØRMAT(13H PARAMETERS =,13F5.2/3H M=,I2,3H F=,1PE14.6)
  GØTØ 1
8 STØP
  END

  SUBRØUTINE F(A,B1,B2,B3,G1,G2,G3,H1,H2,H3,X,Y,Z,M,S)
  S=0.0
  T1=1.0
  DØ 1 N1=1,M
  AN1=FLØAT(N1)
  A1=A+AN1-1.0
  T2=T1
  DØ 2 N2=1,M
  AN2=FLØAT(N2)
  A2=A1+AN2-1.0
  T3=T2
  DØ 3 N3=1,M
  AN3=FLØAT(N3)
  A3=A2+AN3-1.0
  S=S+T3
  T3=T3*(A3/(G3+AN3-1.0))*((B3+AN3-1.0)/(H3+AN3-1.0))*(Z/AN3)
3 CØNTINUE
  T2=T2*(A2/(G2+AN2-1.0))*((B2+AN2-1.0)/(H2+AN2-1.0))*(Y/AN2)
2 CØNTINUE
  D1=(G1+H1-1.0)/2.0
  E1=(G1+H1)/2.0
  T1=T1*(A1/(D1+AN1-1.0))*((B1+AN1-1.0)/(E1+AN1-1.0))*(X/AN1)
1 CØNTINUE
  RETURN
  END
```

B.5 Programs for the Evaluation of Laplace Integrals

B.5.1 $$\frac{p^a}{\Gamma(a)} \int_0^\infty e^{-pt}\, t^{a-1}\; {}_2F_2(c_1,c_2;d_1,d_2;xt)\, dt,$$

$$a,p > 0, \quad -p < x < p.$$

```
1 READ(m,6)A,C1,C2,D1,D2,P,X
  IF(A)8,8,9
6 FØRMAT(7F4.2)
9 READ(m,7)M
7 FØRMAT(I2)
  CALL F(A,C1,C2,D1,D2,P,X,M,S)
  WRITE(n,5)A,C1,C2,D1,D2,P,X,M,S)
5 FØRMAT(13H PARAMETERS =, 7F 5.2/3H M=,I2,3H F=,1PE14.6)
  GØTØ 1
8 STØP
  END

  SUBRØUTINE F(A,C1,C2,D1,D2,P,X,M,S)
  S=0.0
  T=1.0
  DØ 1 N=1,M
  AN=FLØAT(N)-1.0
  S=S+T
  T=T*((A+AN)/P)*((C1+AN)/(D1+AN))*((C2+AN)/(D2+AN))*(X/(AN+1.0))
1 CØNTINUE
  RETURN
  END
```

B.5.2 $$\frac{p^a}{\Gamma(a)} \int_0^\infty e^{-pt}\, t^{a-1}\; F^{0:2;2}_{1:1;1}\left[\begin{matrix} -:b_1,c_1;b_2,c_2; \\ d:\ g_1\ ;\ g_2\ ; \end{matrix} xt,yt\right] dt,$$

$$a,p > 0, \ -p < x,y < p.$$

```
1 READ(m,6)A,B1,B2,C1,C2,D,G1,G2,P,X,Y
  IF(A)8,8,9
6 FØRMAT(11F4.2)
9 READ(m,7)M
7 FØRMAT(I2)
  CALL F(A,B1,B2,C1,C2,D,G1,G2,P,X,Y,M,S)
  WRITE(n,5)A,B1,B2,C1,C2,D,G1,G2,P,X,Y,M,S
5 FØRMAT(13H PARAMETERS =,11F5.2/3H M=,I2,3H F=,1PE14.6)
  GØTØ 1
8 STØP
  END
```

```
      SUBRØUTINE F(A,B1,B2,C1,C2,D,G1,G2,P,X,Y,M,S)
      S=0.0
      T1=1.0
      DØ 1 N1=1,M
      AN1=FLØAT(N1)
      A1=A+AN1-1.0
      D1=D+AN1-1.0
      T2=T1
      DØ 2 N2=1,M
      AN2=FLØAT(N2)
      A2=A1+AN2-1.0
      D2=D1+AN2-1.0
      S=S+T2
      T2=T2*(A2/P)*((B2+AN2-1.0)/D2)*((C2+AN2-1.0)/(G2+AN2-1.0))
      T2=T2*(Y/AN2)
    2 CØNTINUE
      T1=T1*(A1/P)*((B1+AN1-1.0)/D1)*((C1+AN1-1.0)/(G1+AN1-1.0))
      T1=T1*(X/AN1)
    1 CØNTINUE
      RETURN
      END
```

B.5.3 $$\frac{p^a}{\Gamma(a)} \int_0^\infty e^{-pt}\, t^{a-1}\, {}_2F_2(c_1,d_1;g_1,h_1;xt) \times {}_2F_2(c_2,d_2;g_2,h_2;yt)\, {}_2F_2(c_3,d_3;g_3,h_3;zt)\, dt,$$

$$a,p > 0, \quad -p < x+y+z < p.$$

```
    1 READ(m,6)A,C1,C2,C3,D1,D2,D3,G1,G2,G3,H1,H2,H3,P,X,Y,Z
      IF(A)8,8,9
    6 FØRMAT(17F4.2)
    9 READ(m,7)M
    7 FØRMAT(I2)
      CALL F(A,C1,C2,C3,D1,D2,D3,G1,G2,G3,H1,H2,H3,P,X,Y,Z,M,S)
      WRITE(n,5)A,C1,C2,C3,D1,D2,D3,G1,G2,G3,H1,H2,H3,P,X,Y,Z,M,S
    5 FØRMAT(13H PARAMETERS =,9F5.2/2X,8F5.2/3H M=,I2,3H F=,1PE14.6)
      GØTØ 1
    8 STØP
      END

      SUBRØUTINE F(A,C1,C2,C3,D1,D2,D3,G1,G2,G3,H1,H2,H3,P,X,Y,Z,M,S)
      S=0.0
      T1=1.0
      DØ 1 N1=1,M
      AN1=FLØAT(N1)
      A1=A+AN1-1.0
      T2=T1
      DØ 2 N2=1,M
```

```
  AN2=FLØAT (N2)
  A2=A1+AN2-1.0
  T3=T2
  DØ 3 N3=1,M
  AN3=FLØAT(N3)
  A3=A2+AN3-1.0
  S=S+T3
  T3=T3*(A3/P)*((C3+AN3-1.0)/(G3+AN3-1.0))
  T3=T3*((D3+AN3-1.0)/(H3+AN3-1.0))*(Z/AN3)
3 CØNTINUE
  T2=T2*(A2/P)*((C2+AN2-1.0)/(G2+AN2-1.0))
  T2=T2*((D2+AN2-1.0)/(H2+AN2-1.0))*(Y/AN2)
2 CØNTINUE
  T1=T1*(A1/P)*((C1+AN1-1.0)/(G1+AN1-1.0))
  T1=T1*((D1+AN1-1.0)/(H1+AN1-1.0))*(X/AN1)
1 CØNTINUE
  RETURN
  END
```

B.5.4 $$\frac{p^a}{\Gamma(a)} \int_0^\infty e^{-pt}\, t^{a-1}\, {}_1F_1(c;d;xt)\, {}_2F_2(g_1,g_2;h_1,h_2;yt)\, dt,$$

$$a,p > 0, \quad -p < x+y < p.$$

```
1 READ(m,6)A,C,D,G1,G2,H1,H2,P,X,Y
  IF(A)8,8,9
6 FØRMAT(10F4.2)
9 READ(m,7)M
7 FØRMAT(I2)
  CALL F(A,C,D,G1,G2,H1,H2,P,X,Y,M,S)
  WRITE(n,5)A,C,D,G1,G2,H1,H2,P,X,Y,M,S
5 FØRMAT(13H PARAMETERS =,10F5.2/3H M=,I2,3H F=,1PE14.6)
  GØTØ 1
8 STØP
  END

  SUBRØUTINE F(A,C,D,G1,G2,H1,H2,P,X,Y,M,S)
  S=0.0
  T1=1.0
  DØ 1 N1=1,M
  AN1=FLØAT(N1)
  A1=A+AN1-1.0
  T2=T1
  DØ 2 N2=1,M
  AN2=FLØAT(N2)
  A2=A1+AN2-1.0
  S=S+T2
  T2=T2*(A2/P)*((G1+AN2-1.0)/(H1+AN2-1.0))
  T2=T2*((G2+AN2-1.0)/(H2+AN2-1.0))*(Y/AN2)
2 CØNTINUE
```

```
  T1=T1*(A1/P)*((C+AN1-1.0)/(D+AN1-1.0))*(X/AN1)
1 CØNTINUE
  RETURN
  END
```

B.5.5 $$\frac{p^a}{\Gamma(a)}\int_0^\infty e^{-pt}\, t^{a-1}\, {}_1F_1(c;d;xt) \times F^{1:1;1}_{1:1;1}\left[\begin{matrix} g:h_1;h_2; \\ e:q_1;q_2; \end{matrix}\, yt, zt\right] dt,$$

$$a,p > 0, \quad -p < x+y+z < p.$$

```
1 READ(m,6)A,C,D,E,G,H1,H2,P,Q1,Q2,X,Y,Z
  IF(A)8,8,9
6 FØRMAT(13F4.2)
9 READ(m,7)M
7 FØRMAT(I2)
  CALL F(A,C,D,E,G,H1,H2,P,Q1,Q2,X,Y,Z,M,S)
  WRITE(n,5)A,C,D,E,G,H1,H2,P,Q1,Q2,X,Y,Z,M,S
5 FØRMAT(13H PARAMETERS =,13F5.2/3H M=,I2,3H F=,1PE14.6)
  GØTØ 1
8 STØP
  END

  SUBRØUTINE F(A,C,D,E,G,H1,H2,P,Q1,Q2,X,Y,Z,M,S)
  S=0.0
  T1=1.0
  DØ 1 N1=1,M
  AN1=FLØAT(N1)
  A1=A+AN1-1.0
  T2=T1
  DØ 2 N2=1,M
  AN2=FLØAT(N2)
  A2=A1+AN2-1.0
  E2=E+AN2-1.0
  G2=G+AN2-1.0
  T3=T2
  DØ 3 N3=1,M
  AN3=FLØAT(N3)
  A3=A2+AN3-1.0
  E3=E2+AN3-1.0
  G3=G2+AN3-1.0
  S=S+T3
  T3=T3*(A3/P)*(G3/E3)*((H2+AN3-1.0)/(Q2+AN3-1.0))*(Z/AN3)
3 CØNTINUE
  T2=T2*(A2/P)*(G2/E2)*((H1+AN2-1.0)/(Q1+AN2-1.0))*(Y/AN2)
2 CØNTINUE
```

```
  T1=T1*(A1/P)*((C+AN1-1.0)/(D+AN1-1.0))*(X/AN1)
1 CØNTINUE
  RETURN
  END
```

B.5.6
$$\frac{p^a}{\Gamma(a)}\int_0^\infty e^{-pt}\, t^{a-1}\, {}_1F_1(c;d;xt) \times F^{0:2}_{2:0}\left[\begin{matrix} - \;:g_1,h_1;g_2,h_2;g_3,h_3; \\ e,q: \qquad\qquad\qquad ; \end{matrix} yt,zt,ut\right] dt,$$

$$a,p > 0, \quad -p < x+y+z+u < p.$$

```
1 READ(m,6)A,C,D,E,G1,G2,G3,H1,H2,H3,P,Q,U,X,Y,Z
  IF(A)8,8,9
6 FØRMAT(16F4.2)
9 READ(m,7)M
7 FØRMAT(I2)
  CALL F(A,C,D,E,G1,G2,G3,H1,H2,H3,P,Q,U,X,Y,Z,M,S)
  WRITE(n,5)A,C,D,E,G1,G2,G3,H1,H2,H3,P,Q,U,X,Y,Z,M,S
5 FØRMAT(13H PARAMETERS =, 8F5.2/2X,8F5.2/3H M=,I2,3H F=,1PE14.6)
  GØTØ 1
8 STØP
  END

  SUBRØUTINE F(A,C,D,E,G1,G2,G3,H1,H2,H3,P,Q,U,X,Y,Z,M,S)
  S=0.0
  T1=1.0
  DØ 1 N1=1,M
  AN1=FLØAT(N1)
  A1=A+AN1-1.0
  T2=T1
  DØ 2 N2=1,M
  AN2=FLØAT(N2)
  A2=A1+AN2-1.0
  E2=E+AN2-1.0
  Q2=Q+AN2-1.0
  T3=T2
  DØ 3 N3=1,M
  AN3=FLØAT(N3)
  A3=A2+AN3-1.0
  E3=E2+AN3-1.0
  Q3=Q2+AN3-1.0
  T4=T3
  DØ 4 N4=1,M
  AN4=FLØAT(N4)
  A4=A3+AN4-1.0
  E4=E3+AN4-1.0
```

```
   Q4=Q3+AN4-1.0
   S=S+T4
   T4=T4*(A4/P)*((G3+AN4-1.0)/E4)*((H3+AN4-1.0)/Q4)*(U/AN4)
4  CØNTINUE
   T3=T3*(A3/P)*((G2+AN3-1.0)/E3)*((H2+AN3-1.0)/Q3)*(Z/AN3)
3  CØNTINUE
   T2=T2*(A2/P)*((G1+AN2-1.0)/E2)*((H1+AN2-1.0)/Q2)*(Y/AN2)
2  CØNTINUE
   T1=T1*(A1/P)*((C+AN1-1.0)/(D+AN1-1.0))*(X/AN1)
1  CØNTINUE
   RETURN
   END
```

B.6 Programs for the Evaluation of Mellin Integrals

B.6.1
$$\frac{x^{-p}\Gamma(a)\Gamma(b_1)\Gamma(c_1-p)\Gamma(d_1-p)}{\Gamma(p)\Gamma(c_1)\Gamma(d_1)\Gamma(a-p)\Gamma(b_1-p)} \times \int_0^\infty t^{p-1}\, F^{1:1;1}_{0:2;2}\left[\begin{matrix} a: b_1 \quad ; b_2 \quad ; \\ -:c_1,d_1;c_2,d_2; \end{matrix} -xt,y\right] dt,$$

$0 < p < a, \quad 0 < x.$ See Note on page 284.

```
1  READ(m,6)A,B2,C2,D2,P,Y
   IF(A)8,8,9
6  FØRMAT(6F4.2)
9  READ(m,7)M
7  FØRMAT(I2)
   CALL F(A,B2,C2,D2,P,Y,M,S)
   WRITE(n,5)A,B2,C2,D2,P,Y,M,S
5  FØRMAT(13H PARAMETERS =, 6F5.2/3H M=,I2,3H F=,1PE14.6)
   GØTØ 1
8  STØP
   END

   SUBRØUTINE F(A,B2,C2,D2,P,Y,M,S)
   S=0.0
   T=1.0
   DØ 1 N=1,M
   AN=FLØAT(N)-1.0
   S=S+T
   T=T*((A-P+AN)/(C2+AN))*((B2+AN)/(D2+AN))*(Y/(AN+1.0))
1  CØNTINUE
   RETURN
   END
```

B.6.2
$$\frac{x^{-p}\Gamma(a)\Gamma(b)\Gamma(c_1-p)\Gamma(d_1-p)}{\Gamma(p)\Gamma(c_1)\Gamma(d_1)\Gamma(a-p)\Gamma(b-p)} \times \int_0^\infty t^{p-1} F^{2:0}_{0:2}\left[\begin{matrix}a,b: & - & : \\ - & :c_1,d_1;c_2,d_2;c_3,d_3; & \end{matrix} -xt,y,z\right] dt,$$

$0 < p < \mathrm{Min}(a,b)$, $0 < x$. See Note on page 284.

```
1 READ(m,6)A,B,C2,C3,D2,D3,P,Y,Z
  IF(A)8,8,9
6 FØRMAT(9F4.2)
9 READ(m,7)M
7 FØRMAT(I2)
  CALL F(A,B,C2,C3,D2,D3,P,Y,Z,M,S)
  WRITE(n,5)A,B,C2,C3,D2,D3,P,Y,Z,M,S
5 FØRMAT(13H PARAMETERS =, 9F5.2/3H M=,I2,3H F=,1PE14.6)
  GØTØ 1
8 STØP
  END

  SUBRØUTINE F(A,B,C2,C3,D2,D3,P,Y,Z,M,S)
  S=0.0
  T1-1.0
  DØ 1 N1=1,M
  AN1=FLØAT(N1)
  A1=A-P+AN1-1.0
  B1=B-P+AN1-1.0
  T2=T1
  DØ 2 N2=1,M
  AN2=FLØAT(N2)
  A2=A1+AN2-1.0
  B2=B1+AN2-1.0
  S=S+T2
  T2=T2*(A2/(C3+AN2-1.0))*(B2/(D3+AN2-1.0))*(Z/AN2)
2 CØNTINUE
  T1=T1*(A1/(C2+AN1-1.0))*(B1/(D2+AN1-1.0))*(Y/AN1)
1 CØNTINUE
  RETURN
  END
```

B.6.3
$$\frac{z^{p}\Gamma(b_3)\Gamma(c-p)}{\Gamma(p)\Gamma(c)\Gamma(b_3-p)} \int_0^\infty t^{p-1}\ \Phi_2^{(3)}(b_1,b_2,b_3;c;-xt,-yt,-zt)\ dt,$$

$0 < p < b_3$, $0 < x,y,z$.

```
1 READ(m,6)B1,B2,B3,P,X,Y,Z
  IF(P)8,8,9
6 FØRMAT(7F4.2)
```

```
9 READ(m,7)M
7 FØRMAT(I2)
  CALL F(B1,B2,B3,P,X,Y,Z,M,S)
  WRITE(n,5)B1,B2,B3,P,X,Y,Z,M,S
5 FØRMAT(13H PARAMETERS =, 7F 5.2/3H M=,I2,3H F=,1PE14.6)
  GØTØ 1
8 STØP
  END

  SUBRØUTINE F(B1,B2,B3,P,X,Y,Z,M,S)
  S=0.0
  T1=1.0
  DØ 1 N1=1.M
  AN1=FLØAT(N1)
  P1=P+AN1-1.0
  Q1=P-B3+AN1
  T2=T1
  DØ 2 N2=1,M
  AN2=FLØAT(N2)
  P2=P1+AN2-1.0
  Q2=Q1+AN2-1.0
  S=S+T2
  T2=T2*(P2/Z)*((B2+AN2-1.0)/Q2)*(Y/AN2)
2 CØNTINUE
  T1=T1*(P1/Z)*((B1+AN1-1.0)/Q1)*(X/AN1)
1 CØNTINUE
  RETURN
  END
```

B.6.4 $$(x_4/2)^p \frac{\Gamma(c_4-p)\Gamma([p+1]/2)}{\Gamma(p)\Gamma(c_4)\Gamma(1/2)} \times \int_0^\infty t^{p-1}\ {}_0F_1(-;c_1;-x_1^2t^2)\ldots{}_0F_1(-;c_4;-x_4^2t^2)\,dt,$$

$$p,x_1,x_2,x_3,x_4 > 0,\quad c_1+c_2+c_3+c_4-p < 4.$$

```
1 READ(m,6)C1,C2,C3,C4,P,X1,X2,X3,X4
  IF(P)8,8,9
6 FØRMAT(9F4.2)
9 READ(m,7)M
7 FØRMAT(I2)
  CALL F(C1,C2,C3,C4,P,X1,X2,X3,X4,M,S)
  WRITE(n,5)C1,C2,C3,C4,P,X1,X2,X3,X4,M,S
5 FØRMAT(13H PARAMETERS =, 9F5.2/3H M=,I2,3H F=,1PE14.6)
  GØTØ 1
8 STØP
  END

  SUBRØUTINE F(C1,C2,C3,C4,P,X1,X2,X3,X4,M,S)
```

```
  S=0.0
  T1=1.0
  DØ 1 N1=1,M
  AN1=FLØAT(N1)
  P1=(P/2.0)+AN1-1.0
  Q1=((P-1.0)/2.0)-C4+AN1
  T2=T1
  DØ 2 N2=1,M
  AN2=FLØAT(N2)
  P2=P1+AN2-1.0
  Q2=Q1+AN2-1.0
  T3=T2
  DØ 3 N3=1,M
  AN3=FLØAT(N3)
  P3=P2+AN3-1.0
  Q3=Q2+AN3-1.0
  S=S+T3
  Z1=(X1/X4)*(X1/X4)
  Z2=(X2/X4)*(X2/X4)
  Z3=(X3/X4)*(X3/X4)
  T3=T3*(P3/(C3+AN3-1.0))*(Q3/AN3)*Z3
3 CØNTINUE
  T2=T2*(P2/(C2+AN2-1.0))*(Q2/AN2)*Z2
2 CØNTINUE
  T1=T1*(P1/(C1+AN1-1.0))*(Q1/AN1)*Z1
1 CØNTINUE
  RETURN
  END
```

B.6.5 $$\frac{(x_1+x_2)^p}{\Gamma(p)}\int_0^\infty t^{p-1}\,{}_1F_1(a_1;c_1;-x_1t)\ldots{}_1F_1(a_4;c_4;-x_4t)\,dt,$$

$$p,x_1,x_2 > 0.$$

```
1 READ(m,6)A1,A2,A3,A4,C1,C2,C3,C4,P,X1,X2,X3,X4
  IF(P)8,8,9
6 FØRMAT(13F4.2)
9 READ(m,7)M
7 FØRMAT(I2)
  CALL F(A1,A2,A3,A4,C1,C2,C3,C4,P,X1,X2,X3,X4,M,S)
  WRITE(n,5)A1,A2,A3,A4,C1,C2,C3,C4,P,X1,X2,X3,X4,M,S
5 FØRMAT(13H PARAMETERS =,13F5.2/3H M=,I2,3H F=,1PE14.6)
  GØTØ 1
8 STØP
  END

  SUBRØUTINE F(A1,A2,A3,A4,C1,C2,C3,C4,P,X1,X2,X3,X4,M,S)
  S=0.0
  T1=1.0
  DØ 1 N1=1,M
```

```
  AN1=FLØAT(N1)
  P1=P+AN1-1.0
  T2=T1
  DØ 2 N2=1,M
  AN2=FLØAT(N2)
  P2=P1+AN2-1.0
  T3=T2
  DØ 3 N3=1,M
  AN3=FLØAT(N3)
  P3=P2+AN3-1.0
  T4=T3
  DØ 4 N4=1,M
  AN4=FLØAT(N4)
  P4=P3+AN4-1.0
  S=S+T4
  Z1=X1/(X1+X2)
  Z2=X2/(X1+X2)
  Z3=X3/(X1+X2)
  Z4=X4/(X1+X2)
  T4=T4*(P4/(C4+AN4-1.0))*(A4+AN4-1.0)/AN4)*Z4
4 CØNTINUE
  T3=T3*(P3/(C3+AN3-1.0))*((A3+AN3-1.0)/AN3)*Z3
3 CØNTINUE
  T2=T2*(P2/(C2+AN2-1.0))*((A2+AN2-1.0)/AN2)*Z2
2 CØNTINUE
  T1=T1*(P1/(C1+AN1-1.0))*((A1+AN1-1.0)/AN1)*Z1
1 CØNTINUE
  RETURN
  END
```

B.7 Programs for the Evaluation of Multiple Integrals

B.7.1
$$\frac{\Gamma(a_1+b_1)\Gamma(a_2+b_2)\Gamma(a_3+b_3)}{\Gamma(a_1)\Gamma(b_1)\Gamma(a_2)\Gamma(b_2)\Gamma(a_3)\Gamma(b_3)}$$
$$\times \int_0^1\int_0^1\int_0^1 u_1^{a_1-1}\, u_2^{a_2-1}\, u_3^{a_3-1}(1-u_1)^{b_1-1}(1-u_2)^{b_2-1}(1-u_3)^{b_3-1}$$
$$\times F^{2:0}_{0:2}\left[\begin{matrix} c,d: & - & ; \\ - & :g_1,h_1;g_2,h_2;g_3,h_3; \end{matrix}\; u_1x, u_2y, u_3z\right] du_1du_2du_3,$$
$$a_1,a_2,a_3,b_1,b_2,b_3 > 0.$$

```
1 READ(m,6)A1,A2,A3,B1,B2,B3,C,D,G1,G2,G3,H1,H2,H3,X,Y,Z
  IF(A1)8,8,9
6 FØRMAT(17F4.2)
9 READ(m,7)M
7 FØRMAT(I2)
```

```
  CALL F(A1,A2,A3,B1,B2,B3,C,D,G1,G2,G3,H1,H2,H3,X,Y,Z,M,S)
  WRITE(n,5)A1,A2,A3,B1,B2,B3,C,D,G1,G2,G3,H1,H2,H3,X,Y,Z,M,S
5 FØRMAT(13H PARAMETERS =, 9F5.2/2X,8F5.2/3H M=,I2,3H F=,1PE14.6)
  GØTØ 1
8 STØP
  END

  SUBRØUTINE F(A1,A2,A3,B1,B2,B3,C,D,G1,G2,G3,H1,H2,H3,X,Y,Z,M,S)
  S=0.0
  T1=1.0
  DØ 1 N1=1,M
  AN1=FLØAT(N1)
  C1=C+AN1-1.0
  D1=D+AN1-1.0
  T2=T1
  DØ 2 N2=1,M
  AN2=FLØAT(N2)
  C2=C1+AN2-1.0
  D2=D1+AN2-1.0
  T3=T2
  DØ 3 N3=1,M
  AN3=FLØAT(N3)
  C3=C2+AN3-1.0
  D3=D2+AN3-1.0
  S=S+T3
  T3=T3*((C3/(A3+B3+AN3-1.0))*(D3/(G3+AN3-1.0))
  T3=T3*((A3+AN3-1.0)/(H3+AN3-1.0))*(Z/AN3)
3 CØNTINUE
  T2=T2*(C2/(A2+B2+AN2-1.0))*(D2/(G2+AN2-1.0))
  T2=T2*((A2+AN2-1.0)/(H2+AN2-1.0))*(Y/AN2)
2 CØNTINUE
  T1=T1*(C1/(A1+B1+AN1-1.0))*(D1/(G1+AN1-1.0))
  T1=T1*((A1+AN1-1.0)/(H1+AN1-1.0))*(X/AN1)
1 CØNTINUE
  RETURN
  END
```

B.7.2 $$\frac{\Gamma(a_1+a_2+a_3+b)}{\Gamma(a_1)\Gamma(a_2)\Gamma(a_3)\Gamma(b)} \iiint u_1^{a_1-1} u_2^{a_2-1} u_3^{a_3-1} (1-u_1-u_2-u_3)^{b-1}$$

$$\times\ {}_1F_1(c_1;d_1;u_1x)\,{}_1F_1(c_2;d_2;u_2y)\,{}_1F_1(c_3;d_3;u_3z)\,du_1du_2du_3,$$

$$a_1,a_2,a_3,b > 0,\ u_1,u_2,u_3 \geq 0,\ u_1+u_2+u_3 \leq 1.$$

```
1 READ(m,6)A1,A2,A3,B,C1,C2,C3,D1,D2,D3,X,Y,Z
  IF(A1)8,8,9
6 FØRMAT(13F4.2)
9 READ(m,7)M
7 FØRMAT(I2)
```

```
  CALL F(A1,A2,A3,B,C1,C2,C3,D1,D2,D3,X,Y,Z,M,S)
  WRITE(n,5)A1,A2,A3,B,C1,C2,C3,D1,D2,D3,X,Y,Z,M,S
5 FØRMAT(13H PARAMETERS =,13F5.2/3H M=,I2,3H F=,1PE14.6)
  GØTØ 1
8 STØP
  END

  SUBRØUTINE F(A1,A2,A3,B,C1,C2,C3,D1,D2,D3,X,Y,Z,M,S)
  S=0.0
  T1=1.0
  DØ 1 N1=1,M
  AN1=FLØAT(N1)
  B1=A1+A2+A3+AN1+B-1.0
  T2=T1
  DØ 2 N2=1,M
  AN2=FLØAT(N2)
  B2=B1+AN2-1.0
  T3=T2
  DØ 3 N3=1,M
  AN3=FLØAT(N3)
  B3=B2+AN3-1.0
  S=S+T3
  T3=T3*((A3+AN3-1.0)/B3)*((C3+AN3-1.0)/(D3+AN3-1.0))*(Z/AN3)
3 CØNTINUE
  T2=T2*((A2+AN2-1.0)/B2)*((C2+AN2-1.0)/(D2+AN2-1.0))*(Y/AN2)
2 CØNTINUE
  T1=T1*((A1+AN1-1.0)/B1)*((C1+AN1-1.0)/(D1+AN1-1.0))*(X/AN1)
1 CØNTINUE
  RETURN
  END
```

B.7.3 $$\frac{\Gamma(a_1+a_2+a_3+b)}{\Gamma(a_1)\Gamma(a_2)\Gamma(a_3)\Gamma(b)} \iiint u_1^{a_1-1} u_2^{a_2-1} u_3^{a_3-1} (1-u_1-u_2-u_3)^{b-1} \times {}_1F_1(c;d;xu_1+yu_2+zu_3)\, du_1 du_2 du_3,$$

$$a_1, a_2, a_3, b > 0,\quad u_1, u_2, u_3 \geqq 0,\quad u_1+u_2+u_3 \leqq 1.$$

```
1 READ(m,6)A1,A2,A3,B,C,D,X,Y,Z
  IF(A1)8,8,9
6 FØRMAT(9F4.2)
9 READ(m,7)M
7 FØRMAT(I2)
  CALL F(A1,A2,A3,B,C,D,X,Y,Z,M,S)
  WRITE(n,5)A1,A2,A3,B,C,D,X,Y,Z,M,S
5 FØRMAT(13H PARAMETERS =,9F5.2/3H M=,I2,3H F=,1PE14.6)
  GØTØ 1
8 STØP
  END

  SUBRØUTINE F(A1,A2,A3,B,C,D,X,Y,Z,M,S)
```

```
  S=0.0
  T1=1.0
  DØ 1 N1=1,M
  AN1=FLØAT(N1)
  B1=A1+A2+A3+AN1-1.0
  C1=C+AN1-1.0
  D1=D+AN1-1.0
  T2=T1
  DØ 2 N2=1,M
  AN2=FLØAT(N2)
  B2=B1+AN2-1.0
  C2=C1+AN2-1.0
  D2=D1+AN2-1.0
  T3=T2
  DØ 3 N3=1,M
  AN3=FLØAT(N3)
  B3=B2+AN3-1.0
  C3=C2+AN3-1.0
  D3=D2+AN3-1.0
  S=S+T3
  T3=T3*(C3/B3)*((A3+AN3-1.0)/D3)*(Z/AN3)
3 CØNTINUE
  T2=T2*(C2/B2)*((A2+AN2-1.0)/D2)*(Y/AN2)
2 CØNTINUE
  T1=T1*(C1/B1)*((A1+AN1-1.0)/D1)*(X/AN1)
1 CØNTINUE
  RETURN
  END
```

B.7.4 $$\frac{p_1^{a_1}p_2^{a_2}p_3^{a_3}}{\Gamma(a_1)\Gamma(a_2)\Gamma(a_3)}\int_0^\infty\int_0^\infty\int_0^\infty e^{-p_1t_1-p_2t_2-p_3t_3}\, t_1^{a_1-1}t_2^{a_2-1}t_3^{a_3-1}$$

$$\times\ {}_1F_1(c;d;xt_1+yt_2+zt_3)\ dt_1dt_2dt_3,$$

$$a_1,a_2,a_3,p_1,p_2,p_3 > 0$$

```
1 READ(m,6)A1,A2,A3,C,D,P1,P2,P3,X,Y,Z
  IF(A1)8,8,9
6 FØRMAT(11F4.2)
9 READ(m,7)M
7 FØRMAT(I2)
  CALL F(A1,A2,A3,C,D,P1,P2,P3,X,Y,Z,M,S)
  WRITE(n,5)A1,A2,A3,C,D,P1,P2,P3,X,Y,Z,M,S
5 FØRMAT(13H PARAMETERS =,11F5.2/3H M=,I2,3H F=,1PE14.6)
  GØTØ 1
8 STØP
  END
```

```
      SUBRØUTINE F(A1,A2,A3,C,D,P1,P2,P3,X,Y,Z,M,S)
      S=0.0
      T1=1.0
      DØ 1 N1=1,M
      AN1=FLØAT(N1)
      C1=C+AN1-1.0
      D1=D+AN1-1.0
      T2=T1
      DØ 2 N2=1,M
      AN2=FLØAT(N2)
      C2=C1+AN2-1.0
      D2=D1+AN2-1.0
      T3=T2
      DØ 3 N3=1,M
      AN3=FLØAT(N3)
      C3=C2+AN3-1.0
      D3=D2+AN3-1.0
      S=S+T3
      T3=T3*(C3/P3)*((A3+AN3-1.0)/D3)*(Z/AN3)
    3 CØNTINUE
      T2=T2*(C2/P2)*((A2+AN2-1.0)/D2)*(Y/AN2)
    2 CØNTINUE
      T1=T1*(C1/P1)*((A1+AN1-1.0)/D1)*(X/AN1)
    1 CØNTINUE
      RETURN
      END
```

B.7.5 $$\frac{p_1^{a_1}p_2^{a_2}p_3^{a_3}}{\Gamma(a_1)\Gamma(a_2)\Gamma(a_3)}\int_0^\infty\int_0^\infty\int_0^\infty e^{-p_1t_1-p_2t_2-p_3t_3} \times F^{1:1}_{0:2}\left[\begin{matrix} c: d_1 \quad ; d_2 \quad ; d_3 \quad ; \\ -: g_1,h_1; g_2,h_2; g_3,h_3; \end{matrix} xt_1, yt_2, zt_3\right] dt_1dt_2dt_3,$$

$$a_1, a_2, a_3, p_1, p_2, p_3 > 0.$$

```
    1 READ(m,6)A1,A2,A3,C,D1,D2,D3,G1,G2,G3,H1,H2,H3,P1,P2,P3,X,Y,Z
      IF(A1)8,8,9
    6 FØRMAT(19F4.2)
    9 READ(m,7)M
    7 FØRMAT(I2)
      CALL F(A1,A2,A3,C,D1,D2,D3,G1,G2,G3,H1,H2,H3,P1,P2,P3,X,Y,Z,M,
     1S)
      WRITE(n,5)A1,A2,A3,C,D1,D2,D3,G1,G2,G3,H1,H2,H3,P1,P2,P3,X,Y,Z,
     1M,S
    5 FØRMAT(13H PARAMETERS =,10F5.2/2X,9F5.2/3H M=,I2,3H F=,1PE14.6)
      GØTØ 1
    8 STØP
      END
```

```
      SUBRØUTINE F(A1,A2,A3,C,D1,D2,D3,G1,G2,G3,H1,H2,H3,P1,P2,P3,X,
     1Y,Z,M,S)
      S=0.0
      T1=1.0
      DØ 1 N1=1,M
      AN1=FLØAT(N1)
      C1=C+AN1-1.0
      T2=T1
      DØ 2 N2=1,M
      AN2=FLØAT(N2)
      C2=C1+AN2-1.0
      T3=T2
      DØ 3 N3=1,M
      AN3=FLØAT(N3)
      C3=C2+AN3-1.0
      S=S+T3
      T3=T3*(C3/(G3+AN3-1.0))*((A3+AN3-1.0)/(H3+AN3-1.0))
      T3=T3*((D3+AN3-1.0)/AN3)*Z
    3 CØNTINUE
      T2=T2*(C2/(G2+AN2-1.0))*((A2+AN2-1.0)/(H2+AN2-1.0))
      T2=T2*((D2+AN2-1.0)/AN2)*Y
    2 CØNTINUE
      T1=T1*(C1/(G1+AN1-1.0))*((A1+AN1-1.0)/(H1+AN1-1.0))
      T1=T1*((D1+AN1-1.0)/AN1)*X
    1 CØNTINUE
      RETURN
      END
```

B.8 Programs for the Evaluation of Integrals involving the Modified Bessel Function of the Second Kind

B.8.1 $$\frac{x^{a+1}}{2^{a-1}\Gamma([a+b+1]/2)\Gamma([a-b+1]/2)} \times \int_0^\infty t^a\, K_b(xt)\, {}_1F_2(c;d_1,d_2;y^2t^2)\, dt, \quad a+b > -1.$$

```
    1 READ(m,6)A,B,C,D1,D2,X,Y
      IF(A)8,8,9
    6 FØRMAT(7F4.2)
    9 READ(m,7)M
    7 FØRMAT(I2)
      CALL F(A,B,C,D1,D2,X,Y,M,S)
      WRITE(n,5)A,B,C,D1,D2,X,Y,M,S
    5 FØRMAT(13H PARAMETERS =,7F5.2/3H M=,I2,3H F=,1PE14.6)
      GØTØ 1
    8 STØP
      END
```

```
   SUBRØUTINE F(A,B,C,D1,D2,X,Y,M,S)
   S=0.0
   T=1.0
   DØ 1 N=1,M
   AN=FLØAT(N)-1.0
   S=S+T
   P=(A+B+1.0)/2.0
   Q=(A-B+1.0)/2.0
   Z=4.0*(Y/X)*(Y/X)
   T=T*((C+AN)/(D1+AN))*((P+AN)/(D2+AN))*((Q+AN)/(AN+1.0))*Z
 1 CØNTINUE
   RETURN
   END
```

B.8.2 $\dfrac{x^{a+1}}{2^{a-1}\Gamma([a+b+1]/2)\Gamma([a-b+1]/2)}$

$$\times \int_0^\infty t^a K_b(xt) F^{1:0;0}_{1:1;1}\left[\begin{matrix} c: & - & ; \\ g: h_1 ; h_2 ; \end{matrix} y^2t^2, z^2t^2\right] dt, \quad a+b > -1.$$

```
 1 READ(m,6)A,B,C,G,H1,H2,X,Y,Z
   IF(A)8,8,9
 6 FØRMAT(9F4.2)
 9 READ(m,7)M
 7 FØRMAT(I2)
   CALL F(A,B,C,G,H1,H2,X,Y,Z,M,S)
   WRITE(n,5)A,B,C,G,H1,H2,X,Y,Z,M,S
 5 FØRMAT(13H PARAMETERS =,9F5.2/3H M=,I2,3H F=,1PE14.6)
   GØTØ 1
 8 STØP
   END

   SUBRØUTINE F(A,B,C,G,H1,H2,X,Y,Z,M,S)
   S=0.0
   T1=1.0
   DØ 1 N1=1,M
   AN1=FLØAT(N1)
   C1=C+AN1-1.0
   D=(A+B+1)/2.0
   E=(A-B+1)/2.0
   D1=D+AN1-1.0
   E1=E+AN1-1.0
   G1=G+AN1-1.0
   T2=T1
   DØ 2 N2=1,M
   AN2=FLØAT(N2)
```

```
  C2=C1+AN2-1.0
  D2=D1+AN2-1.0
  E2=E1+AN2-1.0
  G2=G1+AN2-1.0
  Z1=4.0*(Y/X)*(Y/X)
  Z2=4.0*(Z/X)*(Z/X)
  S=S+T2
  T2=T2*(C2/G2)*(D2/(H2+AN2-1.0))*(E2/AN2)*Z2
2 CØNTINUE
  T1=T1*(C1/G1)*(D1/(H1+AN1-1.0))*(E1/AN1)*Z1
1 CØNTINUE
  RETURN
  END
```

B.8.3 $$\frac{x^{a+1}}{2^{a-1}\Gamma([a+b+1])/2)\Gamma([a-b+1]/2)}$$

$$\times \int_0^\infty t^a\, K_b(xt)\,{}_0F_1(-;g_1;y^2t^2)\,{}_0F_1(-;g_2;z^2t^2)\,{}_0F_1(-;g_3:u^2t^2)\,dt,$$

$$a+b > -1.$$

```
1 READ(m,6)A,B,G1,G2,G3,U,X,Y,Z
  IF(A)8,8,9
6 FØRMAT(9F4.2)
9 READ(m,7)M
7 FØRMAT(I2)
  CALL F(A,B,G1,G2,G3,U,X,Y,Z,M,S)
  WRITE(n,5)A,B,G1,G2,G3,U,X,Y,Z,M,S
5 FØRMAT(13H PARAMETERS =, 9F5.2/3H M=,I2,3H F=,1PE14.6)
  GØTØ 1
8 STØP
  END

  SUBRØUTINE F(A,B,G1,G2,G3,U,X,Y,Z,M,S)
  S=0.0
  T1=1.0
  DØ 1 N1=1,M
  AN1=FLØAT(N1)
  D=(A+B+1.0)/2.0
  E=(A-B+1.0)/2.0
  D1=D+AN1-1.0
  E1=E+AN1-1.0
  T2=T1
  DØ 2 N2=1,M
  AN2=FLOAT(N2)
  D2=D1+AN2-1.0
  E2=E1+AN2-1.0
  T3=T2
```

```
      DØ 3 N3=1,M
      AN3=FLØAT(N3)
      D3=D2+AN3-1.0
      E3=E2+AN3-1.0
      Z1=4.0*(Y/X)*(Y/X)
      Z2=4.0*(Z/X)*(Z/X)
      Z3=4.0*(U/X)*(U/X)
      S=S+T3
      T3=T3*(D3/(G3+AN3-1.0))*(E3/AN3)*Z3
    3 CØNTINUE
      T2=T2*(D2/(G2+AN2-1.0))*(E2/AN2)*Z2
    2 CØNTINUE
      T1=T1*(D1/(G1+AN1-1.0))*(E1/AN1)*Z1
    1 CØNTINUE
      RETURN
      END
```

B.9 Programs for the Evaluation of Convolution Integrals

B.9.1 $$p^{1-a-b}\frac{\Gamma(a+b)}{\Gamma(a)\Gamma(b)}\int_0^p v^{a-1}(p-v)^{b-1}{}_1F_1(g;a;vx)\times{}_1F_1(h;b;[p-v]y)\,dv,\quad a,b>0.$$

```
    1 READ(m,6)A,B,G,H,P,X,Y
      IF(A)8,8,9
    6 FØRMAT(7F4.2)
    9 READ(m,7)M
    7 FØRMAT(I2)
      CALL F(A,B,G,H,P,X,Y,M,S)
      WRITE(n,5)A,B,G,H,P,X,Y,M,S
    5 FORMAT(13H PARAMETERS =, 7F5.2/3H M=,I2,3H F=,1PE14.6)
      GØTØ 1
    8 STØP
      END

      SUBRØUTINE F(A,B,G,H,P,X,Y,M,S)
      S=0.0
      T=1.0
      DØ 1 N=1.M
      AN=FLØAT(N)-1.0
      S=S+T
      T=T*((G+H+AN)/(A+B+AN))*(P/(AN+1.0))*(X+Y)
    1 CØNTINUE
      RETURN
      END
```

B.9.2 $$p^{1-a-b}\frac{\Gamma(a+b)}{\Gamma(a)\Gamma(b)}\int_0^p v^{a-1}(p-v)^{b-1} \times {}_1F_1(a+b;c;[x-y]v+yp)\,dv, \quad a,b > 0.$$

```
1 READ(m,6)A,B,C,P,X,Y
  IF(A)8,8,9
6 FØRMAT(6F4.2)
9 READ(m,7)M
7 FØRMAT(I2)
  CALL F(A,B,C,P,X,Y,M,S)
  WRITE(n,5)A,B,C,P,X,Y,M,S
5 FØRMAT(13H PARAMETERS =, 6F5.2/3H M=,I2,3H F=,1PE14.6)
  GØTØ 1
8 STØP
  END

  SUBRØUTINE F(A,B,C,P,X,Y,M,S)
  S=0.0
  T1=1.0
  DØ 1 N1=1,M
  AN1=FLØAT(N1)
  C1=C+AN1-1.0
  T2=T1
  DØ 2 N2=1,M
  AN2=FLØAT(N2)
  C2=C1+AN2-1.0
  S=S+T2
  T2=T2*((B+AN2-1.0)/C2)*(Y/AN2)*P
2 CØNTINUE
  T1=T1*((A+AN1-1.0)/C1)*(X/AN1)*P
1 CØNTINUE
  RETURN
  END
```

B.9.3 $$p^{1-a-b}\frac{\Gamma(a+b)}{\Gamma(a)\Gamma(b)}\int_0^p v^{a-1}(p-v)^{b-1} \times F^{2:0;0}_{0:2;2}\left[\begin{matrix} c_1,c_2: & - & ; \\ - & :g_1,h_1;g_2,h_2;g_3,h_3; \end{matrix} xv,(p-v)y\right]dv,$$

$$a,b > 0.$$

```
1 READ(m,6)A,B,C,D,G1,G2,H1,H2,P,X,Y
  IF(A)8,8,9
```

```
 6 FØRMAT(11F4.2)
 9 READ(m,7)M
 7 FØRMAT(I2)
   CALL F(A,B,C,D,G1,G2,H1,H2,P,X,Y,M,S)
   WRITE(m,5)A,B,C,D,G1,G2,H1,H2,P,X,Y,M,S
 5 FØRMAT(13H PARAMETERS =,11F5.2/3H M=,I2,3H F=,1PE14.6)
   GØTØ 1
 8 STØP
   END

   SUBRØUTINE F(A,B,C,D,G1,G2,H1,H2,P,X,Y,M,S)
   S=0.0
   T1=1.0
   DØ 1 N1=1,M
   AN1=FLØAT(N1)
   A1=A+B+AN1-1.0
   C1=C+AN1-1.0
   D1=D+AN1-1.0
   T2=T1
   DØ 2 N2=1,M
   AN2=FLØAT(N2)
   A2=A1+AN2-1.0
   C2=C1+AN2-1.0
   D2=D1+AN2-1.0
   S=S+T2
   T2=T2*(C2/A2)*(D2/(G2+AN2-1.0))*((B+AN2-1.0)/(H2+AN2-1.0))
   T2=T2*(Y/AN2)*P
 2 CØNTINUE
   T1=T1*(C1/A1)*(D1/(G1+AN1-1.0))*((A+AN1-1.0)/(H1+AN1-1.0))
   T1=T1*(X/AN1)*P
 1 CØNTINUE
   RETURN
   END
```

B.9.4 $$p^{1-a-b}\frac{\Gamma(a+b)}{\Gamma(a)\Gamma(b)}\int_0^p v^{a-1}(p-v)^{b-1}\,{}_1F_1(g_1;h_1;xv)$$

$$\times\ {}_1F_1(g_2;h_2;y[p-v])\,{}_1F_1(g_3;h_3;z[p-v])\,dv,\quad a,b > 0.$$

```
 1 READ(m,6)A,B,G1,G2,G3,H1,H2,H3,P,X,Y,Z
   IF(A)8,8,9
 6 FØRMAT(12F4.2)
 9 READ(m,7)M
 7 FØRMAT(I2)
   CALL F(A,B,G1,G2,G3,H1,H2,H3,P,X,Y,Z,M,S)
   WRITE(n,5)A,B,G1,G2,G3,H1,H2,H3,P,X,Y,Z,M,S
 5 FØRMAT(13H PARAMETERS =,12F5.2/3H M=,I2,3H F=,1PE14.6)
   GØTØ 1
```

```
8 STØP
  END

  SUBRØUTINE F(A,B,G1,G2,G3,H1,H2,H3,P,X,Y,Z,M,S)
  S=0.0
  T1=1.0
  DØ 1 N1=1,M
  AN1=FLØAT(N1)
  A1=A+B+AN1-1.0
  T2=T1
  DØ 2 N2=1,M
  AN2=FLØAT(N2)
  A2=A1+AN2-1.0
  B2=B+AN2-1.0
  T3=T2
  DØ 3 N3=1,M
  AN3=FLØAT(N3)
  A3=A2+AN3-1.0
  B3=B2+AN3-1.0
  S=S+T3
  T3=T3*(B3/A3)*((G3+AN3-1.0)/(H3+AN3-1.0))*(Z/AN3)*P
3 CØNTINUE
  T2=T2*(B2/A2)*((G2+AN2-1.0)/(H2+AN2-1.0))*(Y/AN2)*P
2 CØNTINUE
  T1=T1*(B1/A1)*((G1+AN1-1.0)/(H1+AN1-1.0))*(X/AN1)*P
1 CØNTINUE
  RETURN
  END
```

NOTE. The integrals B.6.1 and B.6.2 are such that when the integrations are carried out, a number of the quantities involved are accounted for in the gamma-multipliers of the integrals in question. For this reason, it is unnecessary to include these quantities explicitly in the two programs concerned.

Selected Bibliography

Abbott, W.R. (1949). Evaluation of an integral of a Bessel Function. *J. Math. Physics* 28 192-194.

Abrahmanov, M.A. and Abdikerimov, I.A. (1974). A certain Riemann-Mellin integral. *Isv. Akad. Nauk. Razah S.S.R. Ser. Fiz.-Mat.* 89 1-5.

Abramowitz, M. et al. (1965). *Handbook of Mathematical Functions.* Dover, New York.

Afshar, R. and Mueller, F.M. (1973). Hilbert transformation of densities of states using Hermite functions. *J.Computational Phys.* 11 190-209.

Agahanov, S.A. and Natanson, G.I. (1968). The Lebesque function of Fourier-Jacobi sums. *Vestnik Leningrad Univ.* 23 11-23.

Agarwal, R.P. (1963). *Generalised Hypergeometric Series.* Asia Publishing House, Bombay, New York.

Agostinelli, C. (1942). Sopra alcuni integrali delle funzioni cilindriche. *Boll. Un. Mat. Ital.* 4 25-28.

Agrest, M.M. and Maksimov, M.S. (1971). *Theory of Incomplete Cylindrical Functions and their Applications.* Springer.

Agrest, M.M. and Rikenglaz, M.M. (1967). Incomplete Lipshitz-Hankel integrals. *Z. Vycisl. Mat. i Mat. Fiz.* 7 1370-1374.

Ainsworth, O.R. and Liu, C.K. (1975). An application of Legendre functions in the inversion of a Hilbert matrix. *J. Franklin Inst.* 299 297-299.

Akhaury, S.K. (1972). Abelian theorems for a distributional generalised Laplace transform. *Ranchi. Univ. Math. J.* 3 82-89.

d'Angelo, I.G. and Kalla, S.L. (1973). Some results that involve the H-function of Fox. *Univ. Nac. Tucumán Rev. Ser. A* 23 83-87.

de Anguino, M.E.F. (1975). On an integral transform involving a kernel of Mellin-Barnes type integral. *Kyungpook Math. J.* 15 175-181.

de Anguino, M.E.F.; de Gomez Lopez, A.M.M. and Kalla, S.L. (1972). Integrals that involve the H-function of two variables. *Acta Mexicana Ci. Tecn.* 6 30-41.

Appell, P. (1925). *Sur les Fonctions Hypergéometriques de Plusieurs Variables.* Gautier Villars, Paris.

Appell, P. and Kampé de Fériet, J. (1926). *Fonctions Hypergéometriques et Hypersphériques. Polynomes d'Hermite.* Gautier Villars, Paris.

Askey, R.A. (Ed.). (1976). *Theory and Applications of Special Functions.* Publication No. 35, Mathematical Research Center, University of Wisconsin. Academic Press, New York and London.

Askey, R.A. and Fitch, J. (1969). Integral representations for Jacobi polynomials and some applications. *J. Math. Anal. Appl.* 26 411-437.

Askey, R.A. and Gasper, G. (1971). Jacobi polynomial expansions of Jacobi polynomials with non-negative coefficients. *Proc. Cambridge Philos. Soc.* 70 243-255.

Askey, R.A. and Ismail, M.E.H. (1976). Permutation problems and special functions. *Can. J. Math.* 28 853-874.

Askey, R.A. and Razban, B. (1972). An integral for Jacobi polynomials. *Simon Stevin.* 46 165-169.

Baber, T.D.H. and Mirsky, L. (1944). Note on certain integrals involving Hermite's polynomials. *Philos. Mag.* 35 532-537.

Babeshko, V.A. (1970). Asymptotic properties of the solutions of a class of integral equations of the theory of elasticity and mathematical physics. *Dokl. Akad. Nauk S.S.S.R.* 193 557-560.

Babister, A.W. (1967). *Transcendental Functions Satisfying Nonhomogeneous Linear Differential Equations.* Collier-Macmillan, London.

Bailey, W.N. (1929a). Some definite integrals involving Bessel functions. *Proc. London Math. Soc.* 30 200-208.

Bailey, W.N. (1929b). Some definite integrals allied to an integral of Jacobi. *Proc. London Math. Soc.* 30 417-424.

Bailey, W.N. (1929c). Some integrals of Kapteyn's type involving Bessel functions. *Proc. London Math. Soc.* 30 422-424.

Bailey, W.N. (1930). Some definite integrals involving Legendre functions. *Proc. Cambridge Philos. Soc.* 26 475-479.

Bailey, W.N. (1931). Some series and integrals involving associated Legendre functions. I and II. *Proc. Cambridge Philos. Soc.* 27 184-189 and 381-386.

Bailey, W.N. (1934). Some infinite integrals involving Bessel functions. *Proc. London Math. Soc.* 40 37-48.

Bailey, W.N. (1935). *Generalised Hypergeometric Series.* Cambridge Tracts in Mathematics and Mathematical Physics No.32 Cambridge University Press.

Bailey, W.N. (1936). Some infinite integrals involving Bessel functions. *J. London Math. Soc.* 11 16-20.

Bailey, W.N. (1938a). Some integral formulae involving associated Legendre functions. *J. London Math. Soc.* 13 167-169.

Bailey, W.N. (1938b). Some integrals involving Bessel functions. *Quart. J. Math. Oxford Ser.* 9 141-147.

Bailey, W.N. (1939). On Hermite polynomials and associated Legendre functions. *J. London Math. Soc.* 14 281-286.

Bailey, W.N. (1940). On the product of two associated Legendre functions. *Quart. J. Math. Oxford Ser.* 11 30-35.

Bailey, W.N. (1941). On the double-integral representation of Appell's function F_4. *Quart. J. Math. Oxford Ser.* 12 12-14.

Bailey, W.N. (1948). Some integrals involving Hermite polynomials. *J. London Math. Soc.* 23 291-297.

Bajpai, S.D. (1971). Meijer's G-function and angular displacement in a shaft. *Proc. Nat. Acad. Sci. India* 41(A) 230-232.

Barnett, M.P. and Coulson, C.A. (1951). The evaluation of integrals occurring in the theory of molecular structure. *Philos. Trans. Roy. Soc. London Ser. A.* 243 221-249. I and II.

Barr, C.E. (1969). The integration of generalised hypergeometric functions. *Proc. Cambridge Philos. Soc.* 65 591-595.

Barrucand, P. and Dickenson, D. (1967). On the associated Legendre polynomials. Orthogonal expansions and their continuous analogues. *Proc. Conf. Edwardsville, Ill.* 1967 43-50.

Bateman, H. (1941). Some definite integrals occurring in aerodynamics. *Theodore von Karman Anniversity Vol., Caltec., Pasadena.* 1-7.

Bateman, H. (1949). Some definite integrals occurring in Havelock's work on the wave resistance of ships. *Math. Mag.* 23 1-4.

de Battig, N.E.F. (1974). A theorem on the Meijer transform. *Math. Notae* 24 33-38.

de Battig, N.E.F. and Kalla, S.L. (1971). Some results involving generalised hypergeometric functions of two variables. *Rev. Ci. Mat. Univ. Lourenco Marques, Ser. A* 2 47-53.

Baudoux, P. (1945). Sur les fonctions de Weber et Lommel. *Acad. Roy. Belgique Bull. Cl. Sci.* 31 669-681.

Beck, G. (1947). An application of Poisson's integral. *Math. Notae* 7 191-204.

Belward, J.A. (1969). The solution of an integral equation of the first kind on a finite interval. *Quart. Appl. Math.* 27 313-321.

Ben-Menahem, A. (1975). Properties and applications of a certain operator associated with the Kontrovich-Lebedev transform. *Glasgow Math. J.* 16 109-122.

Bennett, W.R. (1948). Distribution of the sum of randomly phased components. *J. Math. Phys.* 5 385-393.

Benton, T.C. (1975). Concerning

$$\int_0^\infty e^{-at} J_\mu(bt) J_\nu(ct) t^{\mu-\nu} dt.$$

SIAM J. Math. Anal. 6 761-765.

Bingham, N.H. (1972). Integral representations for ultraspherical polynomials. *J. London Math. Soc.* 6 1-11

Blanusa, D. (1945). Eine Klasse von Integraltheoremen der Besselschen Funktionen. *Bull. Intern. Acad. Croate Cl. Sci. Mat. Nat.* 35 103-139.

Blanusha, D. (1950). A type of integral theorem for Bessel functions. *Rad. Jugoslav Akad. Znan. Umjet. Odjel Mat. Fiz. Tehn. Nauke.* 277 55-128.

Blinov, V.S. (1971). A certain class of improper integrals in filtration theory problems. *Izv. Vyss. Ucebn. Zaved. Math.* 1971 33-41.

Bock, P. (1940). Über einige Integrale aus der Theorie der Besselschen, Whittakerschen und verwandter Funktionen. *Niewe Arch. Wiskde.* 20 163-170.

Bonham, R.A. (1966). Some properties of the integrals

$$\int_0^1 P_{2n}(t) \sin xt \, dt \text{ and } \int_0^1 P_{2n+1}(t) \cos xt \, dt.$$

J. Math. and Phys. 23 331-334.

Bouman, J. (1968). On some integrals containing the product of two Bessel functions of order zero or one. *Niewe Arch. Wisk.* 16 186-193.

Bouwkamp, C.J. (1947). A study of Bessel functions in connection with the problem of two mutually attracting circular discs. *Nederl. Akad. Wetensch. Proc. A.* 50 1071-1083.

Bouwkamp, C.J. (1949). On the evaluation of certain integrals occurring the theory of the freely vibrating circular disc and related problems. *Nederl. Akad. Wetensch. Proc. A.* 52 987-994.

Bouwkamp, C.J. (1950). On integrals occurring in the theory of diffraction of electromagnetic waves by a circular disc. *Nederl. Akad. Wetensch. Proc. A.* 53 654-661.

Bouwkamp, C.J. (1976). On some Bessel-function integral equations. *Ann. Mat. Pura Appl.* IV ser. 108 63-67.

Bowie, P.C. (1971). Uncertainty inequalities for Hankel transforms. *SIAM J. Math. Anal.* 2 601-606.

Bragard, L. (1952). Sur quelques formules relatives aux harmoniques sphériques. *Bull. Soc. Roy. Sci. Liège.* 21 46-70, 158-178.

Bragg, L.R. (1967). Dual functions and heat expansions. *Proc.Amer. Math. Soc.* 18 402-407.

de Branges, L. (1964). Self-reciprocal functions. *J. Math. Anal. Appl.* 9 433-457.

Brennan, R.O. and Mulligan, J.L. (1952). Two-center hetero-nuclear hybrid Coulomb exchange integrals. *J. Chem. Phys.* 20 1635-1644.

Braodbent, D. and Jánossy, L. (1948). Production of penetrating particles in extensive air showers. *Proc. Roy. Soc. London Ser. A.* 192 364-382.

Bromwich, T.J.I'A. (1931). *An Introduction to the Theory of Infinite Series.* Second Edition. Macmillan, London.

Buchholz, H. (1958). Ein besonderes uneigentliches Integral über das Produckt zweier regulärer Coulombscher Wellenfunktionen. *Z. Angew. Math. Mech.* 38 115-120.

Burchnall, J.L. and Chaundy, T.W. (1941). Expansions of Appell's double hypergeometric functions. II. *Quart. J. Math. Oxford Ser.* 12 112-128.

Burlak, J. (1962). A pair of dual integral equations occurring in diffraction theory. *Proc. Edin. Math. Soc.* 13 179-187.

Burlak, J. (1963). On the solution of certain dual integral equations. *Proc. Glasgow Math. Assoc.* 6 39-44.

Busbridge, I.W. (1939). The evaluation of certain integrals involving products of Hermite polynomials. *J. London Math. Soc.* 14 93-97.

Busbridge, I.W. (1948). Some integrals involving Hermite polynomials. *J. London Math. Soc.* 23 135-141.

Busbridge, I.W. (1950). On the integro-exponential function and the evaluation of some integrals involving it. *Quart. J. Math. Oxford Ser.* (2) 1 176-184.

Buschman, R.G. (1963). An inversion theorem for a general Legendre transformation. *SIAM Rev.* 5 232-233.
Buschman, R.G. (1965). Integrals of hypergeometric functions. *Math. Z.* 89 74-76.
Buschman, R.G. (1974). Partial derivatives of the H-function with respect to parameters expressed as finite sums and as integrals. *Univ. Nac. Tucumán Rev. Ser. A.* 24 149-155.
Buschman, R.G., Gupta, O.P. and Rathie, P.N. (1973). Integrals involving generalisations of the hypergeometric function. *Univ. Nac. Tucumán Rev. Ser. A.* 23 89-94.
Carlitz, L. (1961). Some integrals containing products of Legendre polynomials. *Arch. Math.* 12 334-340.
Carlitz, L. (1962a). Some integral formulas for the complete elliptic integrals of the first and second kinds. *Proc. Amer. Math. Soc.* 13 913-917.
Carlitz, L. (1962b). A characterisation of the Laguerre polynomials. *Monatsch. Math.* 66 389-392.
Carlitz, L. (1962c). An integral for the product of two Laguerre polynomials. *Boll. Un. Mat. Ital.* 17 25-28.
Carlson, B.C. (1961). Some series and bounds for incomplete elliptical integrals. *J. Math. and Physics.* 40 125-134.
Carlson, B.C. (1974). Inequalities for Jacobi polynomials and Dirichlet averages. *SIAM J. Math. Anal.* 5 586-596.
Cerrillo, M. (1949). Inversion of Laplace transforms of type $F(S,\surd[S^2+B^2])\exp(-A\surd[S^2+B^2])$. *Comisión Impulsora y Coordinatora de la Investigación Cientitura (Mexico).* 1947 31-86.
Ceschino, F. (1950). Les fonctions hypérgéometriques d'ordre supérieur à deux variables. *Ann. Soc. Sci. Bruxelles Ser.* I. 64 13-21.
Chak,A.M. (1955). A generalisation of Whittaker's integral. *Ann. Univ. Lyon Sect. A.* 18 27-33.
Chako, N. (1963). On the evaluation of certain integrals and their application to diffraction theory. *Acta Phys. Polon.* 24 611-620.
Champeney, D.C. (1973). *Fourier Transforms and their Physical Applications*. Academic Press, London, New York.
Cholewinski, F.M. and Haimo, D.T. (1971). Inversion of the Hankel potential transform. *Pacific J. Math.* 37 319-330.
Clemmow, P.C. and Senior, T.B.A. (1953). A note on a generalised Fresnel integral. *Proc. Cambridge Philos. Soc.* 49 570-572.
Collins, W.D. (1958). Some integrals involving Legendre functions. *Proc. Edinburgh Math. Soc.* 11 161-165.
Conte, S.D. (1956). The operational calculus of the Gegenbauer transform. *Z. Angew. Math. Mech.* 36 148-180.
Colombo, S. and Lavoine, J. (1972). *Transformations de Laplace et de Mellin*. Mémorial des Science Mathématiques CLXIX. Gautier Villars, Paris.
Conolly, B.W. (1951). An application of the "Faltung" formula. *Ganita* 2 50-52.

Conolly, B.W. (1955). Two integrals involving modified Bessel functions of the second kind. *Proc. Glasgow Math. Assoc.* 2 147-148.

Cooke, J.C. (1953). Some properties of Legendre functions. *Proc. Cambridge Philos. Soc.* 49 162-164.

Cooke, J.C. (1954). Note on some integrals of Bessel functions with respect to their order. *Monatsch Math.* 58 1-4.

Cooke, J.C. (1956). Some relations between Bessel and Legendre functions. *Monatsch. Math.* 60 322-328.

Cooke, J.C. (1962). Some further triple integral equation solutions. *Proc. Edin. Math. Soc.* 13 303-316.

Cooke, J.C. (1963). Triple integral equations. *Quart. J. Mech. Appl. Math.* 16 193-203.

Cooke, J.C. (1972). The solution of triple and quadruple integral equations and Fourier-Bessel series. *Quart. J. Mech. Appl. Math.* 25 247-263.

Copson, E.T. (1941). On an infinite integral connected with the theory of Bessel functions. *Proc. Cambridge Philos. Soc.* 37 102-104.

Copson, E.T. (1961). On certain dual integral equations. *Proc. Glasgow Math. Assoc.* 5 21-24.

Crothers, D.S.F. (1972). Analytical continuation of hypergeometric functions by complex single-loop Euler integrals. *J. Physics.* 5 256-262.

Cumakov, F.V. (1970). The Abel equation with a hypergeometric kernel. *Proceedings of a Conference on Boundary Value Problems. Izdat. Kazan. Univ., Kazan.* 267-271.

Cumakov, F.V. and Elesevic, T.A. (1972). A generalised Abel equation with a hypergeometric kernel. *Vesci Akad. Navuk BSSR Ser. Fiz.-Mat.* 1972 42-48.

Curtiss, C.F. (1964). Expansions of integrals of Bessel functions of large order. *J. Mathematical Physics.* 5 561-564.

Dekanosidze, E.N. (1955). Some properties of Lommel functions of two variables. *Vycisl. Mat. Vycisl. Tehn.* 2 97-107.

Denis, R.Y. (1972a). On certain infinite integrals involving a generalised hypergeometric function. *Bul. Inst. Politehn. Iasi (N.S.).* 18 61-64.

Denis, R.Y. (1972b). An integral involving a generalised hypergeometric function. *Math. Student* 40A 79-81.

Denman, H.H. and Howard, J.E. (1964). Application of ultraspherical polynomials to non-linear oscillations. *Quart. Appl. Math.* 21 325-330.

Denman, H.H. and Liu, Y.K. (1965). Application of ultraspherical polynomials to non-linear oscillations. *Quart. Appl. Math.* 22 273-292.

Deshpande, V.L. (1972). An integral involving the Lauricella function F_A. *Vijnana Parishad Anusandhan Patrika.* 15 197-204.

Deutsch, E. (1962). Evaluation of certain integrals involving Bessel functions. *Proc. Edin. Math. Soc.* 13 285-290.

Diaz, M. (1972). Integral transforms suitable to solve boundary value problems with radial symmetry. *Rev. Roumanie Math. Pures Appl.* 17 1599-1610.
Dijksma, A. and Koornwinder, T.H. (1971). Spherical harmonics and the product of two Jacobi polynomials. *Nederl. Akad. Wetensch. Proc. A.* 74 191-196.
Dijksma, A. and Kuipers, L. (1970). Generalised Legendre polynomials (an integral theorem). *Z. Angew. Math. Mech.* 50 498-499.
Dimovski, I.H. (1973). An explicit expression for the convolution of the Meijer transformation. *C.R. Acad. Bulgare Sci.* 26 1293-1296.
Dimovski, I.H. (1974). On a Bessel-type integral transformation due to N. Obrechkoff. *C.R. Acad. Bulgare Sci.* 27 23-26.
Ditkin, V.A. and Prudnikov, A.P. (1962). *Operational Calculus in Two Variables and its Applications.* English Edition.Pergamon Press, New York.
Doetsch, G. (1971). *Handbuch der Laplace-Transformation. Band I und II.* Birkhauser Verlag. Basel-Stuttgart.
Donaldson, J.A. (1967). Integral representations of the extended Airy integral type for the modified Bessel function. *J. Math. and Phys.* 46 111-114.
Dörr, J. (1953). Untersuchung einiger Integrale mit Bessel-Funktionen, die für die Elastizitäts theorie von Bedentung sind. *Z. Angew. Math. Physik.* 4 122-127.
Dougall, J. (1919). A theorem of Sonine on Bessel functions, with two extensions to spherical harmonics. *Proc. Edin. Math. Soc.* 37 33-47.
Dzrbasjan, V.A. (1965). Integrals of Bessel functions. *Isv. Akad. Nauk Armjan. SSR Ser. Fiz-Mat. Nauk* 18 3-20.
Eason, G.; Noble, B. and Sneddon, I.N. (1955). On certain integrals of Lipschitz-Hankel type involving products of Bessel functions. *Philos. Trans. Roy. Soc. London Ser. A* 247 529-551.
Erdélyi, A. (1939a). Integral representations for Whittaker functions. *Proc. Benares Math. Soc.* 1 39-53.
Erdélyi, A. (1939b). Two infinite integrals. *Proc. Edin. Math. Soc.* (2) 6 94-104.
Erdélyi, A. (1939c). Integraldarstellungen für Produkte Whittakerschen Funktionen. *Niewe Arch. Wiskde.* 20 1-38.
Erdélyi, A. (1939d). Transformations of hypergeometric integrals by means of fractional integration by parts. *Quart. J. Math. Oxford Ser.* 10 176-189.
Erdélyi, A. (1941a). Integration of the differential equations of Appell's function F_4. *Quart. J. Math. Oxford Ser.* 12 68-77.
Erdélyi, A. (1941b). A note on Heine's integral representation of associated Legendre functions. *Philos. Mag.* (7) 32 351-352.
Erdélyi, A. (1941c). Generating functions of certain continuous orthogonal systems. *Proc. Roy. Soc. Edin. Sect. A* 61 61-70.
Erdélyi, A. (1950). Hypergeometric functions of two variables. *Acta Math.* 83 131-164.

Erdélyi, A. (1963). An integral equation involving Legendre's polynomial. *Amer. Math. Monthly* 70 651-652.
Erdélyi, A. (1964). An integral equation involving Legendre functions. *J. Soc. Indust. Appl. Math.* 12 15-30.
Erdélyi, A. et al. (1953). *Higher Transcendental Functions.* 3 vols. McGraw-Hill, New York.
Erdélyi, A. et al. (1954). *Tables of Integral Transforms.* 2 vols. McGraw-Hill, New York.
Erdélyi, A. and Sneddon, I.N. (1962). Fractional integration and dual integral equations. *Canadian J. Math.* 14 685-693.
Eriksson, H A. (1944). Some applications and properties of hyperspherical harmonics with three polar angles. *Ark. Mat. Astr. Fys.* 30B 8-12.
Eruhimovic, J.A. and Pimenov, J.V. (1964). The evaluation of a definite integral containing a Bessel function. *Z. Vycisl. Mat. i Mat. Fiz.* 4 596-599.
Exton, H. (1972a). On two multiple hypergeometric functions related to Lauricella's F_D. *Jñānabha Sect. A* 2 59-73.
Exton, H. (1972b). Certain hypergeometric functions of four variables. *Bull. Soc. Math. Grèce (N.S.)* 13 104-113.
Exton, H. (1973a). Some integral representations and transfomations of hypergeometric functions of four variables. *Bull. Soc. Math. Grèce (N.S.)* 14 132-140.
Exton, H. (1973b). A note on the Humbert functions. *Jñānabha Sect. A* 3 1-5.
Exton, H. (1975). Incomplete hypergeometric functions. *Nanta Math.* 8 73-78.
Exton, H. (1976). *Multiple Hypergeometric Functions and Applications.* Ellis Horwood, Ltd., Chichester, U.K.
Fasenmayer, M.C. (1947). Some generalised hypergeometric polynomials. *Bull. Amer. Math. Soc.* 53 806-812.
Feldheim, E. (1939). Sur les fonctions génératrices des polynomes de Laguerre et d'Hermite. *Bull. Soc. Math.* 63 307-329.
Feldheim, E. (1940a). Développments en série de polynomes d'Hermite et de Laguerre à l'aide des transformations de Gauss et de Hankel. I et II. *Nederl. Akad. Wetensch. Proc. A.* 43 224-248.
Feldheim, E. (1940b). Développments en série de polynomes d'Hermite et de Laguerre à l'aide des transformations de Gauss et de Hankel. III. *Nederl. Akad. Wetensch. Proc. A.* 43 379-386.
Feldheim, E. (1940c). Expansions and integral transforms for products of Legendre and Hermite polynomials. *Quart. J. Math. Oxford Ser.* 11 18-29.
Feldheim, E. (1941a). Contributi alla teoria della funzioni ipergeometrichi di più variabili. *Ann. Sci. Norm. Sup. Pisa.* 17 17-59.
Feldheim, E. (1941b). Transformata di Hankel di funzioni di Whittaker. *Ann. Sci. Norm. Sup. Pisa.* 17 103-114.

Feldheim, E. (1941c). Alcuni risultati sulle funzioni di Whittaker e del cilindrico parabolico. *Fiz. Mat. Nat.* 76 541-555.

Feldheim, E. (1963). On the positivity of certain sums of ultraspherical polynomials. *J. Analyse Math.* 11 275-284.

Felson, L.B. (1956). Some definite integrals involving conical functions. *J. Math. Phys.* 35 177-178.

Fenyö, I. (1954). Einige Folgerungen aus dem Additionsetz für Zylinderfunktionen. *Magyar Tud. Akad. Alkalm. Mat. Int. Közl.* 2 345-360.

Fettis, H.E. (1957). Lommel type integrals involving three Bessel functions. *J. Math. Phys.* 36 88-95.

Fettis, H.E. (1960). A note on

$$\int_0^\infty e^{-x} J_0(\eta x/\xi) J_1(x/\xi) x^{-n}\, dx.$$

Math. Comp. 14 372-374.

Fettis, H.E. (1968). Generalisation of a Bessel function integral. *SIAM Rev.* 10 214-215.

Fiala, J. (1966). A note on the integrals involving a product of Hermite's polynomials. *Casopis Pest. Mat.* 91 217-220.

Fock, V.A. (1943). On the representation of an arbitrary function by an integral involving Legendre's function of complex index. *C.R. (Doklady) Acad. Sci. URSS (N.S.)* 39 253-256.

Ford, F.A.J. (1958). On certain definite integrals involving Bessel functions. *J. Math. Phys.* 37 157-161.

Ford, F.A.J. (1966). Some infinite integrals involving products of Bessel functions. *J. London Math. Soc.* 41 728-730.

Fox, C. (1965a). An inversion formula for the kernel $K_\nu(x)$. *Proc. Cambridge Philos. Soc.* 61 457-467.

Fox, C. (1965b). A formal solution of certain dual integral equations. *Trans. Amer. Math. Soc.* 119 389-398.

Fox, C. (1972) Applications of Laplace transforms and their inverses. *Proc. Amer. Math. Soc.* 35 193-200.

Gandel, J.V. (1970). On the theory of dual Fourier-Bessel series. *Teor. Funkcii Funkcional Anal. i Prilozen.* 12 59-69.

Garde, R.M. (1965). Applications of Gegenbauer polynomials to non-linear oscillations. Forced and free oscillations without damping. *Indian J. Math.* 7 111-117.

Gasfor, G. (1975). Positive integrals of Bessel functions. *SIAM J. Math. Anal.* 6 868-881.

Geller, M. and Ng, E.W. (1969). A table of integrals of the exponential integral. *J. Res. Nat. Bur. Standards Sect. B.* 73 191-210.

George, D.L. (1962). Numerical values of some integrals involving Bessel functions. *Proc. Edin. Math. Soc.* 13 87-113.

Germain, P. (1955). Remarks on transforms and boundary value problems. *J. Rational Mech. Anal.* 4 925-941.

Gerstl, S.A.W. (1969). Ueber einige neue summierbare Legendre-Reihen. *J. Reine Angew. Math.* 236 131-144.

Gillis, J. (1975). Integrals of products of Laguerre polynomials. *SIAM J. Math. Anal.* 6 318-339.

Gjellestad, G. (1955). Note on the definite integral over products of three Legendre functions. *Proc. Nat. Acad. Sci. USA.* 41 954-956.

Glaeske, H-J. and Sebastian, H-J. (1975). Ueber eine Verallgemeinerung der Besselfunktionen. *Math. Nachr.* 67 41-52.

Glasser, M.L. (1973). Some Bessel function integrals. *Kyungpook Math. J.* 13 171-174.

Glasser, M.L. (1974a). Some integrals involving the modified Bessel function K_0. *Math. Comp.* 28 5-8.

Glasser, M.L. (1974b). Some definite integrals of the product of two Bessel functions of the second kind of order zero. *Math. Comp.* 28 613-615.

Gonzalez, M.O. (1960). Application of the method of the Laplace transform to the evaluation of certain integrals containing non-elementary transcendental functions. *Rev. Un. Mat. Argentina.* 19 146-150.

Gradshteyn, I.S. and Ryzhik, I.M. (1965). *Tables of Integrals, Series and Products.* Fourth edition. Academic Press, New York.

Grant, J.A. and Ludwig, O.G. (1963). Note on the integrals of products of associated Legendre functions. *Comput. J.* 6 356-357.

Griffith, J.L. (1970). On a classical pair of equations. *Proc. Roy. Soc. New South Wales.* 103 31-34.

Grosjean, C.C. (1966). Some new integrals arising from mathematical physics. *Simon Stevin.* 40 49-72.

Grosjean, C.C. (1972). Note on a remark of R. Askey and B.Razban concerning an interesting integral involving the Legendre polynomials. *Simon Stevin.* 46 171-173.

Gudz, V.I. Asymptotic behaviour of the integral kernel of the apparent restistance in the method of vertical electrical sounding. *Isv. Akad. Nauk. SSSR Ser. Fiz. Zemli.* 1974 107-109.

Gutierrez-Suarez, J.J. (1963). Characterisation of functions representable by the generalised Whittaker transform. *Collect. Math.* 15 179-191.

Haimo, D.T. and Cholewinski, F.M. (1968). The Poisson-Laguerre transform. *Bull. Amer. Math. Soc.* 74 137-139.

Haimo, D.T. and Cholewinski, F.M. (1972). Inversion of the reduced Poisson-Hankel transform. *J. Analyse Math.* 25 323-343.

Hallén, E. (1955). Further investigations into iterated sine and cosine integrals and their amplitude functions with reference to antenna theory. *Kungl. Tekn. Högsk. Handl. Stockholm.* 89 1-44.

Hansen, E. (1966). On some sums and integrals involving Bessel functions. *Amer. Math. Monthly.* 73 143-150.

Heatley, A.H. (1939). Some integrals, differential equations and series related to the modified Bessel function of the first kind. *University of Toronto Studies.* 7 1-32.

Henderson, D.J.; Davidson, S.G. and Glasser, M.L. (1974). Indefinite integrals of products of Bessel functions. *Utilitas Math.* 5 227-237.

Henrici, P. (1955). On certain series expansions involving Whittaker functions and Jacobi polynomials. *Pacific J. Math.* 5 725-744.
Henrici, P. (1957). On the representation of a certain integral involving Bessel functions by hypergeometric series. *J. Math. Phys.* 36 151-156.
Higgins, T.P. (1963). An inversion integral for a Gegenbauer transformation. *J. Soc. Indust. Appl. Math.* 11 886-893.
Higgins, T.P. (1964). A hypergeometric function transform. *J. Soc. Indust. Appl. Math.* 12 601-612.
Hirschmann, I.I. (1957). Projections associated with Jacobi polynomials. *Proc. Amer. Math. Soc.* 8 286-290.
Hirschmann, I.I. (1963). Laguerre transforms. *Duke Math. J.* 30 495-510.
Hochstadt, H. (1956). Addition theroems for the functions of the paraboloid of revolution. *Div. Electromag. Res. Inst. Math. Sci. New York Univ.* 18 1-22.
Höfinger, F. (1952). Zur Theorie der hypergeometriche Funktionen. *Monatsh. Math.* 56 126-136.
de Hoop, A.T. (1955). On integrals occurring in the variational formulation of diffraction problems. *Nederl. Akad. Wetensch. Proc. B.* 58 325-330.
Horton, C.W. (1950). On the extension of some Lommel integrals to Struve functions with an application to acoustical *J. Math. Physics.* 29 31-37.
Howell, W.T. (1939a). A note on Laguerre polynomials. *Philos. Mag.* 28 287-288.
Howell, W.T. (1939b). Some formulae for the product of two Whittaker functions with different arguments. *Philos. Mag.* 28 493-495.
Hsü, H-Y. (1949). On Sonine's integral formula and its generalisation. *Bull. Amer. Math. Soc.* 55 370-378.
Igari, S. (1972). On the multipliers of Hankel transforms. *Tohoku Math. J.* 24 201-206.
Isaeva, L.V. (1970). Certain properties of the Kontrovic-Lebedev integral transform. *Tul. Gos. Ped. Inst. Ucen. Zap. Mat. Kaf. Vyp.* 3 82-86.
Ismail, M.H. (1975). Dual and triple sequence equations involving orthogonal polynomials. *Nederl. Akad. Wetensch. Proc. A.* 78 164-169.
Ivanov, V.K. and Melnikova, I.V. (1969). On finding the density of the distributions from a finite number of moments. *Ural. Gos. Univ. Mat. Zap.* 7 34-37.
Ivanov, V.K. and Mel'nikova, I.V. (1972). The problem of finding the density of mass distributions from a finite number of prescribed moments inside a disc or ball. *Ural. Gos. Univ. Mat. Zap.* 8 35-39.
Jackson, A.D. and Maximon, L.C. (1972). Integrals of products of Bessel functions. *SIAM J. Math. Anal.* 3 446-460.

Jaeger, J.C. (1946). On the repeated integrals of Bessel functions. *Quart. Appl. Math.* 4 302-305.
Jaeger, J.C. (1948). Repeated integrals of Bessel functions and the theory of transients in filter circuits. *J. Math. Physics.* 27 210-219.
Jaffé, H.H. (1953). Some overlap integrals involving d-orbitals. *J. Chem. Phys.* 21 258-263.
Jepson, D.W.; Haugh, E.F. and Hirschfelder, J.O. (1955). The integral of the associated Legendre function. *Proc. Nat. Acad. Sci. USA.* 41 645-647.
Jerri, A.J. and Kreisler, D.W. (1975). Sampling expansions with derivatives for finite Hankel and other transforms. *SIAM J. Math. Anal.* 6 262-267.
Jones, R.C. (1940). On the theory of fluctuation in the decay of sound. *J. Acoust. Soc. Amer.* 11 324-332.
Jucys, A.: Perkalskis, B.; Sugurovas, V. and Uspalis, K. (1955). Calculation of integrals of products of several spherical functions. *Vilniaus Valst. Univ. Mosklo Darbai Mat. Fiz. Chem. Mokslu.* 3 35-62.
Kapilevic, M.B. (1966a). Confluent hypergeometric Horn functions. *Differencial'nye Urovnenija.* 2 1239-1254.
Kaplievic, M.B. (1966b). Degenerate hypergeometric functions of Humbert and Horn. *Dokl. Akad. Nauk SSSR.* 170 1251-1254.
Kapilevic, M.B. (1968a). A certain class of hypergeometric function of Horn. *Differencial'nye Uravnenija.* 4 1465-1483.
Kapilevic, M.B. (1968b). Certain properties of Humbert's hypergeometric functions. *Studia Sci. Math. Hungar.* 3 81-91.
Kaplan, E.L. (1950). Multiple elliptic integrals. *J. Math. Phys.* 29 69-75.
Karasev, I.M. (1965). On an integral transformation. *Differencial'nye Uravnenija.* 1 1406-1410.
Karlsson, P.W. (1973). Reduction of certain generalised Kampé de Fériet functions. *Math. Scand.* 32 265-268.
Karlsson, P.W. (1976). A class of hypergeometric functions. *Nederl. Akad. Wetensch. Proc. A.* 79 36-40.
Karpyceva, Z.F. (1972). A certain property of the exponential integral function and its repeated integrals. *Isv. Vyss. Ucebn. Zaved. Mathematika.* 1972 33-41.
Katsura, S.; Inone, Y.; Yamashita, S. and Kilpatrick, J.E. (1965). Tables of integrals of threefold and fourfold products of associated Legendre functions. *Tech. Rep. Tohoku Univ.* 30 93-163.
Klamkin, M.S. (1957). An application of the Gauss multiplication theorem. *Amer. Math. Monthly.* 64 661-663.
Koh, E.L. (1975). The n-dimensional distributional Hankel transformation. *Canadian J. Math.* 27 423-433.
Koornwinder, T. (1975). A new proof of a Paley-Weiner type theorem for the Jabobi transform. *Ark. Mat.* 13 145-159.
Kornhauser, E.T. (1961). Further extensions of Schuster's integral. *Quart. Appl. Math.* 19 153-155.

Koschmeider, L. (1947). Integrale mit hypergeometrischen Integranden. *Acta Math.* 79 241-254.
Koschmeider, L. (1949). Funktionales Rechnen mit allgemeinen Ableitungen. *Anz. Oster. Akad. Wiss. Math. Nat.* 1949 241-244.
Koschmeider, L. (1952). Integrals with hypergeometric integrands. *Univ. Nac. Tucuman Rev. A.* 9 63-78.
Koschmeider, L. (1962). Transzendente Summensätze der Funktion F_D von Lauricella. *Monatsch. Math.* 66 299-305.
Kourganoff, V. (1947). Sur les fonctions $K_n(x)$ et certaines intégrales qui s'y rattachent. I. *Ann. Astrophysique.* 10 282-299.
Kourganoff, V. (1948). Sur les transformées par les operateurs Λ et Φ des fonctions integro-exponentielles. *C.R. Acad. Sci. Paris.* 224 920-922.
Kourganoff, V. (1949). Sur les intégrales

$$\int_0^\infty e^{-px} x^s K_n(x)\, dx.$$

C.R. Acad. Sci. Paris. 225 430-431 and 451-453.
Kozulin, J.N. and Maklovic, S.T. (1967). Some integral transformations for Bessel functions. *Kisinev. Gos. Univ. Ucen. Zap.* 90 63-69.
Knudsen, H.L. (1952). On the calculation of some definite integrals in antenna theory. *Appl. Sci. Research. B.* 3 51-68.
Kreysig, E. (1953). Der allgemeine Integralkosinus $Ci(z,\alpha)$. *Acta Math.* 89 107-131.
Kuipers, L. (1959). Integral transforms in the theory of Jacobi polynomials and generalised Legendre associated functions. I and II. *Nederl. Akad. Wetensch. Proc. A.* 62 148-152 and 63 406-408.
Kummer, E.E. (1836). Ueber die hypergeometrische Reihe $F(\alpha,\beta,x)$. *J. für Math.* 15 39-83 and 127-172.
Lauricella, G. (1893). Sulle funzioni ipergeometriche a più variabili. *Rend. Circ. Mat. Palermo.* 7 111-158.
Lebedev, N.N. (1947). Solution of Dirichlet's problem for hyperboloids of revolution. *Akad. Nauk SSSR. Prikl. Mat. Mech.* 11 251-258.
Lebedev, N.N. (1950). Some integral representations for products of sphere functions. *Doklady Akad. Nauk SSSR (N.S.)* 73 449-451.
Lebedev, N.N. (1962). On an integral representation of an arbitrary function in terms of squares of MacDonald functions with imaginary index. *Sibirsk. Mat. Z.* 3 213-222.
Lebedev, N.N. (1967). Expansion of an arbitrary function in an integral with respect to the squares of Legendre functions with complex index. *Differencial'nye Uravnenija.* 3 422-435.
Lebedev, N.N. and Skalskaia, I.P. (1966). Integral expansion of an arbitrary function in terms of spherical functions. *Prikl. Mat. Meh.* 30 252-258.

Lee, P.A. (1967). On integrals involving parabolic cylinder functions. *J. Math. and Phys.* 46 215-219.

Lew, J.S. (1974). On some integral relations between the Laplace and Mellin transforms. *IBM J. Res. Develop.* 19 582-586.

Lew, J.S. and Ghez, R. (1973). Problem 74-8, an integral equation for crystal growth. *SIAM Rev.* 17 687-690.

Lewin, L. (1953). Theory of wave-guide-fed slots radiating into parallel-plate regions. *J. Appl. Phys.* 24 232-236.

Lindsey, W.C. (1964). Infinite integrals containing Bessel function products. *J. Soc. Indust. Appl. Math.* 12 458-464.

Lindtstrom, F.T. and Oberhettinger, F. (1975). A note on a Laplace transform pair associated with mass transport in porous media and heat transfer problems. *SIAM J. App. Math.* 29 288-292.

Lorch, L. and Szego, P. (1962). A singular integral whose kernel involves a Bessel function. *Acta Math. Acad. Sci. Hungar.* 13 203-217.

Lord, R.D. (1949). Some integrals involving Hermite polynomials. *J. London Math. Soc.* 24 101-112.

Lord, R.D. (1960). Integrals of products of Laguerre polynomials. *Math. Comp.* 14 375-376.

Love, E.R. (1963). Dual integral equations. *Canadian J. Math.* 15 631-640.

Love, E.R. (1965). Franz Neumann's integral of 1848. *Proc. Cambridge Philos. Soc.* 61 445-456.

Love, E.R. (1967a). Some integral equations involving hypergeometric functions. *Proc. Edin. Math. Soc.* 15 169-198.

Love, E.R. (1967b). Two more hypergeometric integral equations. *Proc. Cambridge Philos. Soc.* 63 1055-1076.

Lowndes, J.S. (1964a). Integrals involving Bessel and Legendre functions. *Proc. Edin. Math. Soc.* 14 269-272.

Lowndes, J.S. (1964b). Note on the generalised Mehler transform. *Proc. Cambridge Philos. Soc.* 60 57-59.

Lowndes, J.S. (1973). Simultaneous dual integral equations. *Glasgow Math. J.* 14 73-76.

Lowdon, T.A. (1970). Integral representations of the Hankel function in terms of parabolic cylinder functions. *Quart. J. Mech. Appl. Math.* 23 315-327.

Lowengrub, M. and Sneddon, I.N. (1963). The solution of a pair of dual integral equations. *Proc. Glasgow Math. Assoc.* 6 14-18.

Luke, Y.L. (1950). Some integrals involving Bessel functions. *J. Math. Physics.* 29 27-30.

Luke, Y.L. (1952). An associated Bessel function. *J. Math. Physics.* 31 131-138.

Luke, Y.L. (1962). *Integrals of Bessel Functions*. McGraw-Hill, New York.

Luke, Y.L. (1969). *The Special Functions and their Approximations.* 2 volumes. Academic Press, New York.

MacKinnon, R.F. (1972). The asymptotic expansion of Hankel transforms and related integrals. *Math. Comp.* 26 515-527.

MacRobert, T.M. (1939). Some formulae for the associated Legendre functions of the first kind. *Phil. Mag.* 27 703-705.

MacRobert, T.M. (1940). Some integrals involving Legendre and Bessel functions. *Quart. J. Math. Oxford Ser.* 11 95-100.

MacRobert, T.M. (1941). Some formulae for the E-function. *Phil. Mag.* 31 254-260.

MacRobert, T.M. (1942). Some integrals involving E-functions and confluent hypergeometric functions. *Quart. J. Math. Oxford Ser.* 13 65-68.

MacRobert, T.M. (1943). Proofs of some formulae for the hypergeometric function and the E-function. *Phil. Mag.* 34 422-426.

MacRobert, T.M. (1952). On Neumann's formula for the Legendre function. *Proc. Glasgow Math. Assoc.* 1 10-12.

MacRobert, T.M. (1953a). An integral involving an E-function and an associated Legendre function of the first kind. *Proc. Glasgow Math. Assoc.* 1 111-114.

MacRobert, T.M. (1953b). Integral of an E-function expressed as a sum of two E-functions. *Proc. Glasgow Math. Assoc.* 1 118.

MacRobert, T.M. (1956). Some Bessel function integrals. *Proc. Glasgow Math. Assoc.* 2 183-184.

MacRobert, T.M. (1958). Integrals involving hypergeometric functions and E-functions. *Proc. Glasgow Math. Assoc.* 3 196-198.

MacRobert, T.M. (1959). Integration of E-functions with respect to their parameters. *Proc. Glasgow Math. Assoc.* 4 84-87.

MacRobert, T.M. (1962). *Functions of a Complex Variable.* Fifth Edition. Macmillan, London.

MacRobert, T.M. (1963). Integrals involving E-functions. *Proc. Glasgow Math. Assoc.* 6 31-33.

Magnus, W. (1946). Ueber eine Bezeihung zwischen Whittakerschen Funktionen. *Nachr. Akad. Wiss. Göttingen Math-Phys. Kl.* 1946 4-5.

Makarenko, L.G. (1975). A certain generalisation of dual and triple integral equations. *Vycizl. Prikl. Mat. (Kiev).* 25 72-79.

Maklovic, S.T. (1965). Investigation of integrals containing Bessel and elementary functions. *Kisinev Gos. Univ. Ucen. Zap.* 82 75-81.

Markusevic, G.A. (1969). Determination of the essential upper boundary of a region of finitness of a function with respect to its Fock-Mehler transform. *Sibirsk. Mat. Z.* 10 386-397.

Martic, B. (1973a). The connection between the Riemann-Liouville fractional integral, the Maijer and the Hankel transforms. *Akad. Nauka i Umjet. Bosne i Hercegov. Rad. Kn.* 45 145-148.

Martic, B. (1973b). The Eulerian integrals of the first and second kind that are associated with various functions. *Akad. Nauka i Umjet. Bosne i Hercegov. Rad. Kn.* 45 149-154.

Martic, B. (1973c). On Eulerian integrals of the first and second kind associated with multiple hypergeometric series. *Mat. Vestnik.* 10 123-126.

Mathai, A.M. and Saxena, R.K. (1973). *Generalised Hypergeometric Functions with Applications in Statistics and Physical Sciences.* Springer, Berlin, Heidelberg, New York.

Maximon, L.C. (1956). On the representation of indefinite integrals containing Bessel functions by simple Neumann series. *Proc. Amer. Math. Soc.* 7 1054-1062.

Maximon, L.C. and Morgan, G.W. (1955). On the evaluation of indefinite integrals involving the special functions. *Quart. App. Math.* 13 79-93.

McBride, A.C. (1974). Solution of hypergeometric integral equations involving generalised functions. *Proc. Edin. Math. Soc. (2).* 19 265-285.

Meijer, C.S. (1939a). Ueber Besselsche, Lommelsche und Whittakersche Funktionen. *Nederl. Akad. Wetensch. Proc. A.* 42 872-879.

Meijer, C.S. (1939b). Ueber Produkte Legendreschen Funktionen. *Nederl. Akad. Wetensch. Proc A.* 42 930-937 and 938-947.

Meijer, C.S. (1940). Ueber Besselschen, Struveschen und Lommelschen Funktionen. *Nederl. Akad. Wetensch. Proc. A.* 43 198-210 and366-378.

Meijer, C.S. (1941). Neue Integraldarstellungen für Whittakerschen Finktionen. *Nederl. Akad. Wetensch. Proc. A.* 44 81-92, 186-194, 298-307, 442-451 and 590-598.

Meijer, C.S. (1946). On the G-function. I-VIII. *Nederl. Akad. Wetensch. Proc. A.* 49 3-86.

Meijer, C.S. (1950). Neue Integraldarstellungen für Besselschen Funktionen. *Compositio Math.* 8 49-60.

Meller, N.A. (1961). An application of the operational calculus to the evaluation of certain integrals. *Vycisl. Mat.*7 170-180.

Meulenbeld, B. and de Snoo, H.S.V. (1967). Integrals involving generalised Legendre functions. *J. Engrg. Math.* 1 285-291.

Mihailov, M.D. (1967). The finite hypergeometric integral transform. *C.R. Acad. Bulgare Sci.* 20 887-889.

Miller, J. (1963). Formulas for integrals of products of associated Legendre and Laguerre functions. *Math. Comp.* 17 84-87.

Miller, W. (1971). On Lie algebras of difference operators and the special functions of mathematical physics. *SIAM J. Math. Anal.* 2 307-327.

Minn, H. (1969). A note on certain integrals related with the generalised hypergeometric function. *J. Korean Math. Soc.* 6 51-54.

Mitrinovic, D.S. (1956). Quelques formules concernant les polynomes de Legendre. *Univ. Beograd. Publ. Elektrotehn. Fak. Ser. Mat. Fiz.* 1 1-20.

Muller, G.M. (1955). On certain infinite integrals involving Bessel functions. *J. Math. Phys.* 34 179-181.

Muller, R. (1955). Spezielle Integrale mit Zylinder Funktionen. *Z. Angew. Math. Mech.* 35 62-64.

Müller, R. (1967). Integrals and series involving Bessel functions. *Bul. Inst. Politehn. Iasi (N.S.)* 13 43-45.
Murai, T. and Araki, G. (1952). Calculation of heteronuclear molecular integrals. *Progress Theoret. Physics* 8 615-638.
Mursi, Z. (1948). On the relation of the Airy and allied integrals to the Bessel functions. *Proc. Math. Phys. Soc. Egypt.* 3 23-38.
Nagaraja, K.S. (1964). A note on certain integrals. *J. Math. Phys.* 43 55-59.
Nagel, B. (1963). Some integrals and expansions relating to the hypergeometric functions F_1 and Φ_2 of two variables. *Ark. Fys.* 24 479-493.
Nassim, C. (1975). An inversion formula for the Hankel transform. *Pacific J. Math.* 57 255-258.
Naurzbaev, Z. (1972) Certain integrals that contain Bessel functions. *Izv. Akad. Nauk Kazah. SSR Ser. Fiz.-Mat.* 1972 69-72.
Naylor, D. (1963). On a finite Lebedev transform. *J. Math. Mech.* 12 375-383.
Naylor, D. (1964). On a Lebedev expansion theorem. *J. Math. Mech.* 13 353-363.
Naylor, D. (1966). On a finite Lebedev transform. *J. Math. Mech.* 15 455-464.
Naylor, D. (1972). On an integral transform associated with a certain condition of radiation. *Proc. Cambridge Philos. Soc.* 71 369-379.
Naylor, D. (1975). On an integral transform associated with a certain condition of radition. II. *Math. Proc. Cambridge Philos. Soc.* 77 188-197.
Newman, J.N. and Frank, W. (1963). An integral containing the square of a Bessel function. *Math. Comp.* 17 64-70.
Ng, E.W. and Geller, M. (1970). On some indefinite integrals of confluent hypergeometric functions. *J. Res. Nat. Bur. Stand. Sect. B.* 74 85-98.
Noble, B. (1963). The solution of Bessel function dual integral equations by a multiplying-factor method. *Proc. Cambridge Philos. Soc.* 59 351-362.
Nörlund, N.E. (1956). Sur les fonctions hypergéometriques d'ordre superieur. *Mat.-Fts. Skr. Danske Vid. Selsk.* 1 1-47.
Nuttall, A.H. (1975). Some integrals involving the Q_M function. *IEEE Trans. Information Theory.* 21 95-96.
Oberhettinger, F. (1957). On some expansions for Bessel integral functions. *J. Res. Nat. Bur. Stand.* 59 197-201.
Oberhettinger, F. (1972). *Tables of Bessel Transforms.* Springer, New York, Heidelberg.
Oberhettinger, F. (1973a). *Fourier Expansions. A Collection of Formulas.* Academic Press, New York.
Oberhettinger, F. (1973b). *Fourier Transforms of Distributions and their Inverses.* Academic Press, New York.

Oberhettinger, F. (1974). *Tables of Mellin Transforms.* Springer. New York, Heidelberg.

Oberhettinger, F. and Badii, L. (1973). *Tables of Laplace Transforms.* Springer. New York, Heidelberg.

Offerhaus, M.J. (1971). On integrals of Bessel functions related to Weber's second exponential integral. *Appl. Sci. Res.* 26 374-382.

Okui, S. (1974). Complete elliptic integrals resulting from infinite integrals of Bessel functions. *J. Res. Nat. Bur. Stand. Sect. B.* 78 113-135.

Okui, S. (1976). Complete elliptic integrals resulting from infinite integrals of Bessel functions. *J. Res. Nat. Bur. Stand. Sect. B.* 79 137-170.

Olsson, P.O.M. (1977). On the integration of the differential equations of five-parameter double hypergeometric functions of the second order. *J. Mathematical Physics.* 18 1285-1296.

Ossicini, A. (1962). Sui coefficienti di Gegenbauer-Stieltjes di una funzione non decrescente. *Atti. Accad. Naz. Lincei Rend. Cl. Sci. Fis. Mat. Nat.* 32 313-319.

Palamà G. (1939). Sui delle relazioni integrali relative ai polinomi di Laguerre e d'Hermite. *Rend. Sem. Mat. Univ. Padova.* 10 46-54.

Panda, R. (1973). Some integrals associated with the generalised Lauricella functions. *Publ. Inst. Math. Beograd (N.S.).* 16 115-122.

Panda, R. (1977). Certain dual series equations involving Laguerre polynomials. *Nederl. Akad. Wetensch. Proc. A.* 80 122-127.

Parodi, M. (1954). Sur une propriété des fonctions de Bessel. *C.R. Acad. Sci. Paris.* 238 195-196.

Parodi, M. (1968). Application du produit de convolution à l'obtenir de formules sommatoires. *C.R. Acad. Sci. Paris.* 267 465-468.

Parodi, M. (1969a). Analyse symbolique et formules sommatoires pour les fonctions de Bessel. *C.R. Acad. Sci. Paris.* 268 1099-1102.

Parodi, M. (1969b). Sur certaines intégrales qui interviennent dans l'étude des polynomes de Tchebichiff et d'Hermite. *C.R. Acad. Sci. Paris.* 268 1185-1188.

Patil, K.R. and Thakare N.K. (1977). On dual series equations involving Konhauser biorthogonal polynomials.*J. Mathematical Physics.* 18 1724-1726.

Penrod, D.D. and Farell, C. (1971). On the evaluation of certain integrals containing products of Bessel functions. *Indust. Math.* 21 85-89.

Pevnyi, B.G. (1940). Some functional equations for generalised hypergeometric series and Whitteraker's functions. *C.R. Acad. Sci. URSS (N.S.).* 28 310-312.

Piessene,R. and Haegemans, A. (1973). Inversion of some irrational Laplace transforms. *Comp.(Arch. Elek. Rechen).* 11 39-43.

Pinney, E. (1946). Laguerre functions in the mathematical formulation of the electromagnetic theory of the paraboloidal reflector. *J. Math. Physics.* 25 47-79.

Pinney, E. (1958). Some discontinous Bessel integrals. *J. Math. Physics.* 36 362-370.

Poli, L. (1947). Sur les équations intégrales dont le noyau est une fonction K_n de Bessel. *Ann. Soc. Sci. Bruxelles.* 61 191-198.

Popov, B.S. (1959a). On some integrals involving Legendre polynomials. *Bull. Soc. Roy. Sci. Liège.* 28 188-191.

Popov, B.S. (1959b). Sur les polynomes de Legendre. *Mathesis.* 68 239-242.

Popov, G.J. (1960). On an integro-differential equation.*Ukrain. Mat. Z.* 12 45-54.

Popov, G.J. (1966). Application of the Jacobi polynomials to the solution of integral equations. *Isv. Vyss. Ucebn. Zaved. Mathematika.* 53 77-85.

Popov, G.J. (1967). Some new relations for Jacobi polynomials. *Sibirsk. Mat. Z.* 8 1399-1404.

Pritchard, R.L. (1951). Discussion of papers by Pachner and by Stenzel on radiation from a circular emitter. *J. Acoustical Soc. Amer.* 23 591-593.

Pye, W. (1972). Further generalisations of Neumann's integral. *J. Austral. Math. Soc.* 14 368-374.

Rahman, M. (1976a). Construction of a family of positive kernels from Jacobi polynomials. *SIAM J. Math. Anal.* 7 92-116.

Rahman, M. (1976b). A five-parameter family of positive kernels from Jacobi polynomials. *SIAM J. Math. Anal.* 7 386-413.

Rahman, M. (1976c). Some positive kernels and bilinear sums for Hahn ploynomials. *SIAM J. Math. Anal.* 7 414-435.

Rainville, E.D. (1965). *Special Functions.* Macmillan, New York.

Reissner, E. (1952). Reihen entwicklung eines Integrales aus der Theorie der elastischen Schwingungen. *Math. Nachr.* 8 149-153.

Riekstyn's, E.Y. (1954). On a polynomial applicable to the solution of the telegraph equations. *Prikl. Mat. Meh.* 18 738-744.

Roothaan, C.C.J. (1951). A study of two-center integrals useful in calculations on molecular structure. *J. Chem. Phys.* 19 1445-1477.

Rosati, F. (1962a). Sui coefficienti di Laguerre-Stieltjes di una funzione non decrescente. *Atti. Acad. Naz. Lincei Rend. Cl. Sci. Fis. Mat. Nat.* 32 299-305.

Rosati, F. (1962b). Sui coefficienti di Hermite-Stieltjes di una funzione non decrescente. *Atti. Acad. Naz. Lincei Rend. Cl. Sci. Fis. Mat. Nat.* 32 857-862.

Rosenberg, R.L. (1949). The loss of energy of slow negative mesons in matter. *Philos. Mag.* 40 759-769.

Rosenthal, P. (1974). On an inversion theorem for the general Mehler-Fock transform pair. *Pacific J. Math.* 52 539-545.

Rösler, R. (1968). Eine Integralbeziehung Legendrescher Funktionen. *Z. Angew. Math. Mech.* 48 420-421.

Rousseau, C.C.; Askey, R. and Gasper, G. (1972). Problem 71-13: "A positive integral" by L. Carlitz. *SIAM Rev.* 14 372-378.

Rovnyak, V.G. (1966). Self-reciprocal functions. *Duke Math. J.* 33 363-378.

Rukhovets, A.N. and Ufliand, I.S. (1966). On a class of dual integral equations and their applications to the theory of elasticity. *Prikl. Mat. Meh.* 30 271-277.

Rutgers, J.G. (1941) . Sur des séries et des intégrales définies contenantes les fonctions de Bessel. *Nederl. Akad. Wetensch. Proc. A.* 44 464-474 and 636-647.

Rutgers, J.G. (1942). On series and definite integrals involving Bessel functions. *Nederl. Akad. Wetensch. Proc. A.* 45 376-379 and 484-489.

Ryko, V.S. (1974). The use of a certain connection between integral transforms for the calculation of integrals. *Mat. Zametki.* 15 129-137.

Ryko, V.S. (1975). Certain theorems on integral transforms. *Mat. Zametki.* 18 825-830.

Saaty, T.L. (1958). A note on the Fourier transform of Bessel functions. *Atti Acad. Sci. Torino Cl. Sci. Fis. Mat. Nat.* 93 284-287.

Saxena, V.P. and Nageria, A.R. (1974). Linear flow of heat in an anisotropic finite solid moving in a conducting medium. *Jñānabha Sect. A.* 4 1-6.

Schechter, D. (1976). A method of evaluating integrals of $r^m e^{-ar} j_n(Q_r)$. *J. Physics.* A9 335-336.

Schmidt, P.W. (1964). Asymptotic expansion of certain integrals containing the Bessel function of order zero. *J. Mathematical Physics.* 5 1183-1184.

Schwartz, L. (1944). Untersuchung einiger mit den Zylinderfunktion nulter Ordnung verwalter Funktionen. *Luftfarhtforschung.* 20 341-372.

Shah, M.(1973a). Some results connected with extended Fox's H-functions and their applications. *Vijnana Parishad Anusandhan Patrika* 16 47-66.

Shah, M. (1973b). Fourier expansion formulas for generalised Meijer functions. *Vijnana Parishad Anusandhan Patrika.* 16 97-114.

Shah, M. (1973c). On generalised functions involving classical polynomials. *Indian J. Pure Appl. Math.* 4 473-480.

Shah, M. (1973d). A new formula on generalised functions. *Indian J. Pure Appl. Math.* 4 889-897.

Shah, M. (1973e). On some applications relating to Fox's H-functions of two variables. *Publ. Inst. Math. Beograd.(N.S.)* 16 123-133.

Shah. M. (1973f). A theorem on generalised Meijer functions of two variables. *Istanbul Tek. Univ. Bul.* 26 30-38.

Shah. M. (1973g). Some multiplication theorems involving Fox's H-functions of two variables. *Univ. Nac. Tucuman Rev. A.* 23 143-151.

Shah. M. (1973h). On applications of the inversion theorems of Laplace and Mellin transforms associated with Fox's H-function of two variables. *Univ. Nac. Tucuman Rev. A.* 23 153-163.

Shah, M. (1973j). Several properties of generalised Fox's H-functions and their applications. *Portugal Math.* 32 179-199.

Shah.M. (1973k). A solution of the boundary value problem of heat conduction. *Math. Vestnik.* 10 117-122.

Shah, M. (1973ℓ). On some results involving Kampé de Fériet functions. *Bull. Calcutta Math. Soc.* 65 175-181.

Shah, M. (1974a). On generalised functions of two variables connected with orthogonal polynomials. *Vijnana Parishad Anusandhan Patrika.* 17 97-104.

Shah, M. (1974b). A note on the Mellin inversion formlua. *Indian J. Pure Appl. Math.* 5 464-469.

Shah, M. (1974c). On a unified result involving generalised functions of two variables. *Nanta Math.* 7 1-7.

Shah, M. (1975). An extension of Rice's result on an integral equation. *Publ. Inst. Math. Beograd. (N.S.)* 18 173-179.

Shah, M. (1976). An expansion theorem for the H-function. *J. Reine Angew. Math.* 258 1-5.

Simary, M.A. (1968). Integral representations of the modified Bessel function of the second kind. *J. Natur. Sci. and Math.* 8 133-139.

Simary, M.A. (1972a). Definite integrals involving generalised E-functions. *Math. Student.* 40A 119-123.

Simary, M.A. (1972b). Integrals involving generalised hypergeometric functions. *Rev. Roumaine Math. Pures Appl.* 17 281-286.

Singh Chandel, R.C. (1971). Fractional integration and integral representations of certain generalised hypergeometric functions of several variables. *Jñānabha Sect. A.* 1 45-56.

Singh Chandel, R.C. (1973). On some multiple hypergeometric functions related to Lauricella functions. *Jñānabha Sect. A.* 3 119-136.

Slater, L.J. (1952a). Integrals representing general hypergeometric transformations. *Quart. J. Math. Oxford Ser. (2).* 3 206-216.

Slater, L.J. (1952b). An integral of hypergeometric type. *Proc. Cambridge Philos. Soc.* 48 578-582.

Slater, L.J. (1953). Two double hypergeometric integrals. *Quart. J. Math. Oxford Ser. (2).* 4 127-131.

Slater, L.J. (1955a). The integration of hypergeometric functions. *Proc. Cambridge Philos. Soc.* 51 288-296.

Slater, L.J. (1955b). Integrals for asymptotic expansions of hypergeometric functions. *Proc. Amer. Math. Soc.* 6 226-231.

Slater, L.J. (1955c). Hypergeometric Mellin transforms. *Proc. Cambridge Philos. Soc.* 51 577-589.

Slater, L.J. (1960). *Confluent Hypergeometric Functions.* Cambridge University Press.

Slater, L.J. (1966). *Generalised Hypergeometric Functions.* Cambridge University Press.

Sneddon, I.N. (1951). *Fourier Transforms.* McGraw-Hill, New York.

Sneddon, I.N. (1961). *Special Functions of Mathematical Physics and Chemistry.* Second Edition. Oliver and Boyd. Edinburgh and London.

Sneddon, I.N. (1965). A relation involving Hankel transforms with applications to boundary value problems in potential theory. *J. Math. Mech.* 14 33-40.

Sneddon, I.N. (1966). *Mixed Boundary Value Problems in Potential Theory.* North Holland, Amsterdam.

Sneddon, I.N. (1967). The evaluation of an integral involving the product of two Gegenbauer polynomials. *SIAM Rev.* 9 569-572.

Sneddon, I.N. (1972). *The use of Integral Transforms.* McGraw-Hill, New York.

Srivastava, H.M. (1965). Some integrals involving products of Bessel and Legendre functions. *Rend. Sem. Mat. Univ. Padova.* 35 418-423.

Srivastava, H.M. (1966a). A relation between Meijer and generalised Hankel transforms. *Math. Japon.* 11 11-13.

Srivastava, H.M. (1966b). The products of certain classical polynomials. *Math. Japon.* 11 67-76.

Srivastava, H.M. (1966c). The integration of generalised hypergeometric functions. *Proc. Cambridge Philos. Soc.* 62 761-764.

Srivastava, H.M. (1967a). Some integrals involving products of Bessel and Legendre functions II. *Rend. Sem. Mat. Univ.Padova.* 37 1-10.

Srivastava, H.M. (1967b). Generalised Neumann expansions involving hypergeometric functions. *Proc. Cambridge Philos. Soc.* 63 425-429.

Srivastava, H.M. (1967c). Some integrals representing triple hypergeometric functions. *Rend. Circ. Mat. Palermo (2).* 16 99-115.

Srivastava, H.M. (1968a). Integration of certain products containing Jacobi polynomials. *Collect. Math.* 19 3-9.

Srivastava, H.M. (1968b). A note on a generalisation of Sonine's first finite integral. *Mathematische (Catania).* 23 1-6.

Srivastava, H.M. (1969). On a generalised integral transform. *Math. Z.* 108 197-201.

Srivastava, H.M. (1970). A note on the evaluation of a definite integral. *J. Korean Math. Soc.* 7 29-32.

Srivastava, H.M. (1971). Certain double integrals involving hypergeometric functions. *Jñānabha Sect.A.* 1 1-10.

Srivastava, H.M. (1972a). A contour integral involving Fox's H-function. *Indian J. Math.* 14 1-6.

Srivastava, H.M. (1972b). A class of integral equations involving the H-function as kernel. *Nederl. Akad. Wetensch. Proc. A.* 75 212-220.

Srivastava, H.M. (1972c). Remark on some integrals involving products of Whittaker functions. *Proc. Amer. Math. Soc.* 31 133-134.

Srivastava, H.M. (1972d). On reciprocal functions. *Indian J. Pure Appl. Math.* 3 704-713.

Srivastava, H.M. (1973a). A further note on certain dual equations involving Fourier-Laguerre series. *Nederl. Akad. Wetensch. Proc. A.* 76 137-141.

Srivastava, H.M. (1973b). A note on the integral representation for the product of two generalised Rice polynomials. *Collect. Math.* 24 117-121.

Srivastava, H.M. (1974). Analytical continuation of Appell's function F_4. *Rev. Mat. Hisp.-Amer.* 34 151-156.

Srivastava, H.M. (1975). An integral equation involving the confluent hypergeometric function of several variables. *Applicable Anal.* 5 251-256.

Srivastava, H.M. and Buschman, R.G. (1973). Composition of fractional integral operators involving Fox's H-function. *Acta Mexicana Ci. Tecn.* 7 21-28.

Srivastava, H.M. and Buschman, R.G. (1974). Some convolution integral equations. *Nederl. Akad. Wetensch. Proc. A.* 77 211-216.

Srivastava, H.M. and Buschman, R.G. (1976). Mellin convolutions and H-function transformations. *Rocky Mountain J. Math.* 6 331-343.

Srivastava, H.M. and Buschman, R.G. (1977). *Convolution Integral Equations with Special Function Kernels*. Wiley Eastern, Ltd., New Delhi, Bangalore.

Srivastava, H.M. and Daoust, M.C. (1969), On Eulerian integrals associated with Kampé de Fériet functions. *Publ. Inst. Mat. Beograd. (N.S.)* 9 199-202.

Srivastava, H.M. and Exton, H. (1977). A generalisation of the Weber-Schafheitlin integral.In course of publication.

Srivastava, H.M.; Gupta, K.C. and Hardy S. (1975). A certain double integral transformation. *Nederl. Akad. Wetensch. Proc. A.* 78 402-406.

Srivastava, H.M. and Joshi, C.M. (1967). Certain double Whittaker transforms of generalised hypergeometric functions. *Yokohama Math. J.* 15 17-32.

Srivastava, H.M. and Joshi, C.M. (1968). Certain integrals involving a generalised Meijer function. *Glasnik Mat. Ser.III.* 3 183-191.

Srivastava, H.M. and Joshi, C.M. (1969). Integration of certain products associated with a generalised Meijer transform. *Proc. Cambridge Philos. Soc.* 65 471-477.

Srivastava, H.M. and Joshi, C.M. (1974). Integral representation for the product of a class of generalised hypergeometric polynomials. *Acad. Roy. Belg. Bull. Cl. Sci.* 60 919-926.

Srivastava, H.M. and Panda, R. (1973). Some operational techniques in the theory of special functions. *Nederl. Akad. Wetensch. Proc. A.* 76 308-319.

Srivastava, H.M. and Panda, R. (1975). An integral representation for the product of two Jacobi polynomials. *J. London Math. Soc.* 12 419-425.

Srivastava, H.M. and Singhal, J.P. (1966). Contour integrals associated with certain generalised hypergeometric functions. *Proc. Nat. Acad. Sci. India Sect. A* 36 824-840.

Srivastava, H.M. and Singhal, J.P. (1968). Double Meijer transformations of certain hypergeometric functions. *Proc. Cambridge Philos. Soc.* 64 425-430.

Srivastava, H.M. and Singhal, J.P. (1969). Certain integrals involving Meijer's G-function of two variables. *Proc. Nat. Acad. Sci. India Sect. A.* 35 64-69.

Srivastava, H.M. and Singhal ,J.P. (1972). On a class of generalised hypergeometric distribution. *Jñānabha Sect. A.* 2 1-9.

Srivastava, H.M. and Srivastava, S.C. (1973). On the formal solutions of dual integral equations of two variables. *Defence Sci. J.* 23 71-78.

Stacy, E.W. (1962). A generalisation of the gamma distribution. *Ann. Math. Stastist.* 33 1187-1192.

Steel, W.H. and Ward, J.Y. (1956). Incomplete Bessel and Struve functions. *Proc. Cambridge Philos. Soc.* 52 431-441.

Straubel, R. (1942). Unbestimmte Integrale mit Produkten von Zylinderfunktionen. *Ing.- Arch.* 13 14-20.

Stolov, E.L. (1969). Asymptotic behaviour of a certain integral that contains Gegenbauer polynomials. *Kazan. Gos. Univ. Ucen. Zap.* 129 139-144.

Stolov, E.L. (1973). Asymptotic behaviour of the Mehler-Fock transform of functions that have a singularity. *Isv. Vyss. Ucebn. Zaved. Matematika.* 10 69-72.

Sura-Bura, M.R. Evaluation of an integral containing a product of Bessel functions. *Doklady Akad. Nauk SSSR (N.S.).* 73 901-903.

Swaroop, R. (1964). On a generalisation of the Laplace and the Stieltjes transformations. *Ann. Soc. Sci. Bruxelles.* 78 105-112.

Talat-Erben, M. (1976). Evaluation of the integral

$$I(n,a,b) = \int_0^\infty r^n e^{-ar} \mathrm{Ei}(br)\, dr.$$

Bull. techn. Univ. Istanbul. 29 78-82.

Tanno, Y. (1967). On a class of convolution transform. *Tohoku Math. J.* 19 168-186.

Tarasov, V.F. and Canysev, S.M. (1970). The symmetry of Appell's functions. *Izv. Akad. Nauk Kazah. SSR Ser. Fiz.-Mat.* 1970 55-59.

Thosar, Y.V. (1954). Generalisation of Neumann's formula for $Q_n(y)$. *Math. Z.* 60 52-60.

Tihanov, A.N. (1959). The asymptotic behaviour of integrals containing Bessel functions. *Dokl. Akad. Nauk SSSR* 125 982-985.

Titchmarsh, E.C. (1948a). *Introduction to the Theory of Fourier Integrals*. Oxford University Press.

Titchmarsh, E.C. (1948b). Some integrals involving Hermite polynomials. *J. London Math. Soc.* 23 15-16.

Toscano, L. (1941). Transformata di Laplace di prodotti di funzioni di Bessel e polinomi di Laguerre. Relazione integrale sui funzioni ipergeometriche più generali della F_A di Lauricella. *Pont. Acad. Sci. Comment.* 5 471-500.

Toscano, L. (1949). Calcolo di un integrale della teoria del potenziale di un ellipsoide. *Ist. Lombardo Sci. Lett. Rend. Cl. Sci. Mat. Nat. (3).* 13 192-200.

Toscano, L. (1953). Relazioni integrali sulla funzioni ipergeometriche di Kummer. *Mathematische (Catania).* 8 51-58.

Toscano, L. (1976). Calcul de deux intégrales définies de polynomes hypérgéometriques à l'aide du théorème de composition de la transformation de Laplace. *Bull. Math. Soc. Sci. Math. R.S. Roumanie (N.S.).* 18 403-414.

Tranter, C.J. (1956). *Integral Transforms in Mathematical Physics*. Second Edition. Methuen, London.

Tranter, C.J. (1963). A generalisation of Sonine's first finite integral. *Proc. Glasgow Math. Assoc.* 6 97-98.

Tranter, C.J. and Cooke, J.C. (1973). A Fourier-Neumann series and its application to the reduction of triple cosine series. *Glasgow Math. J.* 14 198-201.

Tricomi, F.G. (1950). Sulle funzioni gamma incompleta. *Ann. Mat. Pura Appl.* 31 263-279.

Tricomi, F.G. (1951). Generalizzazione di un teorema d'additione per le funzioni ipergeometriche confluenti. *Univ. e Politecnico Torino Rend. Sem. Mat.* 10 211-216.

Tsu, R. (1961). The evaluation of incomplete normalisation integrals and derivatives with respect to the order of associated Legendre polynomials. *J. Math. and Phys.* 40 232-243.

Tsuchikura, T. (1950). On the functions Ci(x,y) and Si(x,y). *Tohoku Math. J. (2).* 2 68-73.

Tuzilin, A.A. (1967). The theory of MacDonald integrals. *Differencial'nye Uravnenija.* 3 1195-1212 and 1751-1765.

Vaisleit, J.V. (1973). Certain integrals and series for the incomplete cylindrical functions. *Isv. Vyss. Ucebn. Zaved. Matematika.* 12 22-27.

Van Tuyl, A.H. (1964). The evaluation of some definite integrals involving Bessel functions which occur in hydrodynamics and elasticity. *Math. Comp.* 18 421-432.

Vilenkin, N.Y. (1958). Some relations for Gegenbauer functions. *Uspehi. Mat. Nauk. (N.S.).* 13 167-172.

Vircenko, N.A. and Makarenko, L.G. (1975). Certain dual integral equations. *Ukrain. Mat. Z.* 27 790-794.

Voelkar, D. and Doetsch, G. (1950). *Die zweidimensionale Laplace-Transformation.* Verlag Birkhauser, Basel.

Vovkodov, I.P. and Ulitko, A.T. (1973). Expansion of an arbitrary function as an integral in the Legendre functions

$$P^{i\lambda}_{-1/2+i\tau}(\pm i \text{ sh } \xi).$$

Dopovidi Akad. Nauk. Ukrain RSR Ser. A 1973 678-683.

Vroelant, C. (1953). Calcul des intégrales intervenant pour certaines formes approchées de la fonction d'onde. *C.R. Acad. Sci. Paris.* 236 2504-2506.

Waldron, R.A. (1968). The integration of Bessel functions. *J. Inst. Math. Appl.* 4 315-319.

Watson, G.N. (1944). *A Treatise on the Theory of Bessel Functions.* Cambridge University Press.

Weinstein, A. (1948). Discontinuous integrals and generalised potential theory. *Trans. Amer. Math. Soc.* 63 342-354.

Westmann, R.A. (1965). Simultaneous pairs of dual integral equations. *SIAM Rev.* 7 341-348.

van der Wetering, R.L. (1968). A generalisation of the Mehler-Dirichlet integral. *Nederl. Akad. Wetensch. Proc. A.* 71 234-238.

van der Wetering, R.L. (1975). Variation diminishing Fourier-Jacobi transforms. *SIAM J. Math. Anal.* 6 774-783.

Wheeler, T.S. (1944). A note on the evaluation of the Schrodinger hydrogenic intensity integral. *Proc. Roy. Irish Acad. Sect. A.* 50 7-12.

Whittaker, E.T. and Watson, G.N. (1952). *A Course in Modern Analysis.* Fourth Edition. Cambridge University Press.

Widder, D.V. (1963). The inversion of a convolution transform whose kernel is a Laguerre polynomial. *Amer. Math. Monthly.* 70 291-293.

Wilkins, J.E. (1948). Nicholson's integral for $J_n^2(z)+Y_n^2(z)$. *Bull. Amer. Math. Soc.* 54 232-234.

Wilks, S.S. (1962). *Mathematical Stastics.* Wiley, New York.

Williams, W.E. (1960). The solution of certain dual integral equations. *Proc. Edin. Math. Soc.* 12 213-216.

Williams, W.E. (1963a). A class of integral equations. *Proc. Cambridge Philos. Soc.* 59 589-597.

Williams, W.E. (1963b). Note on the reduction of dual and triple series equations to dual and triple integral equations. *Proc. Cambridge Philos. Soc.* 59 731-734.

Wimp, J. (1964). A class of integral transform. *Proc. Edin. Math. Soc.* 14 33-40.

Wimp, J. (1965). Two integral transform pairs involving hypergeometric functions. *Proc. Glasgow Math. Assoc.* 7 42-44.

Wimp, J. (1971). Integral transforms involving Whittaker functions. *Glasnik. Mat. Ser. III.* 6 67-70.

Winch, D.E. (1976). Integration formulae for Wigner's 3-; coefficients. *J. Mathematical Physics.* 17 1166-1170.

Witschel, W. (1973). Integral properties of Hermite polynomials by operator methods. *Z. Angew. Math. Phys.* 24 861-870.

Wong, C.F. and Kesarwani, R.N. (1973). On an integral transform involving Meijer's G-function. *Kyungpook Math. J.* 13 281-286.

Wu, A.C.T. (1977). Structure of the Azzarelli-Collas representation for the scattering amplitude and generalisation of the Rice representation and the Euler-Pochhammer representation. *J. Mathematical Physics.* 18 1935-1937.

Yeh, C. (1966). A note on integrals involving parabolic functions. *J. Math. and Phys.* 45 231-232.

Zemanian, A.H. (1968). *Generalised Integral Transforms.* Interscience. New York.

Zemanian, A.H. (1975). The Kontrovich-Lebedev transformation on distributions of compact support and its inversion. *Math. Proc. Cambridge Philos. Soc.* 77 139-143.

Zonneveld, J.A. and Bergius, J. (1955). The asymptotic expansion of a special function and some relations with Bessel functions. *Math. Centrum Amsterdam. Rekenafdeling no.* MR18 1-8.

Index of Symbols

NOTE 1. Indices of multiple summation are only included explicitly in association with the corresponding sign of summation if this is necessary to avoid ambiguity. Otherwise, they are omitted for convenience.

NOTE 2. In some instances, contrary to the usually accepted convention, it may be necessary to place square brackets, [ ], within ordinary brackets, ( ), in the notation. This is on account of the fact that the definition of a particular function may demand a particular kind of bracket in a fixed location.

Subject Index